# Multi-Step Enzyme Catalysis

*Edited by*
*Eduardo Garcia-Junceda*

## *Further Reading*

M. T. Reetz

**Directed Evolution of Selective Enzymes**

**Catalysts for Synthetic Organic Chemistry**

2009

ISBN: 978-3-527-31660-1

V. Gotor, I. Alfonso, E. García-Urdiales (Eds.)

**Asymmetric Organic Synthesis with Enzymes**

2008

ISBN: 978-3-527-31825-4

W. Aehle (Ed.)

**Enzymes in Industry**

**Production and Applications**

2007

ISBN: 978-3-527-31689-2

J.-L. Reymond (Ed.)

**Enzyme Assays**

**High-throughput Screening, Genetic Selection and Fingerprinting**

2006

ISBN: 978-3-527-31095-1

U. T. Bornscheuer, R. J. Kazlauskas

**Hydrolases in Organic Synthesis**

**Regio- and Stereoselective Biotransformations**

2006

ISBN: 978-3-527-31029-6

# Multi-Step Enzyme Catalysis

Biotransformations and Chemoenzymatic Synthesis

*Edited by*

*Eduardo Garcia-Junceda*

WILEY-VCH Verlag GmbH & Co. KGaA

**The Editor**

Dr. Eduardo Garcia-Junceda
Institute of Organic Chemistry (CSIC)
Department of Organic Chemistry
Juan de la Cierva, 3
28006 Madrid
Spanien



**Library of Congress Card No.:** applied for

**British Library Cataloguing-in-Publication Data**
A catalogue record for this book is available from the British Library.

**Bibliographic information published by the Deutsche Nationalbibliothek**
The Deutsche Nationalbibliothek lists this publication in the Deutsche Nationalbibliografie; detailed bibliographic data are available on the Internet at <http://dnb.d-nb.de>.



Printed in the Federal Republic of Germany
Printed on acid-free paper

**Cover Design** Grafik-Design Schulz, Fußgönheim
**Typesetting** SNP Best-set Typesetter Ltd., Hong Kong
**Printing** betz-druck GmbH, Darmstadt
**Binding** Litges & Dopf GmbH, Heppenheim

**ISBN:** 978-3-527-31921-3

# Contents

**Preface** *XI*

**List of Contributors** *XIII*

1 **Asymmetric Transformations by Coupled Enzyme and Metal Catalysis: Dynamic Kinetic Resolution** *1*
*Mahn-Joo Kim, Jaiwook Park, and Yoon Kyung Choi*
1.1 Introduction *1*
1.2 Some Fundamentals for DKR *2*
1.2.1 Enzymes for Kinetic Resolution *2*
1.2.2 Metal Catalysts for Racemization *3*
1.2.3 Enzyme–Metal Combination for DKR *5*
1.2.4 (*R*)- and (*S*)-Selective DKR *5*
1.3 Examples of DKR *6*
1.3.1 First DKR of Secondary Alcohols *6*
1.3.2 DKR of Secondary Alcohols with Racemization Catalyst **1** *6*
1.3.3 DKR of Secondary Alcohols with Racemization Catalyst **2** *8*
1.3.4 DKR of Secondary Alcohols with Racemization Catalyst **3** *9*
1.3.5 DKR of Secondary Alcohols with Racemization Catalyst **4** *10*
1.3.6 DKR of Secondary Alcohols with Racemization Catalyst **5** *10*
1.3.7 DKR of Secondary Alcohols with Racemization Catalyst **6** *11*
1.3.8 DKR of Secondary Alcohols with Racemization Catalyst **7** *12*
1.3.9 DKR of Secondary Alcohols with Air-Stable Racemization Catalysts *13*
1.3.10 DKR of Secondary Alcohols with Racemization Catalyst **10** *14*
1.3.11 DKR of Secondary Alcohols with Aluminum Catalysts *14*
1.3.12 DKR of Secondary Alcohols with Vanadium Catalysts *15*
1.4 Conclusions *16*
References *17*

*Multi-Step Enzyme Catalysis: Biotransformations and Chemoenzymatic Synthesis*
Edited by Eduardo Garcia-Junceda

ISBN: 978-3-527-31921-3

**2 Chemoenzymatic Routes to Enantiomerically Pure Amino Acids and Amines** *21*
*Nicholas J. Turner*
2.1 Introduction *21*
2.2 Amino Acids *23*
2.3 Amines *33*
References *38*

**3 Oxidizing Enzymes in Multi-Step Biotransformation Processes** *41*
*Stephanie G. Burton and Marilize le Roes-Hill*
3.1 Oxidizing Enzymes in Biocatalysis *41*
3.2 Classes of Oxidizing Enzymes *41*
3.3 Mechanisms of Biological Oxidation and Implications for Multi-Enzyme Biocatalysis *44*
3.4 Multi-Step Biotransformation Processes Involving Oxidation *45*
3.5 Design and Development of New Multi-Enzyme Oxidizing Processes *48*
3.5.1 Coupling Redox Enzymes *48*
3.5.2 Cofactor Recycle in Multi-Step Oxidizing Biocatalytic Systems *51*
3.6 Examples of Multi-Enzyme Biotransformation Processes Involving Oxidizing Enzymes *52*
3.6.1 Coupling of Oxidases with Non-Redox Enzymes *53*
3.6.2 Biocatalytic Systems Involving Coupled Oxidizing Enzymes *53*
3.7 Multi-Enzyme Systems in Whole-Cell Biotransformations and Expression of Redox Systems in Recombinant Hosts *55*
3.8 Other Applications of Multi-Enzyme Oxidizing Systems *56*
3.9 Conclusions *58*
References *58*

**4 Dihydroxyacetone Phosphate-Dependent Aldolases in the Core of Multi-Step Processes** *61*
*Laura Iturrate and Eduardo García-Junceda*
4.1 Introduction *61*
4.2 DHAP-Dependent Aldolases *63*
4.2.1 Problem of DHAP Dependence *63*
4.2.2 DHAP-Dependent Aldolases in the Core of Aza Sugar Synthesis *68*
4.2.3 Combined Use of Aldolases and Isomerases for the Synthesis of Natural and Unnatural Sugars *71*
4.2.4 DHAP-Dependent Aldolases in the Synthesis of Natural Products *73*
4.3 Fructose-6-Phosphate Aldolase: An Alternative to DHAP-Dependent Aldolases? *76*
4.4 Conclusions *78*
References *79*

**5 Multi-Enzyme Systems for the Synthesis of Glycoconjugates** *83*
*Birgit Sauerzapfe and Lothar Elling*
5.1 Introduction *83*
5.2 *In Vitro* and *In Vivo* Multi-Enzyme Systems *85*
5.3 Combinatorial Biocatalysis *86*
5.3.1 Synthesis and *In Situ* Regeneration of Nucleotide Sugars *88*
5.3.2 Synthesis of Oligosaccharides, Glycopeptides and Glycolipids Oligosaccharides *94*
5.4 Combinatorial Biosynthesis *97*
5.4.1 Synthesis of Oligosaccharides with Metabolically Engineered Cells *98*
5.5 Conclusions *102*
References *102*

**6 Enzyme-Catalyzed Cascade Reactions** *109*
*Roger A. Sheldon*
6.1 Introduction *109*
6.2 Enzyme Immobilization *110*
6.3 Reaction Types: General Considerations *111*
6.4 Chiral Alcohols *112*
6.5 Chiral Amines *114*
6.6 Chiral Carboxylic Acid Derivatives *121*
6.7 C–C Bond Formation: Aldolases *127*
6.8 Oxidations with $O_2$ and $H_2O_2$ *130*
6.9 Conclusions and Prospects *131*
References *132*

**7 Multi-modular Synthases as Tools of the Synthetic Chemist** *137*
*Michael D. Burkart and Junhua Tao*
7.1 Introduction *137*
7.2 Excised Domains for Chemical Transformations *139*
7.2.1 Function of Individual Domains and Domain Autonomy *139*
7.2.2 Heterocyclization and Aromatization *139*
7.2.3 Macrocyclization *144*
7.2.4 Halogenation *147*
7.2.5 Glycosylation *150*
7.2.6 Methyltransferases *151*
7.2.7 Oxidation *153*
7.3 Conclusions *155*
References *156*

**8 Modifying the Glycosylation Pattern in Actinomycetes by Combinatorial Biosynthesis** *159*
*José A. Salas and Carmen Méndez*
8.1 Bioactive Natural Products in Actinomycetes *159*
8.2 Deoxy Sugar Biosynthesis and Gene Clusters *161*
8.3 Characterization of Sugar Biosynthesis Enzymes *161*
8.4 Strategies for the Generation of Novel Glycosylated Derivatives *165*
8.4.1 Gene Inactivation *165*
8.4.2 Gene Expression *166*
8.4.3 Combining Gene Inactivation and Gene Expression *166*
8.4.4 Endowing a Host with the Capability of Synthesizing Different Sugars *166*
8.5 Generation of Glycosylated Derivatives of Bioactive Compounds *166*
8.5.1 Macrolides *167*
8.5.2 Aureolic Acid Group *175*
8.5.3 Angucyclines *181*
8.5.4 Anthracyclines *186*
8.5.5 Indolocarbazoles *191*
8.5.6 Aminocoumarins *193*
References *194*

**9 Microbial Production of DNA Building Blocks** *199*
*Jun Ogawa, Nobuyuki Horinouchi, and Sakayu Shimizu*
9.1 Introduction *199*
9.2 Screening of Acetaldehyde-Tolerant Deoxyriboaldolase and Its Application for DR5P Synthesis *200*
9.3 Construction of Deoxyriboaldolase-Overexpressing *E. coli* and Metabolic Analysis of the *E. coli* Transformants for DR5P Production from Glucose and Acetaldehyde *201*
9.4 Efficient Production of DR5P from Glucose and Acetaldehyde by Coupling of the Alcoholic Fermentation System of Baker's Yeast and Deoxyriboaldolase-Expressing *E. coli* *203*
9.5 Biochemical Retrosynthesis of 2′-Deoxyribonucleosides from Glucose Acetaldehyde and a Nucleobase: Three-Step Multi-Enzyme-Catalyzed Synthesis *204*
9.6 One-Pot Multi-Step Enzymatic Synthesis of 2′-Deoxyribonucleoside from Glucose, Acetaldehyde and a Nucleobase *206*
9.7 Improvement of the One-Pot Multi-Step Enzymatic Process for Practical Production of 2′-Deoxyribonucleoside from Glucose, Acetaldehyde and a Nucleobase *207*
9.8 Conclusions *208*
References *210*

**10 Combination of Biocatalysis and Chemical Catalysis for the Preparation of Pharmaceuticals Through Multi-Step Syntheses** *213*
*Vicente Gotor-Fernández, Rosario Brieva, and Vicente Gotor*
10.1 Introduction: Biocatalysis and Chemical Catalysis *213*
10.2 Pharmaceuticals with Hydrolases *214*
10.2.1 Enzymatic Hydrolysis *214*
10.2.2 Enzymatic Transesterification *219*
10.2.3 Enzymatic Aminolysis *222*
10.3 Pharmaceuticals with Oxidoreductases *226*
10.4 Pharmaceuticals with Lyases *227*
10.5 Conclusions *230*
References *231*

**Index** *235*

# Preface

> "The abundance of substances of which animals and plants are composed, the remarkable processes whereby they are formed and then broken down again have claimed the attention of mankind of old, and hence from the early days they also persistently captivated the interest of chemists".
>
> *Emil Fischer, Nobel Lecture, 1902*

> "The chemist who designs and completes an original and esthetically pleasing multistep synthesis is like the composer, artist or poet who, with great individuality, fashions new forms of beauty from the interplay of mind and spirit".
>
> *Elias James Corey, Nobel Lecture, 1990*

Nature has always been a permanent source of inspiration for chemists but, as Emil Fischer brightly indicated in his Nobel award acceptance lecture in 1902, it is not only the vast diversity of compounds that living beings are capable of creating, but also the extraordinary strategies of synthesis deployed. Evidently, the catalysts used by living beings – enzymes – are key to Nature's Synthesis Strategies. Emil Fischer himself foretold, a few paragraphs further into his lecture, that chemistry would employ enzymes at large and – to our greatest surprise, bearing in mind that these words were written as early as 1902 – that artificial enzymes would be tailor-made to serve its purposes.

The longing of biocatalysis to transfer to the laboratory the exquisite efficiency shown by enzymes in Nature has begun to become a reality since the late 1980s, with the invention of the polymerase chain reaction (PCR). The level of development and access brought about by the PCR to genetic material handling and transformation, has allowed the number of available enzymes to grow exponentially. Modifying the catalytic properties of enzymes to adapt them to their new environment in a test tube has become a reality. We have learned to imitate the strategies used by Nature to create new enzymes, and to adapt the existing ones to new synthetic needs. Eventually, Emil Fischer's prediction has come true.

*Multi-Step Enzyme Catalysis: Biotransformations and Chemoenzymatic Synthesis*
Edited by Eduardo Garcia-Junceda

ISBN: 978-3-527-31921-3

Living beings do not use enzymes in isolation, however. A large portion of the extraordinary synthetic effectiveness that enzymes display in Nature comes from the fact that living beings apply a multistep synthesis strategy, catalyzed by enzymes acting sequentially. It is the utilization of more- or less-complicated biosynthetic routes that allows living beings to build complex structures from simple elements; to obtain and to store energy; and to know and to communicate with their environment. The jointed action of a sequence of enzymes can make irreversible a reversible process, eliminate inhibition problems caused by product excess, or prevent the lack of substrate scattered on the bulk solution. Evidently, in order to develop, biocatalysis could not look away from these and other synthesis opportunities served by multistep reactions. The level of relevance attained by the development of this synthetic strategy in the field of biocatalysis and biotransformation is evidenced by the celebration in April 2006 of the first Symposium on *Multistep Enzyme-Catalyzed Processes*, organized jointly by the Applied Biocatalysis Research Centre at Gratz and the European Federation of Biotechnology Section of Applied Biocatalysis (ESAB), to which this book is indebted.

The aim of this handbook is to bring together various key aspects to cover the broad field of *multistep enzyme-catalyzed processes*, from the 'simplest' system in which one or a few isolated enzymes are used alone or in combination with non-enzyme-catalyzed steps, to the most 'complex' system in which artificial or natural pathways are created or even whole cells are modified to be used as synthetic factories.

I would like thank all those authors who have participated in this exciting project for their superb work, valuable time and remarkable efforts; and in particular, I thank Elke Maase and Stefanie Volk at Wiley-VCH for their patience, friendliness and precious help in editing.

I hope that you enjoy reading this book, and that it can serve as an inspirational source and stimulus to researchers of all levels – especially the youngest – who are working in the biocatalysis field.

Madrid, July 2008 *Eduardo García-Junceda*

# List of Contributors

***Rosario Brieva***
Universidad de Oviedo
Facultad de Química
Departamento de Química
Orgánica e Inorgánica
Julián Clavería, 6
33006 Oviedo
Spain

***Michael D. Burkart***
University of California,
San Diego
Department of Chemistry and
Biochemistry
9500 Gilman Drive
La Jolla, CA 92093-0358
USA

***Stephanie G. Burton***
University of Cape Town
Department of Chemical
Engineering
Private Bag
7701, Rondebosch
Cape Town
South Africa

***Yoon Kyung Choi***
Pohang University of Science and
Technology
Department of Chemistry
San-31 Hyojadong
Pohang, 790–784
South Korea

***Lothar Elling***
RWTH Aachen University
Institute for Biotechnology and
Helmholtz Institute for Biomedical
Engineering
Laboratory for Biomaterials
Worringer Weg 1
52074 Aachen
Germany

***Eduardo García-Junceda***
Consejo Superior de Investigaciones
Científicas
Instituto de Química Orgánica
General
Juan de la Cierva, 3
28006 Madrid
Spain

*Multi-Step Enzyme Catalysis: Biotransformations and Chemoenzymatic Synthesis*
Edited by Eduardo Garcia-Junceda

ISBN: 978-3-527-31921-3

***Vicente Gotor***
Universidad de Oviedo
Facultad de Química
Departamento de Química
Orgánica e Inorgánica
Julián Clavería, 6
33006 Oviedo
Spain

***Vicente Gotor-Fernández***
Universidad de Oviedo
Facultad de Química
Departamento de Química
Orgánica e Inorgánica
Julián Clavería, 6
33006 Oviedo
Spain

***Nobuyuki Horinouchi***
Kyoto University
Graduate School of Agriculture
Division of Applied Life Sciences
Laboratory of Industrial
Microbiology
Kitashirakawa-oiwakecho
Sakyo-ku
Kyoto 606-8502
Japan

***Laura Iturrate***
Consejo Superior de
Investigaciones Científicas
Instituto de Química Orgánica
General
Juan de la Cierva, 3
28006 Madrid
Spain

***Mahn-Joo Kim***
Pohang University of Science and
Technology
Department of Chemistry
San-31 Hyojadong
Pohang, 790–784
South Korea

***Carmen Méndez***
Universidad de Oviedo
Facultad de Medicina
Departamento de Biología Funcional
Instituto Universitario de Oncología
del Principado de Asturias (IUOPA)
c/Julián Clavería s/n
33006 Oviedo
Spain

***Jun Ogawa***
Kyoto University
Graduate School of Agriculture
Division of Applied Life Sciences
Laboratory of Fermentation Physiology
and Applied Microbiology
Kitashirakawa-oiwakecho
Sakyo-ku
Kyoto 606-8502
Japan

***Jaiwook Park***
Pohang University of Science and
Technology
Department of Chemistry
San-31 Hyojadong
Pohang, 790–784
South Korea

***Marilize le Roes-Hill***
University of Cape Town
Department of Chemical Engineering
Private Bag
7701, Rondebosch
Cape Town
South Africa

**José A. Salas**
Universidad de Oviedo
Facultad de Medicina
Departamento de Biología
Funcional Instituto Universitario
de Oncología del Principado de
Asturias (IUOPA)
c/Julián Clavería s/n
33006 Oviedo
Spain

**Birgit Sauerzapfe**
RWTH Aachen University
Institute for Biotechnology and
Helmholtz Institute for Biomedical
Engineering
Laboratory for Biomaterials
Worringer Weg 1
52074 Aachen
Germany

**Roger A. Sheldon**
Delft University of Technology
Department of Biocatalysis and
Organic Chemistry
Julianalaan 136
2628 BL Delft
The Netherlands

**Sakayu Shimizu**
Kyoto University
Graduate School of Agriculture
Division of Applied Life Sciences
Laboratory of Fermentation Physiology
and Applied Microbiology
Kitashirakawa-oiwakecho
Sakyo-ku
Kyoto 606-8502
Japan

**Junhua Tao**
BioVerdant Inc.
7330 Carroll Road
San Diego, CA 92121
USA

**Nicholas J. Turner**
University of Manchester
School of Chemistry
Manchester Interdisciplinary Biocentre
131 Princess Street
Manchester M1 7DN
United Kingdom

# 1
# Asymmetric Transformations by Coupled Enzyme and Metal Catalysis: Dynamic Kinetic Resolution

*Mahn-Joo Kim, Jaiwook Park, and Yoon Kyung Choi*

## 1.1
## Introduction

The enzymatic resolution of racemic substrates now is a well-established approach for the synthesis of single enantiomers [1, 2]. A representative example is the kinetic resolution of secondary alcohols via lipase-catalyzed transesterification for the preparation of enantiomerically enriched alcohols and esters [3]. The enzymatic resolution in general is straightforward and satisfactory in terms of optical purity, but it has an intrinsic limitation in that the theoretical maximum yield of a desirable enantiomer cannot exceed 50%. Accordingly, additional processes such as isolation, racemization and recycling of unwanted isomers are required to obtain the desirable isomer in a higher yield (Scheme 1.1).

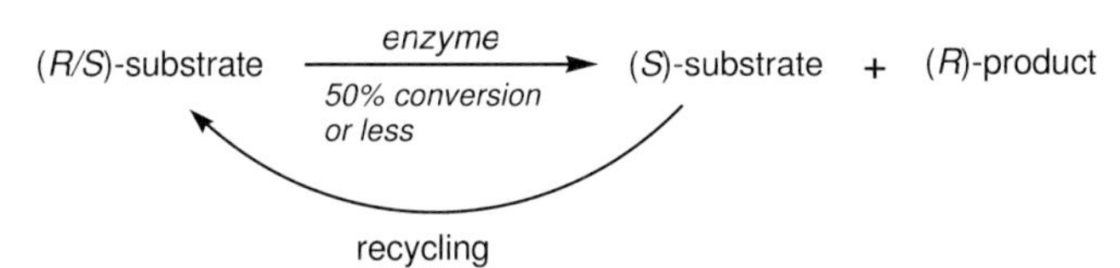


**Scheme 1.1** (*R*)-Selective enzymatic resolution with recycling of unreacted (*S*)-substrate.

The limitation of enzymatic resolution, however, can be overcome by introducing an efficient catalyst for racemization of substrate into the resolution, leading to the process called dynamic kinetic resolution (DKR) [4]. Theoretically, DKR can provide single enantiomeric products [99% enantiomeric excess (e.e.) or greater] in 100% yield in the case where a highly efficient racemization catalyst is combined with a highly enantioselective enzyme. In the last decade, several metal-based catalysts have been developed for the racemization and successfully incorporated into the resolution process [5]. Now a wide range of racemic substrates can be converted to enantiomeric products of high optical purity in good yields via the enzymometallic DKR (Scheme 1.2). This chapter covers these developments with detailed examples.

*Multi-Step Enzyme Catalysis: Biotransformations and Chemoenzymatic Synthesis*
Edited by Eduardo Garcia-Junceda

ISBN: 978-3-527-31921-3

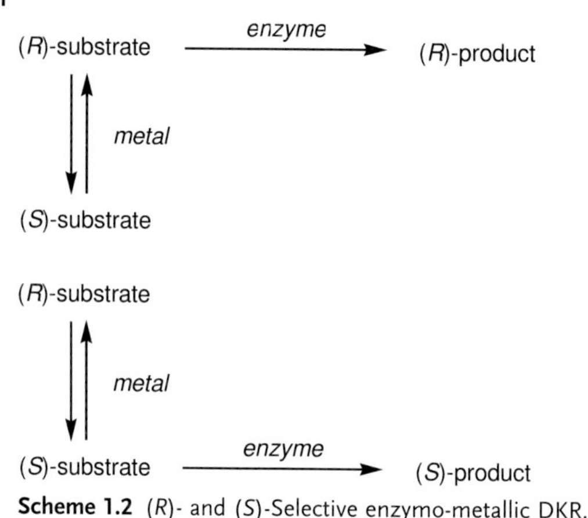


**Scheme 1.2** (*R*)- and (*S*)-Selective enzymo-metallic DKR.

## 1.2
## Some Fundamentals for DKR

### 1.2.1
### Enzymes for Kinetic Resolution

The resolution of a racemic substrate can be achieved with a range of hydrolases including lipases and esterases. Among them, two commercially available lipases, *Candida antarctica* lipase B (CALB; trade name, Novozym-435) and *Pseudomonas cepacia* lipase (PCL; trade name, Lipase PS-C), are particularly useful because they have broad substrate specificity and high enantioselectivity. They display satisfactory activity and good stability in organic media. In particular, CALB is highly thermostable so that it can be used at elevated temperature up to 100 °C.

The lipase-catalyzed resolutions usually are performed with racemic secondary alcohols in the presence of an acyl donor in hydrophobic organic solvents such as toluene and *tert*-butyl methyl ether (Scheme 1.3). In case the enzyme is highly enantioselective ($E = 200$ or greater), the resolution reaction in general is stopped at nearly 50% conversion to obtain both unreacted enantiomers and acylated enantiomers in enantiomerically enriched forms. With a moderately enantioselective enzyme ($E = 20$–$50$), the reaction carries to well over 50% conversion to get unreacted enantiomer of high optical purity at the cost of acylated enantiomer of lower optical purity. The enantioselectivity of lipase is largely dependent on the structure of substrate as formulated by Kazlauskas [6]: most lipases show

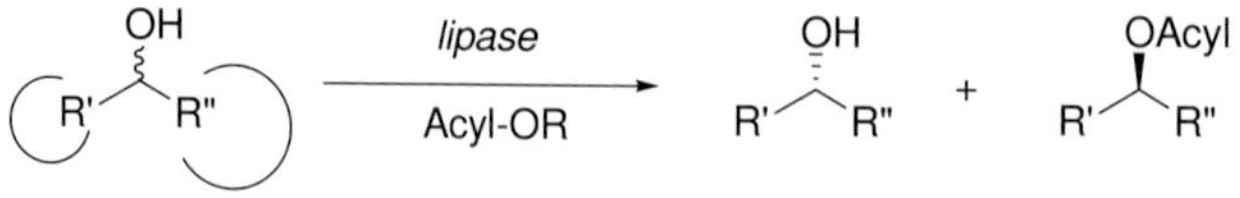

**Scheme 1.3** Lipase-catalyzed resolution of secondary alcohols.

(*R*)-selectivity toward simple secondary alcohols carrying one small and one relatively larger substituent at the hydroxyl methane center, and the selectivity in general increases with an increase in the size difference between two substituents. The size of the small substituent limits the reactivity of substrate toward lipase. If it exceeds a three-carbon unit, the substrate reacts very slowly or does not react at a synthetically useful rate. Accordingly, the Kazlauskas rule is useful as a guideline for predicting substrates that can be efficiently resolved by lipase as well as the stereochemistry of resolved substrates.

Lipase, which is highly useful for kinetic resolution, however, has a limitation for use in DKR in that it cannot be used for (*S*)-configuration products. For this purpose, subtilisin, a protease from *Bacillus licheniformis*, can replace lipase since it provides complementary enantioselectivity (Scheme 1.4). Subtilisin, however, has been much less frequently employed in resolution compared to lipase because it displays poor catalytic performance in organic media. Subtilisin is inferior to lipase in several properties such as activity, enantioselectivity and stability. Accordingly, the use of the enzyme usually requires some special treatments for activation and stabilization before use. For example, the treatment of subtilisin with surfactants has enhanced substantially its activity and stability up to a synthetically useful level.

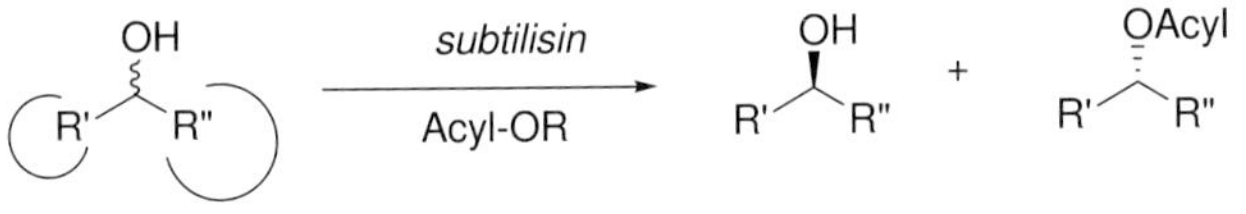

**Scheme 1.4** Subtilisin-catalyzed resolution of secondary alcohols.

## 1.2.2
## Metal Catalysts for Racemization

Many different metal catalysts have been explored for racemization of secondary alcohols. Among them, ruthenium-based organometallic complexes have been most intensively tested as the racemization catalyst (Figure 1.1).

These ruthenium catalysts catalyze the racemization of secondary alcohol through a dehydrogenation/hydrogenation cycle with or without releasing ketone as a byproduct (Scheme 1.5). Catalysts **6–9** display good activities at room temperature, while others show satisfactory activities at elevated temperatures. Catalyst **1**, for example, requires a high temperature (70 °C) for dissociation into two monomeric species (**1a** and **1b**) acting as racemization catalysts (Scheme 1.6).

Most ruthenium catalysts except **8** and **9** are highly sensitive to oxygen or air and must be used under anaerobic conditions. The latter can be used under aerobic conditions. Currently, no rationale is available for explaining the difference in stability between these ruthenium catalysts. In general, racemizations by these catalysts take place more rapidly with benzylic alcohols compared to non-benzylic or aliphatic alcohols.

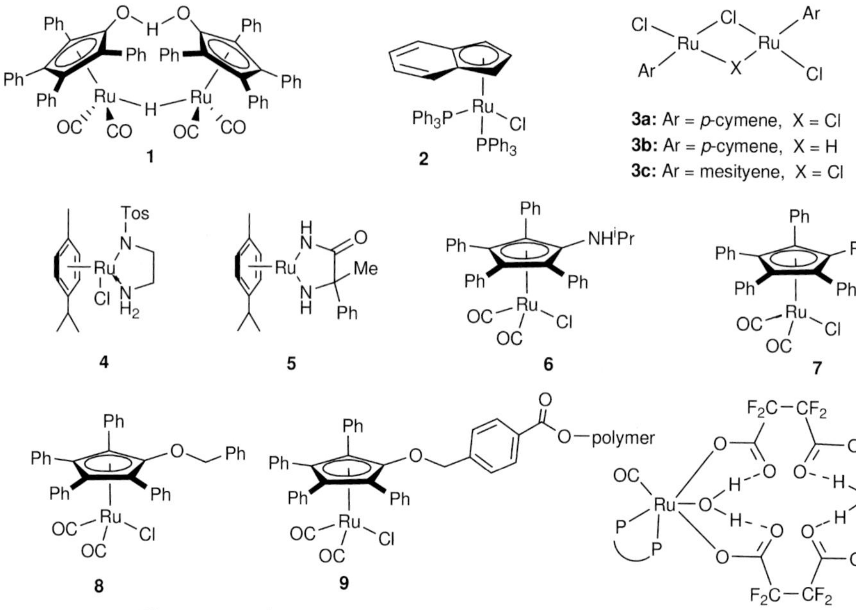
1
2
3a: Ar = p-cymene, X = Cl
3b: Ar = p-cymene, X = H
3c: Ar = mesityene, X = Cl
4
5
6
7
8
9
10

**Figure 1.1** Ruthenium catalysts.

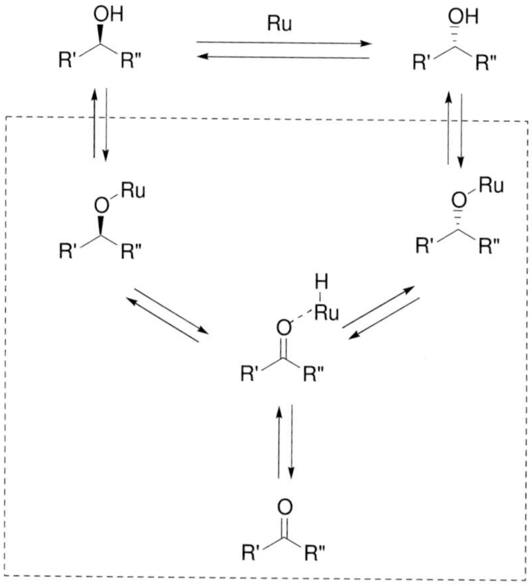
Ru

**Scheme 1.5** A simplified mechanism for ruthenium-catalyzed racemization of *sec*-alcohol.

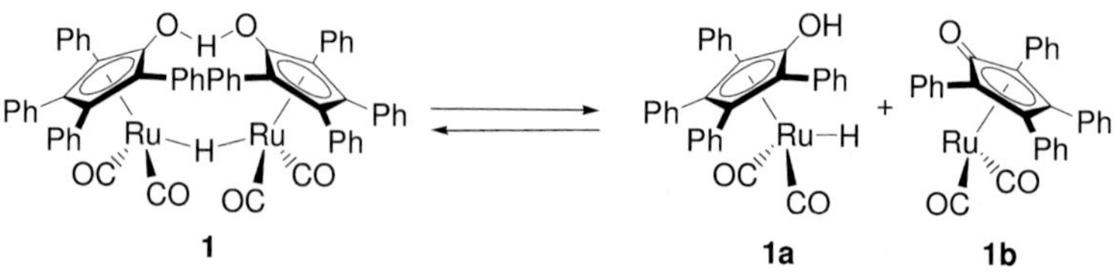
1
1a
1b

**Scheme 1.6** Dissociation of catalyst **1**.

### 1.2.3
### Enzyme–Metal Combination for DKR

DKR of secondary alcohol is achieved by coupling enzyme-catalyzed resolution with metal-catalyzed racemization. For efficient DKR, these catalytic reactions must be compatible with each other. In the case of DKR of secondary alcohol with the lipase–ruthenium combination, the use of a proper acyl donor (required for enzymatic reaction) is particularly crucial because metal catalyst can react with the acyl donor or its deacylated form. Popular vinyl acetate is incompatible with all the ruthenium complexes, while isopropenyl acetate can be used with most monomeric ruthenium complexes. *p*-Chlorophenyl acetate (PCPA) is the best acyl donor for use with dimeric ruthenium complex **1**. On the other hand, reaction temperature is another crucial factor. Many enzymes lose their activities at elevated temperatures. Thus, the racemization catalyst should show good catalytic efficiency at room temperature to be combined with these enzymes. One representative example is subtilisin. This enzyme rapidly loses catalytic activities at elevated temperatures and gradually even at ambient temperature. It therefore is compatible with the racemization catalysts **6–9**, showing good activities at ambient temperature. In case the racemization catalyst requires an elevated temperature, CALB is the best counterpart.

### 1.2.4
### (*R*)- and (*S*)-Selective DKR

Thanks to two complementary enzymes, lipase and subtilisin, both (*R*)- and (*S*)-selective DKR can be performed to obtain the corresponding enantiomeric products.

DKR by the lipase–ruthenium combination provides (*R*)-products, while DKR by the subtilisin–ruthenium combination gives products of the opposite configuration (Schemes 1.7 and 1.8).

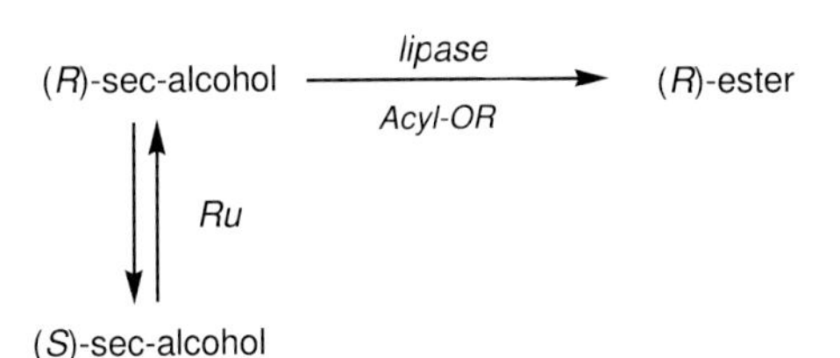


**Scheme 1.7** (*R*)-Selective DKR with the lipase–ruthenium combination.

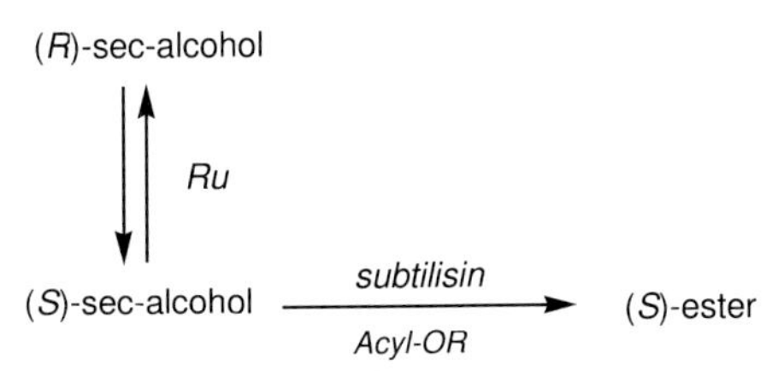


**Scheme 1.8** (*S*)-Selective DKR with the subtilisin–ruthenium combination.

## 1.3
## Examples of DKR

### 1.3.1
### First DKR of Secondary Alcohols

The first use of a metal catalyst in the DKR of secondary alcohols was reported by Williams *et al.* [7]. In this work, various rhodium, iridium, ruthenium and aluminum complexes were tested. Among them, only $Rh_2(OAc)_4$ and $[Rh(cod)Cl]_2$ showed reasonable activity as the racemization catalyst in the DKR of 1-phenylethanol. The racemization occurred through transfer-hydrogenation reactions and required stoichiometric amounts of ketone as hydrogen acceptor. The DKR of 1-phenylethanol performed with $Rh_2(OAc)_4$ and *Pseudomonas fluorescens* lipase gave (*R*)-1-phenylethyl acetate of 98% e.e. at 60% conversion after 72 h.

### 1.3.2
### DKR of Secondary Alcohols with Racemization Catalyst 1

Significantly improved DKR was reported by Bäckvall *et al.* who used diruthenium complex **1** together with CALB [8]. This work demonstrated for the first time the superiority of PCPA as the acyl donor over popular acyl donors such as vinyl and isopropenyl acetate. The DKR of 1-phenylethanol by this procedure afforded optically pure (*R*)-1-phenylethyl acetate in a high yield (Scheme 1.9) [8b].

However, the procedure has some drawbacks to overcome. First, it requires an elevated temperature (70 °C) for the activation of the racemization catalyst. Such a high temperature is unacceptable for thermally less-stable enzymes. Second, the racemization proceeds through a mechanism including the release of ketone as a byproduct and thus the lowering of yield is inevitable. Third, PCPA used in an

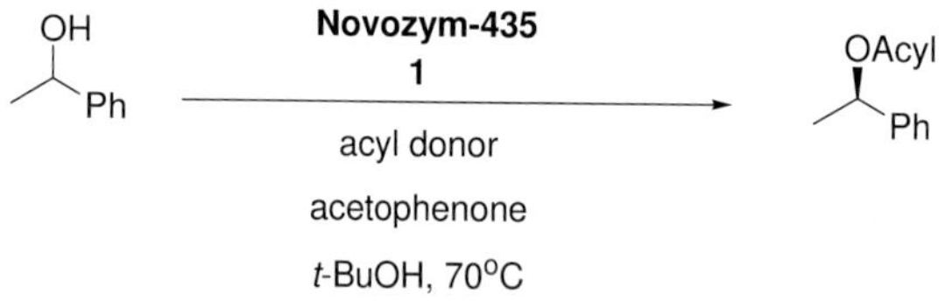

OH Ph → (Novozym-435, **1**, acyl donor, acetophenone, *t*-BuOH, 70°C) → OAcyl Ph

| acyl donor | ee (%) | yield (%) |
|---|---|---|
| vinyl acetate | >99 | 50 |
| Isopropenyl acetate | >99 | 72 |
| *p*-chlorophenyl acetate | >99 | 100 |

**Scheme 1.9** DKR of 1-phenylethanol with ruthenium catalyst **1**.

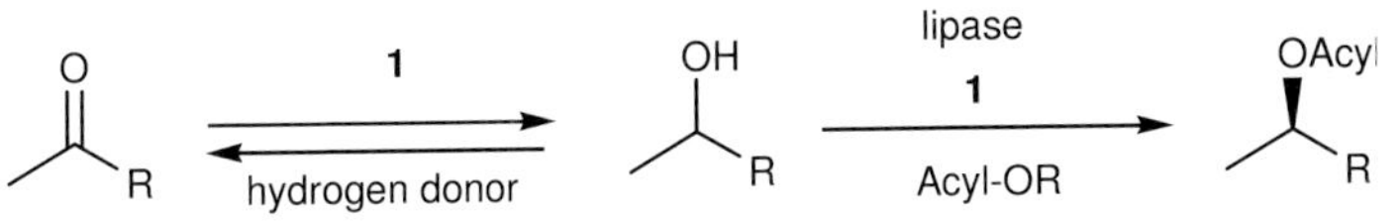

Scheme 1.10 Hydrogenation and DKR of ketones.

excess amount is often difficult to remove from the acylated products during work-up. In spite of these limitations, the procedure with ruthenium catalyst **1** has been successfully applied in the DKR of a variety of simple and functionalized alcohols, including diols [9], hydroxy acid esters [9b, 10], hydroxyl aldehydes [9b], β-azido alcohols [11], β-hydroxyl nitriles [12], β-halo alcohols [13] and hydroxyalkanephosphonates [14].

An interesting application of **1** is the use in the asymmetric reductive acetylation of ketones via DKR of alcohol intermediates. In this transformation, ruthenium-catalyzed hydrogenation of ketone takes place in a concerted fashion with DKR of alcohol to produce the corresponding acyl products (Scheme 1.10). The idea of this process was to take advantage of ketone formation, which is a problem observed in the DKR of secondary alcohols with **1**. A key to this process was the selection of hydrogen donors compatible with the DKR conditions. 2,6-Dimethyl-4-heptanol, which cannot be acylated by lipases, and hydrogen molecules were effective hydrogen donors [15]. Asymmetric reductive acetylation of ketones under 1 atm hydrogen in ethyl acetate gave products in good yields and high optical purities (Scheme 1.11) [15b]. Here, ethyl acetate was used as both acyl donor and solvent.

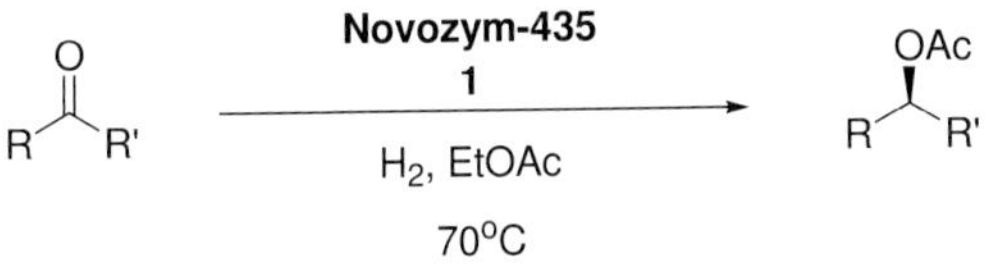

Novozym-435, 1; $H_2$, EtOAc, 70°C

| R | R' | ee (%) | yield (%) |
|---|---|---|---|
| Me | Ph | 96 | 81 |
| Me | 4-MeO-Ph | 99 | 85 |
| Me | 4-Cl-Ph | 97 | 72 |

Scheme 1.11 Asymmetric reductive acetylation of ketones.

Asymmetric reductive acetylation was also applicable to acetoxyphenyl ketones. In this case the substrate itself acts as an acyl donor. For example, *m*-acetoxyacetophenone was transformed to (*R*)-1-(3-hydroxyphenyl)ethyl acetate under 1 atm $H_2$ in 95% yield [16] (Scheme 1.12). The pathway of this reaction is rather complex. It was confirmed that nine catalytic steps are involved: two steps for ruthenium-catalyzed reductions, two steps for ruthenium-catalyzed racemizations, two steps

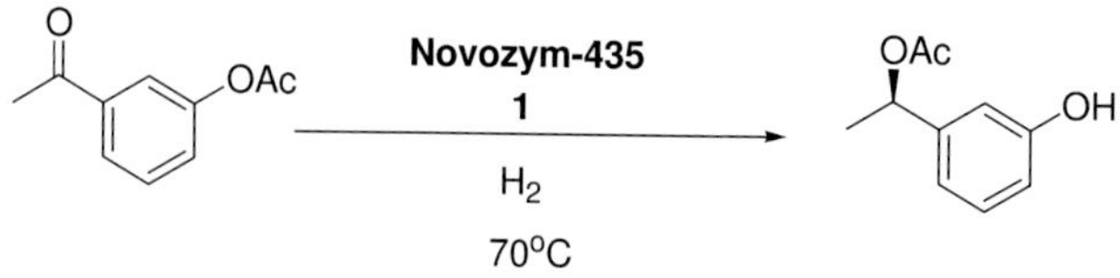

**Scheme 1.12** Asymmetric transformation of *m*-acetoxyacetophenone.

for ruthenium-catalyzed deacylations and three steps for lipase-catalyzed acylations. This process was applicable to a wide range of acyloxyphenyl ketones.

Another example showing the utility of **1** is the asymmetric hydrogenation of vinyl esters which usually are used as acyl donors in enzymatic resolution. In this transformation, vinyl esters are converted to ketones which then undergo asymmetric reductive acylation to give chiral esters as described in Scheme 1.13. The overall reaction thus corresponds to the asymmetric hydrogenation of vinyl ester to the corresponding alkyl esters.

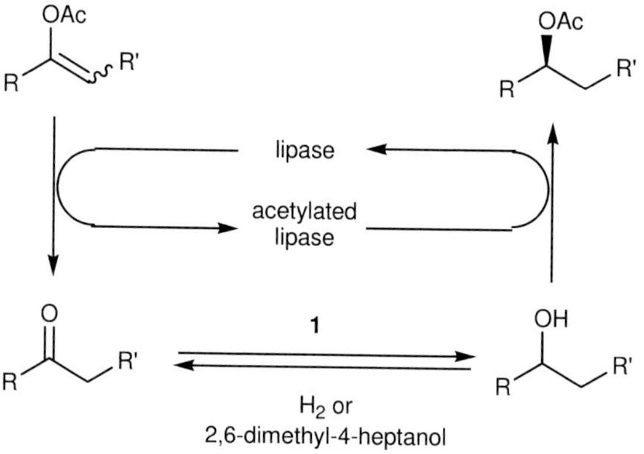

**Scheme 1.13** Asymmetric hydrogenation of vinyl esters.

The reaction was carried out with CALB and diruthenium complex **1** in the presence of 2,6-dimethyl-4-heptanol or molecular hydrogen (1 atm). In the case of 1-phenylvinyl acetate, (*R*)-1-phenylethyl acetate was obtained in 89% yield and 98% ee (Scheme 1.14) [15a].

### 1.3.3 DKR of Secondary Alcohols with Racemization Catalyst 2

We reported the use of an indenyl ruthenium complex **2** as a racemization catalyst which did not produce ketones as the byproducts [17]. The metal catalyst requires a weak base like triethylamine and molecular oxygen to be activated. The DKR with **2** in combination with an immobilized PCL was carried out at a lower temperature (60 °C) to afford good yields and high optical purities (Scheme 1.15). It is noteworthy that **2** does not require ketone as hydrogen mediator for racemization.

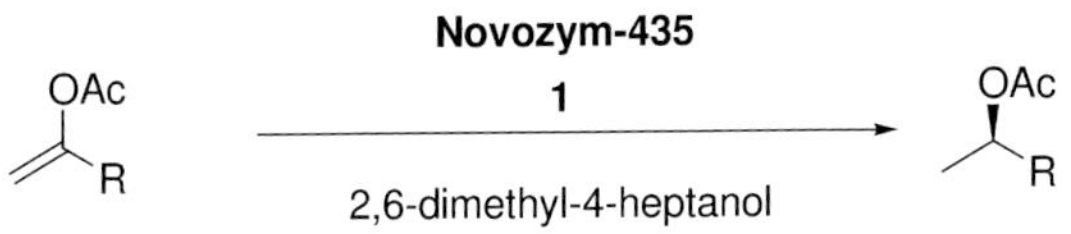

Novozym-435
1
2,6-dimethyl-4-heptanol

| R | ee (%) | yield (%) | R | ee (%) | yield (%) |
|---|---|---|---|---|---|
| Ph | 98 | 89 | $PhCH_2$ | 79 | 90 |
| 4-MeO-Ph | 98 | 80 | $PhCH_2CH_2$ | 94 | 92 |
| 4-Cl-Ph | 97 | 91 | $CH_3(CH_2)_4CH_2$ | 91 | 95 |
| c-$C_6H_{11}$ | 99 | 94 | | | |

**Scheme 1.14** Asymmetric hydrogenation of vinyl acetates.

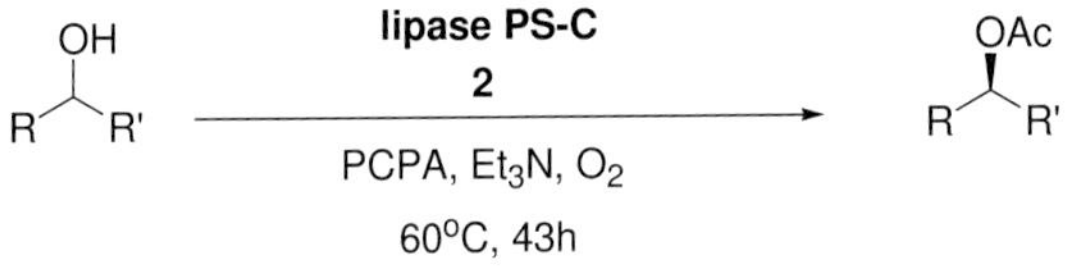

lipase PS-C
2
PCPA, $Et_3N$, $O_2$
60°C, 43h

| R | R' | ee (%) | yield (%) |
|---|---|---|---|
| Me | Ph | 96 | 86 |
| Me | Ph-4-OMe | 99 | 82 |
| Me | Ph-4-Br | 99 | 98 |
| Me | $CH_2Ph$ | 97 | 60 |

**Scheme 1.15** DKR with indenyl ruthenium complex **2**.

This work thus presented the first example for DKR without the formation of ketones as the side products.

## 1.3.4 DKR of Secondary Alcohols with Racemization Catalyst 3

We discovered that cymene–ruthenium catalysts **3a–c** were effective catalyst systems for facile DKR of secondary alcohols at 40 °C. This catalyst system was particularly useful for the DKR of allylic alcohols [18], which underwent smoothly at room temperature to provide the corresponding chiral acetates with excellent optical purities (Scheme 1.16). This work has for the first time demonstrated that DKR can be performed at room temperature.

Interestingly, catalyst **3a** showed higher racemization activities in ionic liquids such as $[EMIm]BF_4$ and $[BMIm]PF_6$ ([EMIm]=1-ethyl-3-methylimidazolium, [BMIm]=1-butyl-3-methyl-imidazolium) [19]. The DKR in ionic liquids has one big

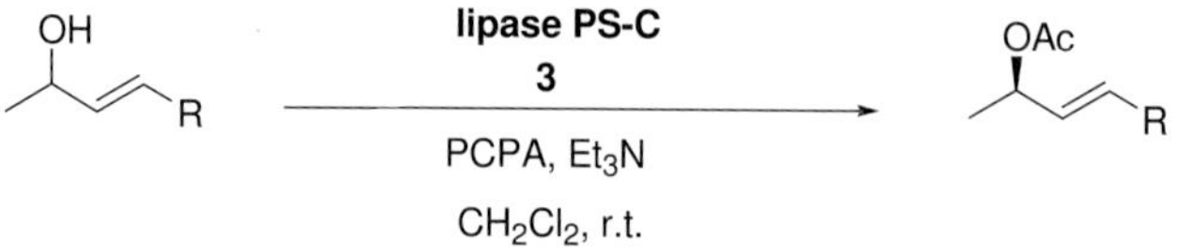

OH R → (lipase PS-C, 3; PCPA, $Et_3N$; $CH_2Cl_2$, r.t.) → OAc R

| R | ee (%) | yield (%) |
|---|---|---|
| Ph | >99 | 84 |
| 4-Cl-Ph | 99 | 91 |
| 4-MeO-Ph | 99 | 85 |
| 2-furyl | 99 | 92 |
| c-$C_6H_{11}$ | 95 | 90 |
| $(CH_3)_3C$ | >99 | 85 |

**Scheme 1.16** DKR of allylic alcohols with cymene–ruthenium catalyst **3**.

advantage that both ruthenium catalyst and enzyme together with ionic liquid can be reused after a simple extraction of the products by ether [20].

### 1.3.5 DKR of Secondary Alcohols with Racemization Catalyst 4

Chiral Ru(II) complexes formed from $[RuCl_2(p\text{-cymene})]_2$ and chiral bidentate nitrogen ligands have been extensively studied as catalysts for the asymmetric transfer hydrogenation of prochiral ketones [21]. Sheldon *et al.* utilized an achiral analog **4** for alcohol racemization. A cocatalyst system of **4** and 2,2,6,6-tetramethyl-1-piperidinyloxyl (TEMPO) racemized benzylic alcohols under the conditions for enzymatic acylation [22]. They suggested that a ruthenium hydride species generated from the mixture of **4**, TEMPO and a benzylic alcohol is the active catalyst for racemization. This DKR process with **4** gave (*R*)-1-phenylethyl acetate (greater than 99% e.e.) in 76% yield after 48 h with the production of acetophenone in 15%.

### 1.3.6 DKR of Secondary Alcohols with Racemization Catalyst 5

Palmans *et al.* prepared **5** by the reaction of $[RuCl_2(p\text{-cymene})]_2$ and 2-phenyl-2-aminopropionamide in the presence of potassium carbonate. They used **5** in an iterative tandem catalysis for the synthesis of chiral oligoesters. The enzymatic ring opening of 6-methyl-ε-caprolactone was combined with ruthenium-catalyzed alcohol racemization to produce optically active oligomers of 6-methyl-ε-caprolactone [23] (Scheme 1.17).

The same catalyst system was applied to the condensation of racemic α,α′-dimethyl-1,4-benzenedimethanol and dimethyl adipate. Optically active polyesters ($M_w$ = 3400 g/mol; $M_n$ = 2100 g/mol) were obtained [24] (Scheme 1.18).

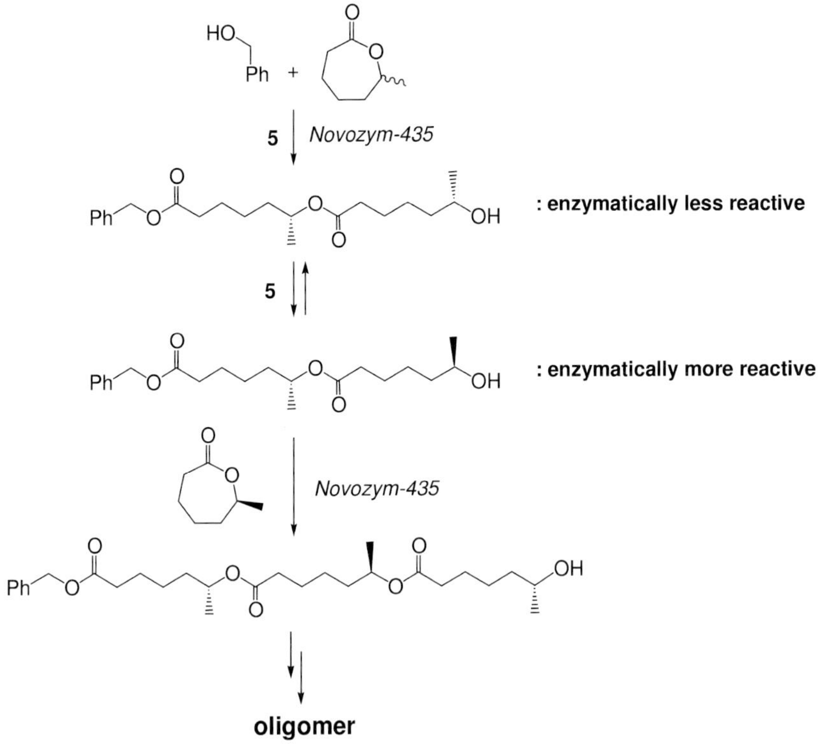

Scheme 1.17 Synthesis of optically active oligoesters by DKR.

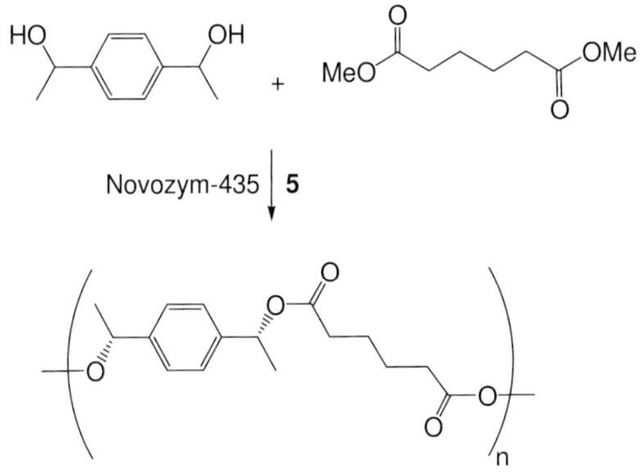

Scheme 1.18 Synthesis of optically active polyester by DKR.

## 1.3.7
## DKR of Secondary Alcohols with Racemization Catalyst 6

In an effort directed at developing a racemization catalyst that works uniformly for a wide range of substrates at room temperature, we designed and synthesized a novel aminocyclopentadienyl ruthenium chloride complex **6** [25]. The DKR of aromatic as well as aliphatic alcohols with **6** was successfully conducted at room temperature. The resulting acylated products were obtained in high yields and in high enantiomeric excess (Scheme 1.19). In the case of aromatic alcohols the substituent effects were found insignificant in the DKR; however, comparatively aromatic alcohols have faster conversion rates than their aliphatic counterparts. These results represent the first successful DKR of secondary alcohols at room temperature.

Novozym-435, **6**; isopropenyl acetate; *t*BuOK, $Na_2CO_3$; toluene, 25°C

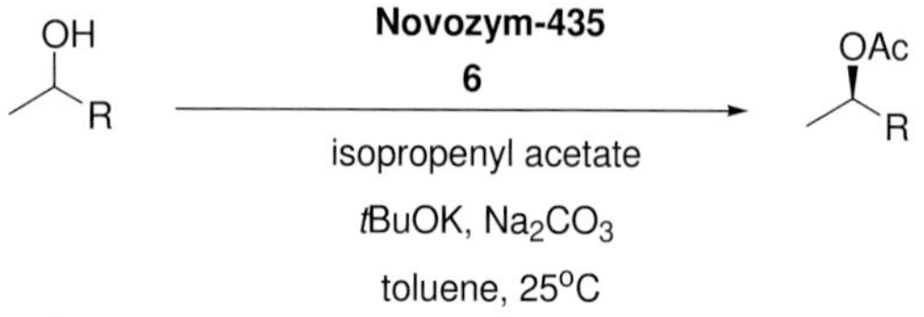

| R | ee (%) | yield (%) | R | ee (%) | yield (%) |
|---|---|---|---|---|---|
| Ph | >99 | 95 | *c*-$C_6H_{11}$ | >99 | 86 |
| Ph-4-Cl | >99 | 94 | $CH_2(CH_2)_4CH_3$ | 91 | 89 |
| Ph-4-OMe | >99 | 90 | $CH_2O(Ph)_3C$ | 99 | 97 |
| Ph-4- $NO_2$ | >99 | 97 | CH=CHPh | 98 | 93 |
| Ph-4-CN | >99 | 95 | | | |

**Scheme 1.19** DKR with aminocyclopentadienyl ruthenium complex **6**.

An additional feature of this DKR is the use of isopropenyl acetate, which is readily available, more active than PCPA and easily separable from the DKR mixtures [26]. Although the mechanism of the catalytic racemization is not clear yet, according to our interpretation, it can be deduced that the amino group in **6** seems to play a crucial role in the racemization.

The high activity of **6** at room temperature allowed us for the first time to combine it with thermally weak subtilisin for the (*S*)-selective DKR. A commercially available form of subtilisin is not practical due to its low activity and instability. However, we succeeded in enhancing its activity and stability by treating it with a surfactant before use. Room temperature DKRs with subtilisin and ruthenium catalyst **6** were performed in the presence of trifluoroethyl butanoate as an acylating agent, and the (*S*)-products were obtained in good yields and high optical purities (Scheme 1.20) [27].

### 1.3.8
### DKR of Secondary Alcohols with Racemization Catalyst 7

Bäckvall *et al.* found that **7** also catalyzes efficiently the racemization of alcohols at room temperature after being activated by potassium *tert*-butoxide [28a]. The racemization of 1-phenylethanol was completed within 10 min by using 0.5 mol%

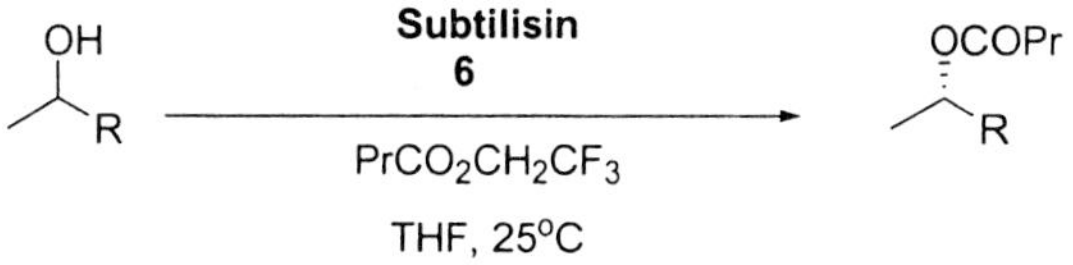

OH, R → (Subtilisin 6; $PrCO_2CH_2CF_3$, THF, 25°C) → OCOPr, R

| R | ee (%) | yield (%) | R | ee (%) | yield (%) |
|---|---|---|---|---|---|
| Ph | 92 | 95 | $CH_2Ph$ | 92 | 77 |
| Ph-4-Cl | 99 | 92 | $CH_2CH_2Ph$ | 98 | 80 |
| Ph-4-MeO | 94 | 93 | $CH_2(CH_2)_4CH_3$ | 98 | 77 |
| $c$-$C_6H_{11}$ | 98 | 80 | CH=CHPh | 95 | 90 |

**Scheme 1.20** (*S*)-Selective DKR by the subtilisin–ruthenium combination.

of **7** at room temperature. They succeeded in alcohol DKR by using **7**, CALB and isopropenyl acetate as acyl donor at room temperature [28b, 28c]. Under optimized conditions the DKR of benzylic alcohols was about one order of magnitude faster than the previous DKR with **6**. For example, the DKR of 1-phenylethanol was completed in 3 h (Scheme 1.21). This catalyst system was also effective for the DKR of aliphatic alcohols. The synthesis of (*S*)-esters was performed by combining **7** and surfactant-treated subtilisin [29].

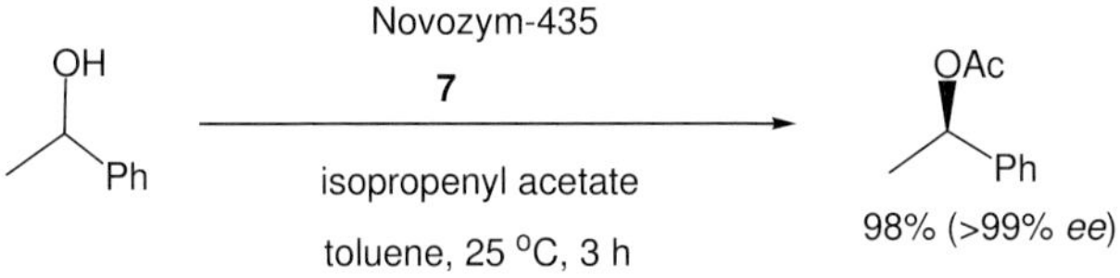

OH, Ph → (Novozym-435, **7**; isopropenyl acetate, toluene, 25 °C, 3 h) → OAc, Ph; 98% (>99% *ee*)

**Scheme 1.21** DKR with racemization catalyst **7**.

## 1.3.9 DKR of Secondary Alcohols with Air-Stable Racemization Catalysts

We synthesized **8** by the one-step reaction of $[Ph_4(\eta^4\text{-}C_4CO)]Ru(CO)_3$ with benzyl chloride. In contrast to previous alcohol racemization catalysts, **8** was stable in the air during racemization [30]. The racemization was performed even under 1 atm of molecular oxygen. Thus, alcohol DKR was for the first time possible with **8** in the air at room temperature; (*R*)-1-phenylethyl acetate (99% yield, greater than 99% e.e.) was obtained from 1-phenylethanol by using 4 mol% of **8**, CALB and isopropenyl acetate in the presence of potassium phosphate (Scheme 1.22). This catalyst system was effective for both benzylic and aliphatic alcohols. The synthetic method for **8** was applied to the preparation of a polymer-bound derivative (**9**). Hydroxymethyl polystyrene was reacted with 4-(chloromethyl)benzoyl chloride to

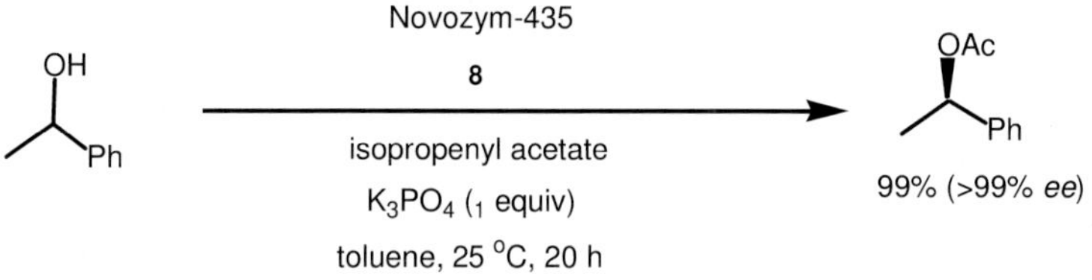

Novozym-435
**8**
isopropenyl acetate
$K_3PO_4$ (1 equiv)
toluene, 25 °C, 20 h
99% (>99% *ee*)

**Scheme 1.22** DKR with racemization catalyst **8**.

attach chlorobenzyl groups. Heating a mixture of the resulting polymer and $[Ph_4(\eta^4\text{-}C_4CO)]Ru(CO)_3$ gave the polymer **9**, which are readily recyclable for practical DKR.

### 1.3.10 DKR of Secondary Alcohols with Racemization Catalyst 10

Hulshof *et al.* introduced **10** as an alcohol racemization catalyst [31]. Alcohol DKR was performed with 0.1 mol% of **10**, CALB, isopropyl butyrate as the acyl donor, potassium carbonate and about 20 mol% of the corresponding ketone at 70 °C (Scheme 1.23). Without the ketone, yield and optical purity of the product ester were decreased significantly. 2-Propanol produced by the acyl transfer reaction was removed at reduced pressure during the DKR to shift the equilibrium to acylated products.

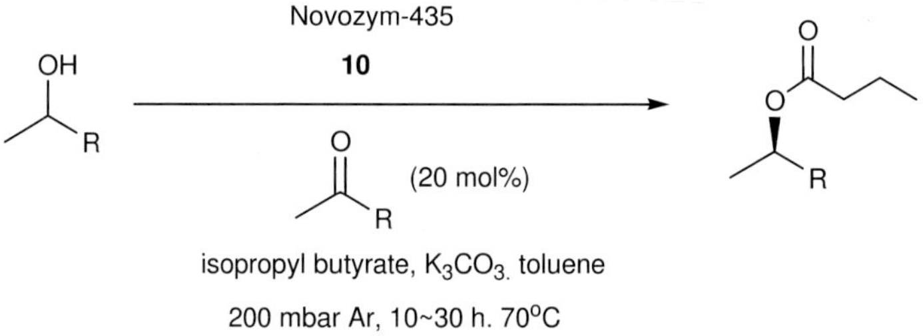

Novozym-435
**10**
(20 mol%)
isopropyl butyrate, $K_3CO_3$, toluene
200 mbar Ar, 10~30 h. 70 °C

**Scheme 1.23** DKR with racemization catalyst **10**.

### 1.3.11 DKR of Secondary Alcohols with Aluminum Catalysts

Based on the catalytic activity of aluminum alkoxides in the Meerwein–Ponndorf–Verley-Oppenauer reaction, Berkessel *et al.* envisioned that aluminum complexes can act as alcohol racemization catalysts [32]. Aluminum alkoxide complexes generated from a 1 : 1 mixture of $AlMe_3$ and a bidentate ligand such as binol or 2,2′-biphenol were effective catalysts for alcohol racemization. At room temperature, 10 mol% of the aluminum catalyst racemized 1-phenylethanol completely within 3 h in the presence of 0.5 equiv. of acetophenone. The aluminum catalysts were applicable for the DKR of various alcohols. However, a specific acylating reagent was required for each of the resolved alcohols. The enol ester derived from ketone corresponding to each alcohol substrate must be employed as the specific acyl donor. For example, (*R*)-1-phenylethyl acetate (96%e.e.) was obtained in 96% yield from a mixture of 1-phenylethanol and 1-phenylvinyl acetate (1.2 equiv.) by using $AlMe_3$ (10 mol%), binol (10 mol%) and CALB at room temperature (Scheme 1.24).

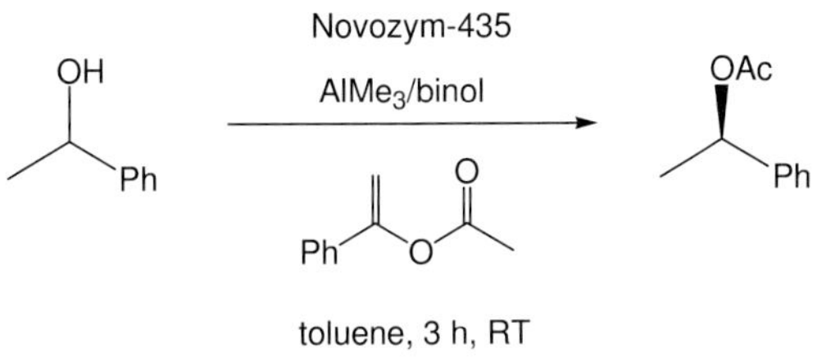

Novozym-435

$AlMe_3$/binol

toluene, 3 h, RT

**Scheme 1.24** Aluminum-catalyzed DKR of 1-phenylethanol.

## 1.3.12 DKR of Secondary Alcohols with Vanadium Catalysts

Oxovanadium(V) complex, $VO(OR)_3$, has been known to catalyze the rearrangement of allylic alcohols through the formation of allyl vanadate intermediates [33] (Scheme 1.25). Akai *et al.* employed vanadium complexes for the DKR of allyic alcohols. They selected $VO(OSiPh_3)_3$ and optimized conditions for the DKR of 1-cyclohexylidenepropan-2-ol. (*R*)-1-Cyclohexylidenepropan-2-yl acetate (98% ee) was obtained in 95% yield by using 10 mol% of $VO(OSiPh_3)_3$, ethoxyvinyl acetate as acyl donor and CALB in acetone at room temperature for 60 h [34] (Scheme 1.26). (*E*)-1-(Prop-1-enyl)cyclohexanol was formed as the byproduct at 5%.

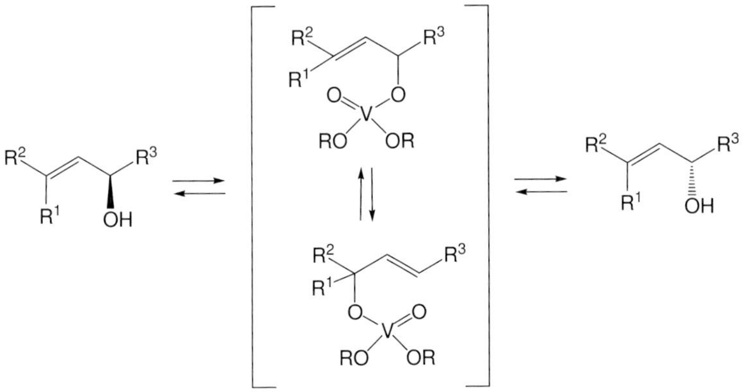

**Scheme 1.25** Vanadium-catalyzed racemization of allylic alcohol.

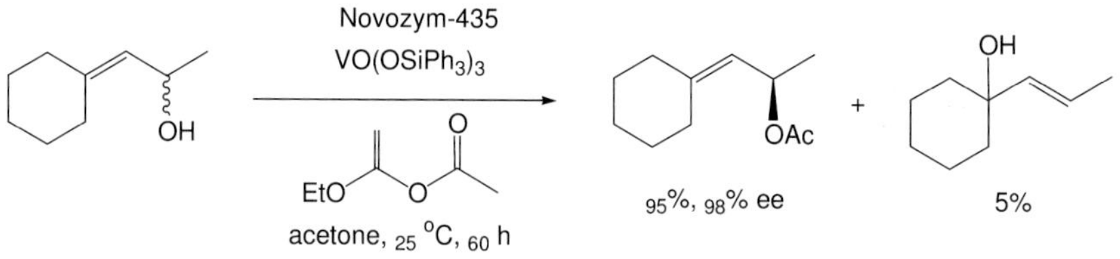

Novozym-435
$VO(OSiPh_3)_3$
OH
EtO O O
acetone, 25 °C, 60 h
OAc
95%, 98% ee
OH
5%

**Scheme 1.26** DKR of allylic alcohols with $VO(OSiPH_3)_3$.

The amount of the byproduct could be minimized to near 0% if more efficient racemization can be achieved.

## 1.4 Conclusions

We have demonstrated that enzyme–metal combinations are of great use as a new class of catalyst systems for the conversion of racemic substrates to single enantiomeric products. A wide range of substrates were converted into their respective enantiopure products under DKR conditions. Two enzyme–metal combination systems, lipase–ruthenium and subtilisin–ruthenium, are particularly useful for the DKR of *sec*-alcohols. They are complementary in stereoselectivity. The former provides (*R*)-products, while the latter gives (*S*)-products. These catalyst systems are readily available or can be prepared without difficulty. The DKR reactions with them can be performed at ambient temperatures, and provide high yields and excellent optical purities in many cases. However, the scope of DKR is limited by the specificity of catalysts, both enzyme and metal. Accordingly, new catalysts should be developed to expand the range of substrates resolved through DKR. These catalysts should be highly active and thermally stable. In the near future, we could see the developments of many different types of enzyme–metal combinations that are applicable for the DKR of substrates other than *sec*-alcohols (Scheme 1.27). Currently, several groups are developing efficient metal catalysts applicable for the DKR of amines and amino acids [35]. We also could see the development of recyclable enzyme–metal catalysts so that DKR can become practically good enough to use in the industrial manufacturing of enantiopure chiral compounds.

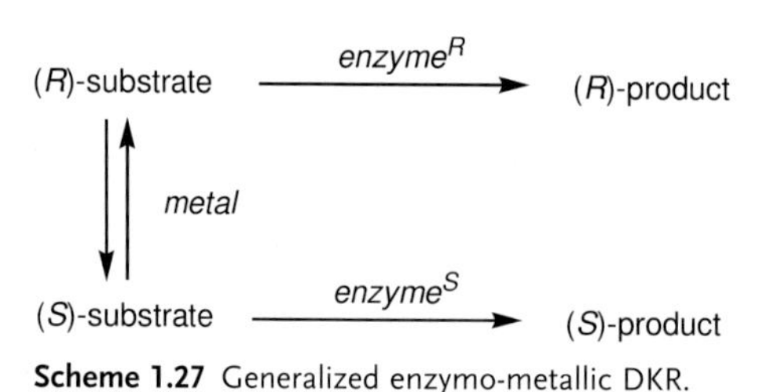


**Scheme 1.27** Generalized enzymo-metallic DKR.

## References


1 Sheldon, R.A. (1993) *Chirotechnology, Industrial Synthesis of Optically Active Compounds*, Dekker, New York.

2 (a) Wong, C.-H. and Whitesides, G.M. (1994) *Enzymes in Synthetic Organic Chemistry*, Pergamon, Oxford.
(b) Koskinen, A.M.P. and Klibanov, A.M. (1996) *Enzymatic Reactions in Organic Media*, Blackie Academic & Professional, Glasgow.
(c) Faber, K. (1997) *Biotransformations in Organic Chemistry*, 3rd edn, Springer, Berlin.
(d) Bornscheuer, U.T. and Kazlauskas, R.J. (1999) *Hydrolases in Organic Synthesis*, Wiley-VCH, Weinheim.
(e) Deauz, K. and Waldmann, H. (2002) *Enzyme Catalysis in Organic Synthesis: A Comprehensive Handbook*, 2nd edn, Vols I–III. Wiley-VCH, Weinheim.

3 (a) Kim, M.-J., Choi, G.-B. and Kim, H.-J. (1995) *Tetrahedron Letters*, **36**, 6253.
(b) Kim, M.-J. and Lim, I.-T. (1996) *Synlett*, **2**, 138.
(c) Kim, M.-J., Lim, I.-T., Choi, G.-B., Whang, S.-Y., Ku, B.-C. and Choi, J.-Y. (1996) *Bioorganic and Medicinal Chemistry Letters*, **6**, 71.
(d) Kim, M.-J., Lim, I.-T., Kim, H.-J. and Wong, C.-H. (1997) *Tetrahedron: Asymmetry*, **8**, 1507.
(e) Lee, D. and Kim, M.-J. (1998) *Tetrahedron Letters*, **39**, 2163.
(f) Chung, S.K., Chang, Y.-T., Lee, E.J., Shin, B.-G., Kwon, Y.-U., Kim, K.-C., Lee, D. and Kim, M.-J. (1998) *Bioorganic and Medicinal Chemistry Letters*, **8**, 1503.
(g) Lee, D. and Kim, M.-J. (1999) *Tetrahedron Letters*, **39**, 9039.
(h) Lee, D. and Kim, M.-J. (1999) *Organic Letters*, **1**, 925.
(i) Im, A.S., Cheong, C.S. and Lee, S.H. (2003) *Bulletin of the Korean Chemical Society*, **24**, 1269.
(j) Kang, H.-Y., Ji, Y., Yu, Y.-K., Yu, J.-Y., Lee, Y. and Lee, S.-J. (2003) *Bulletin of the Korean Chemical Society*, **24**, 1819.

4 Ward, R.S. (1995) *Tetrahedron: Asymmetry*, **6**, 1475.

5 (a) Stürmer, R. (1997) *Angewandte Chemie (International Edition in English)*, **36**, 1173.
(b) Azerad, R. and Buisson, D. (2000) *Current Opinion in Biotechnology*, **11**, 565.
(c) Huerta, F.F., Minidis, A.B.E. and Bäckvall, J.-E. (2001) *Chemical Society Reviews*, **30**, 321.
(d) Kim, M.-J., Ahn, Y. and Park, J. (2002) *Current Opinion in Biotechnology*, **13**, 578.
(e) Pellissier, H. (2004) *Tetrahedron*, **59**, 8291.
(f) Pámies, O. and Bäckvall, J.-E. (2003) *Chemical Society Reviews*, **103**, 3247.
(g) Pámies, O. and Bäckvall, J.-E. (2003) *Current Opinion in Biotechnology*, **14**, 407.
(h) Turner, N.J. (2004) *Current Opinion in Biotechnology*, **8**, 114.

6 (a) Kazlauskas, R.J., Weissfloch, A.N.E., Rappaport, A.T. and Cuccia, L.A. (1991) *Journal of Organic Chemistry*, **56**, 2656.
(b) Kazlauskas, R.J. and Weissfloch, A.N.E. (1997) *Journal of Molecular Catalysis B: Enzymatic*, **3**, 65.

7 Dinh, P.M., Howarth, J.A., Hudnott, A.R., Williams, J.M.J. and Harries, W. (1996) *Tetrahedron Letters*, **37**, 7623.

8 (a) Larsson, A.L.E., Persson, B.A. and Bäckvall, J.-E. (1997) *Angewandte Chemie (International Edition in English)*, **36**, 1211.
(b) Persson, B.A., Larsson, A.L.E., Ray, M. L. and Bäckvall, J.-E. (1999) *Journal of the American Chemical Society*, **121**, 1645.
(c) Lee, H.K. and Ahn, Y. (2004) *Bulletin of the Korean Chemical Society*, **25**, 1471.

9 (a) Persson, B.A., Huerta, F.F. and Bäckvall, J.-E. (1999) *Journal of Organic Chemistry*, **64**, 5237.
(b) Kim, M.-J., Choi, Y.K., Choi, M.Y., Kim, M.J. and Park, J. (2001) *Journal of Organic Chemistry*, **66**, 4736.
(c) Edin, M. and Bäckvall, J.-E. (2003) *Journal of Organic Chemistry*, **68**, 2216.
(d) Martin-Matute, B. and Bäckvall, J.-E. (2004) *Journal of Organic Chemistry*, **69**, 9191.
(e) Fransson, A.-B.L., Xu, Y., Leijondahl, K. and Bäckvall, J.-E. (2006) *Journal of Organic Chemistry*, **71**, 6309.

10 (a) Huerta, F.F., Laxmi, S.Y.R. and Bäckvall, J.-E. (2000) *Organic Letters*, **2**, 1037.

(b) Huerta, F.F. and Bäckvall, J.-E. (2001) *Organic Letters*, **3**, 1209.
(c) Runmo, A.B.L., Pámies, O., Faber, K. and Bäckvall, J.-E. (2002) *Tetrahedron Letters*, **43**, 2983.
(d) Pámies, O. and Bäckvall, J.-E. (2002) *Journal of Organic Chemistry*, **67**, 1261.

**11** Pámies, O. and Bäckvall, J.-E. (2001) *Journal of Organic Chemistry*, **66**, 4022.

**12** Pámies, O. and Bäckvall, J.-E. (2001) *Advanced Synthesis and Catalysis*, **343**, 726.

**13** Pámies, O. and Bäckvall, J.-E. (2002) *Journal of Organic Chemistry*, **67**, 9006.

**14** Pámies, O. and Bäckvall, J.-E. (2003) *Journal of Organic Chemistry*, **68**, 4815.

**15** (a) Jung, H.M., Koh, J.H., Kim, M.-J. and Park, J. (2000) *Organic Letters*, **2**, 409.
(b) Jung, H.M., Koh, J.H., Kim, M.-J. and Park, J. (2000) *Organic Letters*, **2**, 2487.

**16** Kim, M.-J., Choi, M.Y., Han, M.Y., Choi, Y.K., Lee, J.K. and Park, J. (2002) *Journal of Organic Chemistry*, **67**, 9481.

**17** Koh, J.H., Jeong, H.M., Kim, M.-J. and Park, J. (1999) *Tetrahedron Letters*, **40**, 6281.

**18** Lee, D., Huh, E.A., Kim, M.-J., Jung, H.M., Koh, J.H. and Park, J. (2000) *Organic Letters*, **2**, 2377.

**19** (a) Kim, K.W., Song, B., Choi, M.Y. and Kim, M.-J. (2001) *Organic Letters*, **3**, 1507.
(b) Lee, J.K. and Kim, M.-J. (2002) *Journal of Organic Chemistry*, **67**, 6845.
(c) Kim, M.-J., Choi, M.Y., Lee, J.K. and Ahn, Y. (2003) *Journal of Molecular Catalysis B: Enzymatic*, **26**, 115.
(d) Erbeldinger, M., Mesiano, A.J. and Russel, A. (2000) *Biotechnology Progress*, **16**, 1129.
(e) Lau, R.M., Rantwijk, F., Seddon, K.R. and Sheldon, R.A. (2000) *Organic Letters*, **2**, 4189.
(f) Itoh, T., Akasaki, E., Kudo, K. and Shirakami, S. (2001) *Chemistry Letters*, 262.
(g) Schoefer, S.H., Kraftzik, N., Wasserscheid, P. and Kragl, U. (2001) *Chemical Communications*, 425.
(h) Park, S. and Kazlauskas, R. (2001) *Journal of Organic Chemistry*, **66**, 8395.

**20** Kim, M.-J., Kim, D., Kim, H.M., Ahn, Y. and Park, J. (2004) *Green Chemistry*, **6**, 471.

**21** Masutani, K., Uchida, T., Irie, R. and Katsuki, T. (2000) *Tetrahedron Letters*, **41**, 5119.

**22** Dijksman, A., Elzinga, J.M., Li, Y.X., Arends, W.C.E. and Sheldon, R.A. (2002) *Tetrahedron: Asymmetry*, **13**, 879.

**23** van As, B.A.C., van Buijtenen, J., Heise, A., Broxterman, Q.B., Verzijl, G.K.M., Palmans, A.R.A. and Meijer, E.W. (2005) *Journal of the American Chemical Society*, **127**, 9964.

**24** Hilker, I., Rabani, G., Verzijl, G.K.M., Palmans, A.R.A. and Heise, A. (2006) *Angewandte Chemie (International Edition in English)*, **45**, 2130.

**25** Choi, J.H., Kim, Y.H., Nam, S.H., Shin, S.T., Kim, M.-J. and Park, J. (2002) *Angewandte Chemie (International Edition in English)*, **41**, 2373.

**26** Choi, J.H., Choi, Y.K., Kim, Y.H., Park, E.S., Kim, E.J., Kim, M.-J. and Park, J. (2004) *Journal of Organic Chemistry*, **69**, 1972.

**27** Kim, M.-J., Chung, Y.I., Choi, Y.K., Lee, H.K., Kim, D. and Park, J. (2003) *Journal of the American Chemical Society*, **125**, 11494.

**28** (a) Csjernyik, G., Bogár, K. and Bäckvall, J.-E. (2004) *Tetrahedron Letters*, **45**, 6799.
(b) Martin-Matute, B., Edin, M., Bogár, K. and Bäckvall, J.-E. (2004) *Angewandte Chemie (International Edition in English)*, **43**, 6535.
(c) Martin-Matute, B., Edin, M., Bogár, K., Kaynak, F.B. and Bäckvall, J.-E. (2005) *Journal of the American Chemical Society*, **127**, 8817.

**29** Borén, L., Martín-Matute, B., Xu, Y., Córdova, A. and Bäckvall, J.-E. (2006) *Chemistry–A European Journal*, **12**, 225.

**30** Kim, N., Ko, S.-B., Kwon, M.S., Kim, M.-J. and Park, J. (2005) *Organic Letters*, **7**, 4523.

**31** van Nispen, S.F.G., van Buijtenen, J., Vekemans, J.A.J., Meuldijk, J. and Hulshof, L.A. (2006) *Tetrahedron: Asymmetry*, **17**, 2299.

**32** Berkessel, A., Sebastian-Ibarz, M.L. and Müller, T.N. (2006) *Angewandte Chemie (International Edition in English)*, **45**, 6567.

**33** Chabardes, P., Kuntz, E. and Varagnat, J. (1997) *Tetrahedron*, **33**, 1775.

**34** Akai, S., Tanimoto, K., Kanao, Y., Egi, M., Yamamoto, T. and Kita, Y. (2006) *Angewandte Chemie (International Edition in English)*, **45**, 2592.

**35** (a) Reetz, M.T. and Schimossek, K. (1996) *Chimia*, **50**, 668.
(b) Choi, Y.K., Kim, M., Ahn, Y. and Kim, M.-J. (2001) *Organic Letters*, **3**, 4099.
(c) Pámies, O., Ell, A.H., Samec, J.S.M., Hermanns, N. and Bäckvall, J.-E. (2002) *Tetrahedron Letters*, **43**, 4699.
(d) Paetzold, J. and Bäckvall, J.-E. (2005) *Journal of the American Chemical Society*, **127**, 17620.
(e) Parvulescu, A., Vos, D.D. and Jacobs, P. (2005) *Chemical Communications*, 5307.
(f) Kim, M.-J., Kim, W.H., Han, K.-W., Choi, Y.K. and Park, J. (2007) *Organic Letters*, **9**, 1157.
(g) Parvulescu, A., Jacobs, P. and Vos, D.D. (2007) *Chemistry–A European Journal*, **13**, 2034.

# 2
# Chemoenzymatic Routes to Enantiomerically Pure Amino Acids and Amines

*Nicholas J. Turner*

## 2.1
## Introduction

This chapter describes various chemoenzymatic strategies that have been developed for the preparation of chiral amino acids and amines in non-racemic form. Although a wide range of different technologies have been developed for the commercial production of amino acids [1], this class of molecule continues to remain a highly attractive target for biocatalysis [2]. Part of this reason derives from the fact that a large number of different enzymes [e.g. ammonia lyases, amino acid dehydrogenases, amino acid oxidases (AAOs), etc.] are able to produce amino acids directly as substrates from inexpensive starting materials. In addition, a second set of enzymes (e.g. acylases, proteases, esterases, nitrilases, etc.) catalyze transformations on functional groups that are either derived from, or lead to production of, an amino or carboxyl group (e.g. conversion of a nitrile group to a carboxylic acid using a nitrile). Such diversity offers many opportunities for developing biocatalytic routes, sometimes involving more than one enzyme. Furthermore, a wide range of structurally different unnatural amino acids are currently in demand as components of pharmaceuticals that are in development to treat a range of different therapeutic conditions. A prime example is provided by the new generation of hepatitis C virus (HCV) protease inhibitors that are in clinical development, particularly telaprevir (Vertex) and SCH 503034 (Schering-Plough) which contain no less than four and three different non-natural α-amino acids, respectively, including the unusual bicyclic proline analogs highlighted in red (Figure 2.1).

Similarly, chiral amines also feature increasingly prominently as intermediates in new pharmaceuticals (Figure 2.2). Compared to amino acids, amines offer a particular challenge in view of the paucity of different available chiral technologies that have been developed to a point where they are of practical utility [3].

*Multi-Step Enzyme Catalysis: Biotransformations and Chemoenzymatic Synthesis*
Edited by Eduardo Garcia-Junceda

ISBN: 978-3-527-31921-3

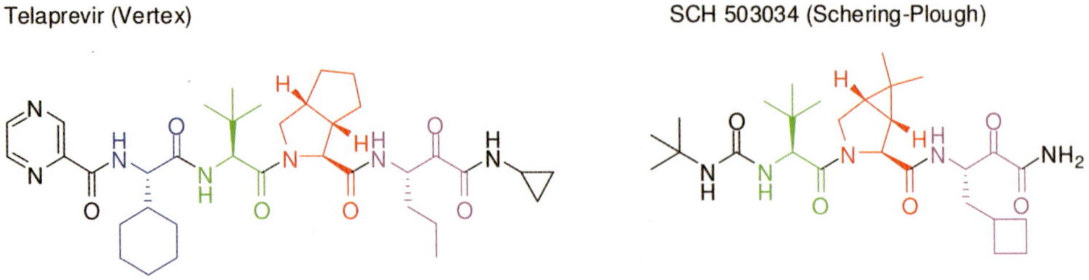

Telaprevir (Vertex)

SCH 503034 (Schering-Plough)

**Figure 2.1** Structure of two different HCV inhibitors highlighting amino acid building blocks.

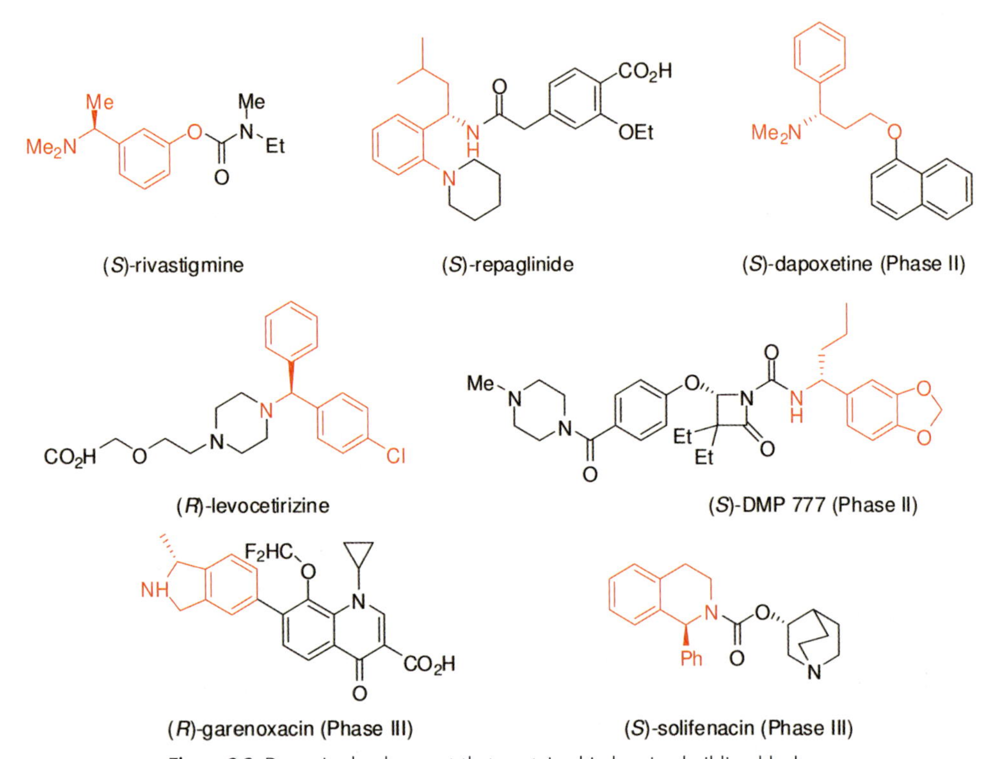


**Figure 2.2** Drugs in development that contain chiral amine building blocks.

This review covers the recent literature (2002–2007) in which either multi-enzyme systems or, alternatively, true chemoenzymatic processes (i.e. those in which an enzyme is combined with a chemical catalysts/reagent in a key step) are employed for the synthesis of chiral amino acids and amines. Not included are those papers in which the term 'chemoenzymatic' refers to the fact that the substrate for the biotransformation has simply been prepared via chemical synthesis.

## 2.2 Amino Acids

One approach to the chemoenzymatic synthesis of α-amino acids which continues to attract attention is the dynamic kinetic resolution (DKR) of a racemic starting material using an enzyme as the enantioselective catalyst. DKR processes are highly attractive in that they are intrinsically more enantioselective (since the reactive enantiomer is constantly being regenerated and presented to the enzyme) and they also offer the possibility of isolating the product in much greater than 50% yield (Scheme 2.1). A number of systems have recently been reported in which the racemization step is effected either by a chemical reagent/catalyst or an additional enzyme (racemase).

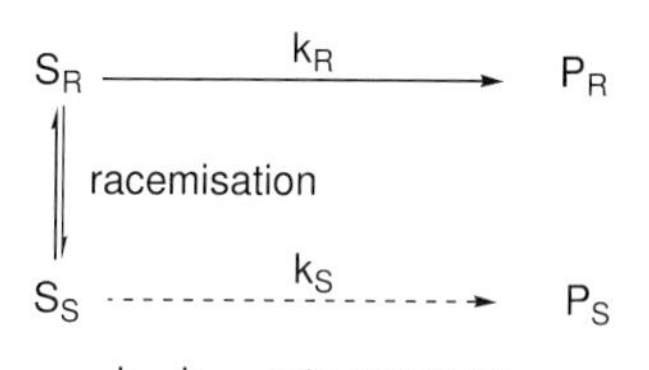


$S_R$, $S_S$ = substrate enantiomers
$P_R$, $P_S$ = product enantiomers
$k_R$»$k_S$ preferably irreversible

**Scheme 2.1** General principle of DKR.

Various strategies were investigated by Kanerva *et al.* for the kinetic resolution of racemic tetrahydroisoquinoline-1-carboxylic acid ethyl ester **1** (Scheme 2.2) [4]. Initial attempts involving the aminolysis of the substrate ester in the presence of *Candida antarctica* lipase B (CALB; trade name, Novozyme-435) were unsuccessful and resulted in formation of hydrolysis product (*R*)-**2** rather than the expected amide. However, racemization of ester **1** was observed under the reaction conditions, raising the prospect of developing a DKR process. Further subsequent studies on the hydrolysis reaction resulted in the development of a successful kinetic resolution, affording both (*R*)-**2** and (*S*)-**1** in 42–46% yield and 92–94% enantiomeric excess (e.e.). Thereafter, a DKR process was developed, using catalytic quantities of Hunig's base (di-iso-propylethylamine) to effect the racemization, delivering (*R*)-**2** in good yield (80%) and high optical purity (96% e.e.).

Thioesters are well known to racemize more easily than their oxygen counterparts, as well as being more labile towards acyl transfer reactions. The L-forms of

NH
$CO_2Et$
*rac*-**1**
*Candida antarctica* lipase B
toluene:acetonitrile (4:1)
*i*-$Pr_2NH$, water
25 °C, 6d
NH
$CO_2H$
(*R*)-**2**
80%; 96% e.e.

**Scheme 2.2** DKR of *rac*-**1** using CALB.

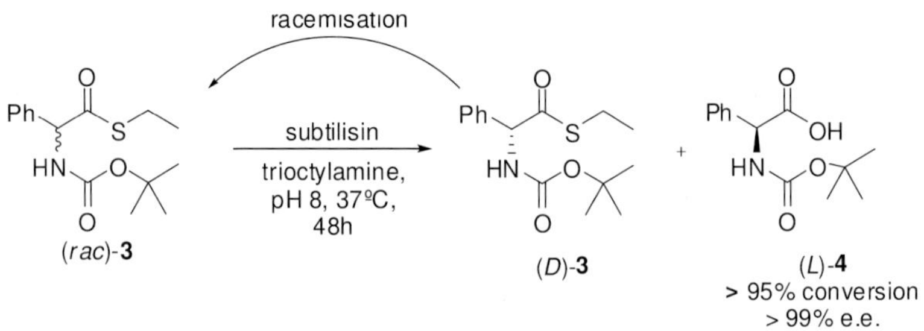

**Scheme 2.3** DKR of thioester **3** using subtilisin as catalyst.

racemic-*N*-protected-β,γ-unsaturated α-amino acid thioesters (e.g. phenylglycine derivative **3**) were shown to be substrates for the subtilisin-catalyzed production of the corresponding acid **4** (Scheme 2.3) [5]. The D-enantiomer (D-**3**) was continuously racemized in the presence of an organic base, thus allowing a DKR to occur with high yields and enantioselectivities.

Another possible mechanism for the racemization of amino acid esters involves the *in situ*, transient, formation of Schiff's bases by reaction of the amine group of an amino acid ester with an aldehyde. Using this approach, DKR of the methyl esters of proline **5** and pipecolic acid **6** was achieved using lipase A from *C. antarctica* as the enantioselective hydrolytic enzyme and acetaldehyde as the racemizing agent (Scheme 2.4). Interestingly, the acetaldehyde was released *in situ* from vinyl butanoate, which acted as the acyl donor, in the presence of triethylamine. The use of other reaction additives was also investigated. Yields of up to 97% and up to 97% e.e. were obtained [6].

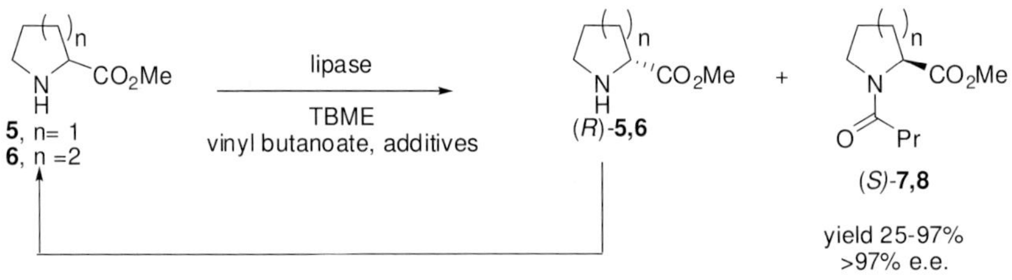

**Scheme 2.4** DKR of proline and pipecolic acid esters.

L-Alanine amide (*S*)-**8** was converted to D-alanine (*R*)-**9** in excellent yield and enantiomeric excess by incubation of the substrate with α-amino-ε-caprolactam racemase from *Achromobacter obae* and D-aminopeptidase from *Ochrobactrum anthropi* (Scheme 2.5) [7].

D-Allo-isoleucine and L-isoleucine derivatives have been prepared from the corresponding mixture of stereoisomers via a diastereoselective hydrolysis reaction catalyzed by the enzyme alcalase (Scheme 2.6). Initially, enantiomerically and diastereomerically pure derivatives of L-isoleucine **10** were submitted to chemical epimerization to yield a 1:1 mixture of stereoisomers at the α-position. Thereafter,

ACL racemase
D-aminopeptidase
pyridoxal phosphate
phosphate buffer
pH 7, 30 °C, 8hr
(*S*)-**8**
(*R*)-**9**
>99.7%
>99.9% e.e.

**Scheme 2.5** Stereoinversion of L-alanine amide **8** to D-alanine **9**.

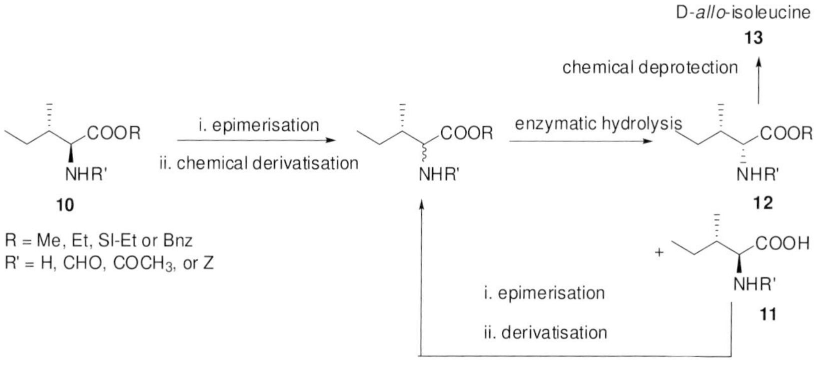


**Scheme 2.6** Preparation of D-allo-isoleucine **13** from L-isoleucine.

treatment with alcalase resulted in hydrolysis of the ester group of L-isoleucine **11** which could then be easily separated from the protected D-allo-isoleucine **12**. The L-isoleucine **11** could be recycled by a subsequent round of epimerization and derivatization [8].

A potential versatile route into α-amino acids and their derivatives is via a combination of (i) nitrile hydratase/amidase-mediated conversion of substituted malononitriles to the corresponding amide/acid followed by (ii) stereospecific Hofmann rearrangement of the amide group to the corresponding amine. Using a series of α,α-disubstituted malononitriles **14**, cyanocarboxamides **15** and bis-carboxamides **16**, the substrate specificity of the nitrile hydratase and amidase from *Rhodococcus rhodochrous* IF015564 was initially examined (Scheme 2.7). The amidase hydrolyzed the diamide **16** to produce (*R*)-**17** with 95% conversion and 98% e.e. Amide **17** was then chemically converted to a precursor of (*S*)-α-methyldopa. It was found

*Rhodococcus rhodochrous* nitrile hydratase
*R. rhodochrous* nitrile hydratase
*R. rhodochrous* amidase
**14** **15** **16** (*R*)-**17**

**Scheme 2.7** Asymmetric hydrolysis of disubstituted malononitrile **14** using a nitrile hydratase.

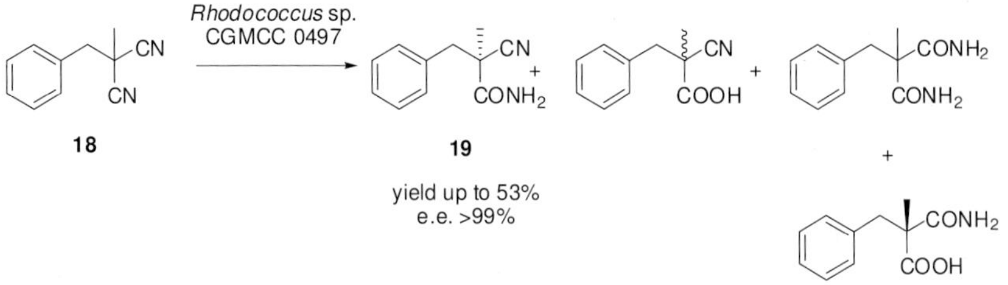

Scheme 2.8 Asymmetric hydrolysis of disubstituted malononitrile 18.

that fluorine substituents at the α-position of the diamides caused inhibition of the amidase enzyme [9].

In a similar vein, Wu *et al.* prepared α?α-disubstituted-α-cyanoacetamides **19** from the corresponding α,α-disubstituted-malononitriles **18** by using whole cells of *Rhodococcus* sp. CGMCC 0497 (Scheme 2.8). The biotransformation proceeded with greater than 99% e.e. and yields up 53% [10].

Racemic α-amino nitriles have also been examined as precursors to α-amino acids via nitrilase-catalyzed enantioselective hydrolysis (Scheme 2.9). A large library of nitrilases was screened for activity against phenylglycinonitrile and 4-fluorophenylglycinonitrile **20**. *In situ* racemization of the starting material **20** occurred at pH 8.5, resulting in a high yield (87%) and high enantiomeric excess (98%) of the product **21** [11].

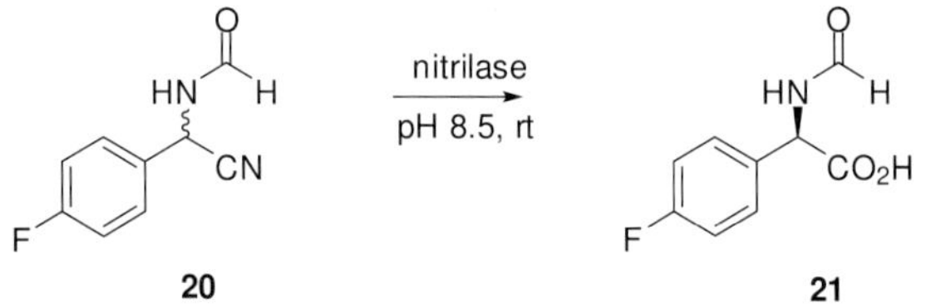

Scheme 2.9 DKR of *N*-formyl amino nitriles.

Hensel *et al.* have also reported the use of α-amino nitriles as substrates (Scheme 2.10). They discovered a number of new bacterial isolates with stereoselective nitrile hydratase activity. A combination of stereoselective nitrile hydratases and amidases was shown to be responsible for the production of phenylglycine **24** from nitrile **22** via amide **23**. By investigating five isolates both (*R*)- and (*S*)-phenylglycine were produced in greater than 99% e.e. [12].

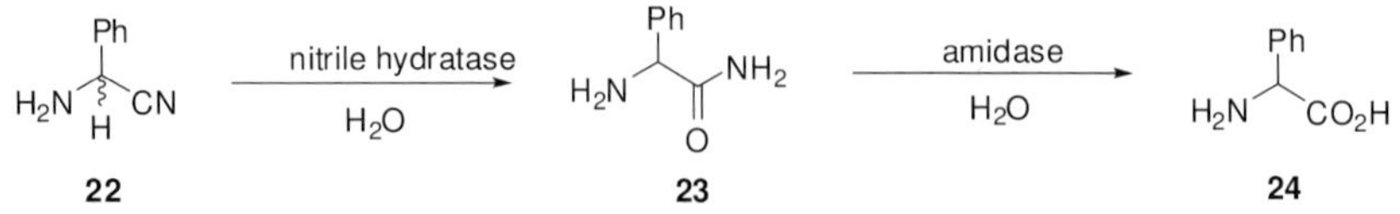

Scheme 2.10 Kinetic resolution of amino nitriles.

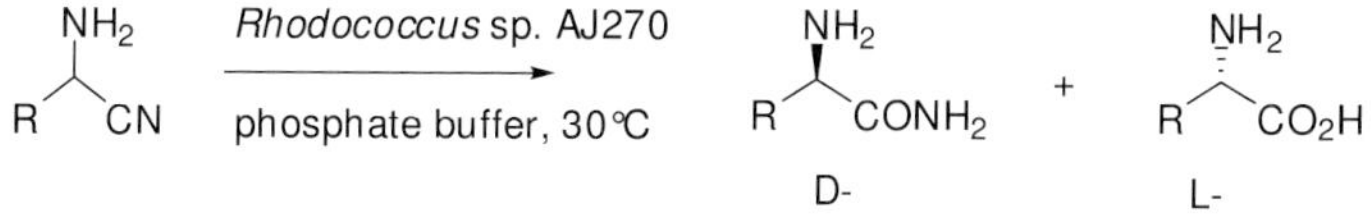

**Scheme 2.11** Synthesis of amino acids from amino nitriles using a nitrile hydratase and amidase.

D-Amino acid amides and L-amino acids were synthesized in generally high yield and high enantiomeric excess utilizing the nitrile hydratase and amidase activities present in *Rhodococcus* sp. AJ270 cells (Scheme 2.11). The amidase was shown to be highly (*S*)-enantioselective and the nitrile hydratase was found to be (*S*)-selective, but to a significantly lower degree. Variations in buffer pH were investigated [13].

A range of 2-aryl-4-pentenenitriles **25** underwent effective hydrolysis with *Rhodococcus* sp. AJ270 microbial cells under mild conditions to afford excellent yields of enantiomerically pure (*R*)-(−)-2-aryl-4-pentenamides **26** and (*S*)-(+)-2-aryl-4-pentenoic acids **27** in most cases (Scheme 2.12). The application of this biotransformation was exemplified by the two-step synthesis of (*R*)-(−)-baclofen **28** [14].

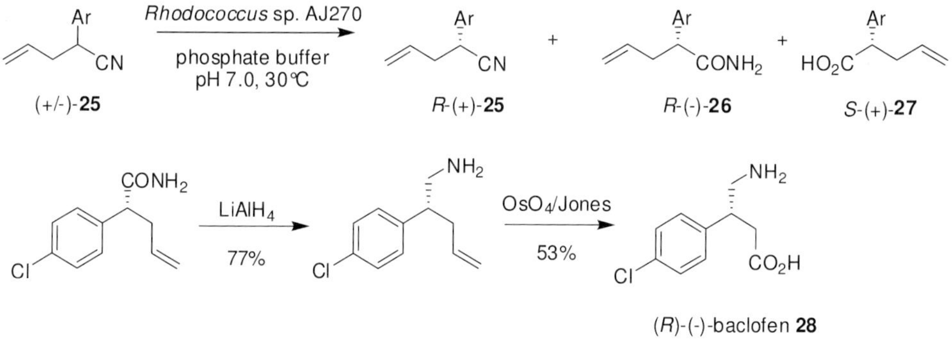

**Scheme 2.12** Synthesis of (*R*)-(−)-baclofen **28**.

Amino acid dehydrogenases catalyze the reductive amination of α-keto acids to yield the corresponding α-amino acids. In order for these reactions to be economically viable it is essential to develop efficient methods for recycling the cofactor. In a paper from the group at Bristol-Myers Squibb, phenylalanine dehydrogenase (PDH) from *Thermoactinomyces intermedius*, expressed in *Escherichia coli* and *Pichia pastoris*, was used for the reduction amination of **29** to the amino acid product **30** (Scheme 2.13). Wet cells, heat-dried cells and cell extracts from *P. pastoris* were examined in the biotransformation of **29**, with the cell extract giving a yield close to 100%. The conditions of expression were investigated, using several constructs, with and without formate dehydrogenase. When the sequence of PDH was compared to the first cloned and sequenced PDH, some amino acid changes and a C-terminal extension were found. Finally, a pilot plant-scale reductive amination of **29** with recombinant *P. pastoris* extract was carried out [15].

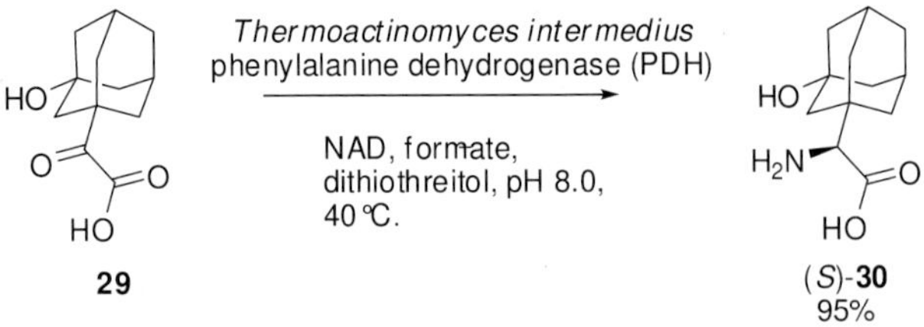

*Thermoactinomyces intermedius*
phenylalanine dehydrogenase (PDH)
NAD, formate,
dithiothreitol, pH 8.0,
40 °C.

**29** → (*S*)-**30** 95%

**Scheme 2.13** Reductive amination of α-keto acid **29** using PDH.

In a related example, the N145A mutant of PDH from *Bacillus sphaericus* was supported on Celite and investigated for the reductive amination of phenylpyruvic acid **31** (Scheme 2.14). The immobilized enzyme and cofactor regeneration system both showed remarkable activity in a wide range of homogeneous and biphasic aqueous/organic solvents, affording (*S*)-phenylalanine **33** with excellent conversion and complete enantioselectivity. The same system gave high yields and enantiomeric excesses of the non-natural amino acid **34** when **32** was used as the substrate [16].

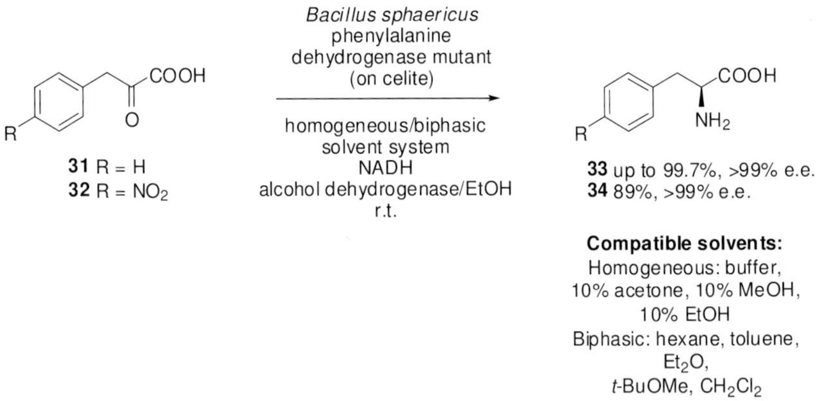

*Bacillus sphaericus*
phenylalanine
dehydrogenase mutant
(on celite)
homogeneous/biphasic
solvent system
NADH
alcohol dehydrogenase/EtOH
r.t.

**31** R = H
**32** R = $NO_2$

**33** up to 99.7%, >99% e.e.
**34** 89%, >99% e.e.

**Compatible solvents:**
Homogeneous: buffer,
10% acetone, 10% MeOH,
10% EtOH
Biphasic: hexane, toluene,
$Et_2O$,
*t*-BuOMe, $CH_2Cl_2$

**Scheme 2.14** Reductive amination of phenylpyruvic acid derivatives under biphasic conditions.

The vast majority of amino acid dehydrogenases use ammonium ions as the amine donor. However, recently a novel *N*-methyl-L-amino acid dehydrogenase (NMAADH), from *Pseudomonas putida*, was isolated and used to synthesize *N*-methyl-L-phenylalanine **36** from phenylpyruvic acid **31** and methylamine **35** in 98% yield and greater than 99% e.e. (Scheme 2.15). The enzyme was shown to accept a number of different ketoacids and also use various amine donors. Glucose dehydrogenase from *Bacillus subtilis* was used to recycle the NADPH cofactor [17].

The same group has exploited this interesting dehydrogenase for the synthesis of cyclic amino acids from linear precursors by developing a one-pot, two-enzyme system (Scheme 2.16). L-Lysine oxidase or L- or D-AAO were initially used to

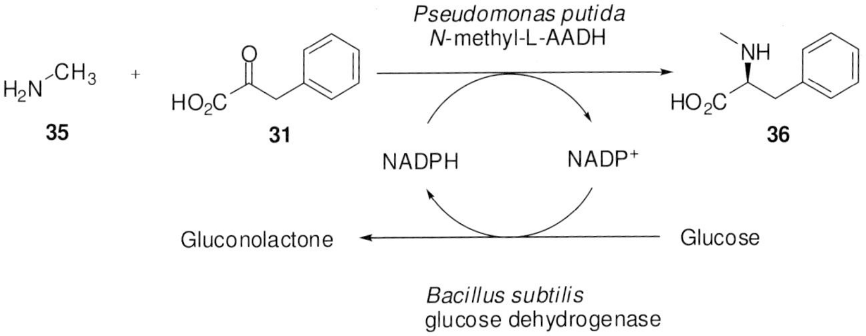

$H_2N-CH_3$ **35** + $HO_2C$ **31** → *Pseudomonas putida* *N*-methyl-L-AADH → $HO_2C$ NH **36**

NADPH NADP+

Gluconolactone ← Glucose

*Bacillus subtilis*
glucose dehydrogenase

**Scheme 2.15** Synthesis of *N*-methyl-L-phenylalanine **36**.

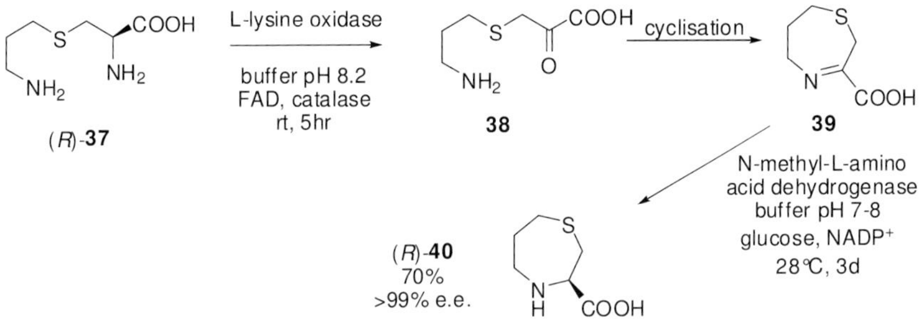

COOH S $NH_2$ $NH_2$ (*R*)-**37** → L-lysine oxidase, buffer pH 8.2, FAD, catalase, rt, 5hr → S COOH $NH_2$ O **38** → cyclisation → S N COOH **39**

N-methyl-L-amino acid dehydrogenase, buffer pH 7-8, glucose, NADP+, 28°C, 3d

(*R*)-**40** 70% >99% e.e. S N H COOH

**Scheme 2.16** Synthesis of cyclic amino acid (*R*)-**40** by combined use of L-lysine oxidase and NMAADH.

convert diamino acids such as (*R*)-**37** to the corresponding α-keto amino acid **38**, which subsequently cyclized spontaneously to the imino acid **39**. Addition of NMAADH then afforded the cyclic amino acid **40** in good yield and excellent optical purity. A deracemization was also accomplished using racemic analogs of **4** in conjunction with D-AAO and NMAADH [18].

In a different, but related, approach a lysine cyclodeaminase gene was identified in the rapamycin gene cluster, and subsequently cloned, overexpressed and the enzyme purified (Scheme 2.17). The enzyme catalyzes the conversion of L-lysine

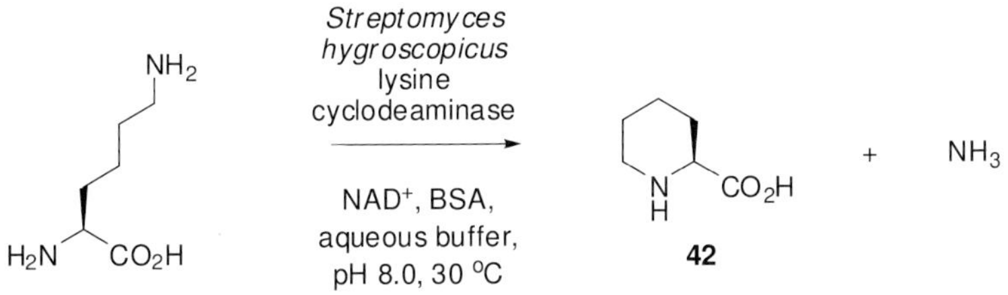

$NH_2$ $H_2N$ $CO_2H$ → *Streptomyces hygroscopicus* lysine cyclodeaminase; $NAD^+$, BSA, aqueous buffer, pH 8.0, 30 °C → N H $CO_2H$ **42** + $NH_3$

**Scheme 2.17** Synthesis of L-pipecolic acid **42** using a lysine cyclodeaminase.

**41** to pipecolic acid **42**, but displayed much lower activity towards L-ornithine and D-lysine was not active. The enzyme mechanism was investigated via study of enzyme cofactor requirements and isotopic substrate labeling [19].

Reduction of β-nitroacrylates **43** by *Saccharomyces carlsbergensis* 'Old Yellow Enzyme' is the key step in the synthesis of optically pure β-2-amino acids (Scheme 2.18). High yields and enantioselectivities of the initially derived nitro-compounds **44** were achieved. The latter can be further chemically elaborated to the corresponding β-amino acids **45** [20].

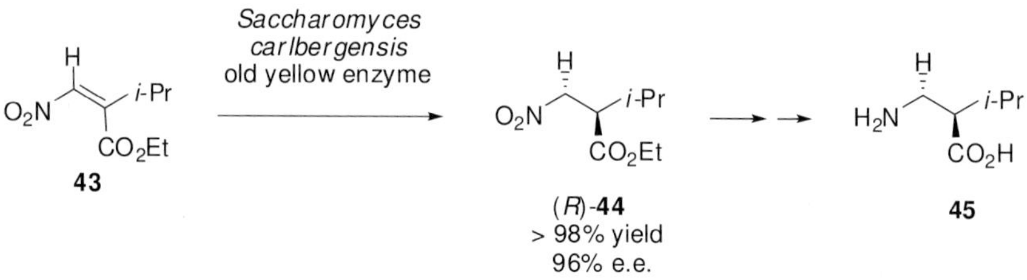

*Saccharomyces carlbergensis* old yellow enzyme

**43** → (*R*)-**44** > 98% yield 96% e.e. → → **45**

**Scheme 2.18** Chemoenzymatic synthesis of β-2-amino acid **45** using 'Old Yellow Enzyme'.

Some researchers have begun to explore the possibility of combining transition metal catalysts with a protein to generate novel synthetic 'chemzymes'. The transition metal can potentially provide access to novel reaction chemistry with the protein providing the asymmetric environment required for stereoselective transformations. In a recent example from Reetz's group, directed evolution techniques were used to improve the enantioselectivity of a biotinylated metal catalyst linked to streptavidin (Scheme 2.19). The Asn49Val mutant of streptavidin was shown to catalyze the enantioselective hydrogenation of α-acetamidoacrylic acid ester **46** with moderate enantiomeric excess [21].

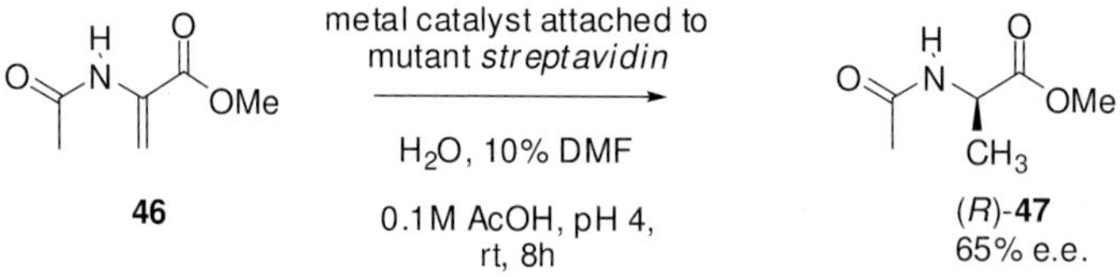

metal catalyst attached to mutant *streptavidin*; $H_2O$, 10% DMF; 0.1 M AcOH, pH 4, rt, 8h

**46** → (*R*)-**47** 65% e.e.

**Scheme 2.19** Asymmetric reduction of alkene **46** using a hybrid transition metal/protein catalyst.

Ammonia lyases catalyze the enantioselective addition of ammonia to an activated double bond. A one-pot, three-step protocol was developed for the enantioselective synthesis of L-arylalanines **50** using phenylalanine ammonia lyase (PAL) in the key step (Scheme 2.20). After formation of the unsaturated esters **48** *in situ* via a Wittig reaction from the corresponding aldehydes, addition of porcine liver esterase and basification of the reaction mixture resulted in hydrolysis to the carboxylic acids **49**. Once this reaction had gone to completion, introduction of PAL and further addition of ammonia generated the amino acids **50** in good yield and excellent optical purity [22].

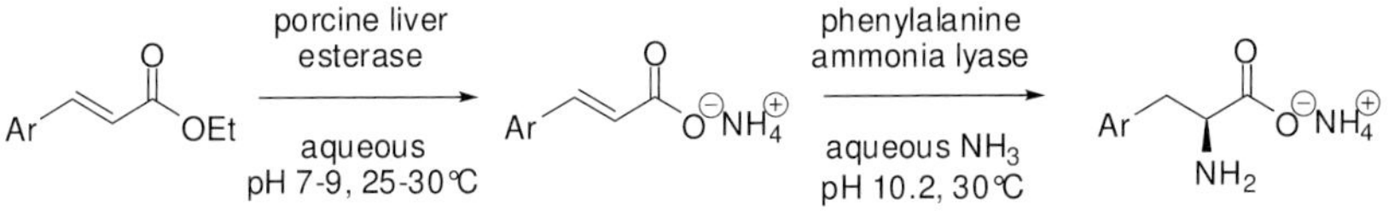

porcine liver esterase, aqueous pH 7-9, 25-30 °C
49 a-d
phenylalanine ammonia lyase, aqueous $NH_3$ pH 10.2, 30 °C

**48a** Ar = Ph
**b** Ar = 4-$ClC_6H_4$
**c** Ar = 3-$FC_6H_4$
**d** Ar = 2-thiophene

**50a** 88%, >98% e.e.
**b** 78%, >98% e.e.
**c** 72%, >98% e.e.
**d** 91%, >98% e.e.

**Scheme 2.20** Synthesis of L-phenylalanine derivatives **50a–d** using a tandem pig liver esterase/PAL combination.

Transition metal catalysts and biocatalysts can be combined in tandem in very effective ways as shown by the following example (Scheme 2.21). An immobilized rhodium complex-catalyzed hydrogenation of **46** was followed by enzymatic hydrolysis of the amide and ester groups of **47** to afford alanine (*S*)-**9** in high conversion and enantiomeric excess. Removal of the hydrogenation catalyst by filtration prior to addition of enzyme led to improved yields when porcine kidney acylase I was used, although the acylase from *Aspergillus melleus* was unaffected by residual catalyst [23].

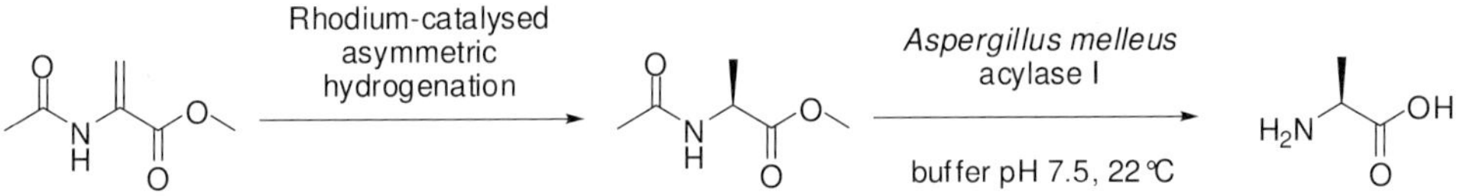

**46** → Rhodium-catalysed asymmetric hydrogenation → (*S*)-**47** → *Aspergillus melleus* acylase I, buffer pH 7.5, 22 °C → (*S*)-**9** 98%, >98% e.e.

**Scheme 2.21** Chemoenzymatic synthesis of L-alanine **9** via a sequential rhodium-catalyzed asymmetric reduction followed by acylase I.

Aminotransferases are useful enzymes for the synthesis and manipulation of amino acids. Young *et al.* investigated the ability of PLP-dependent aminotransferases to catalyze β-substitution of L- and D-β-chloroalanine **51** using β-mercaptoethanol (Scheme 2.22). L-Aspartate aminotransferase and D-aminotransferase catalyzed the substitution of L-**51** and D-**51**, respectively, to **52** with retention of stereochemistry [24].

Turner *et al.* have developed a chemoenzymatic method for the deracemization of α-amino acids in which an enantioselective oxidase enzyme is combined in a one-pot process with a non-selective chemical reducing agent. Since the method is in essence a stereoinversion process it can also be applied to substrates possessing more than one stereogenic center. For example, the interconversion of a number of D- and L-diastereomers of β- and γ- α-amino acids was investigated (Scheme 2.23). Various combinations of L- and D-AAOs with chemical reducing agents (e.g. sodium borohydride) were explored, resulting in generally good yields and high enantiomeric excesses for all the substrates investigated [25].

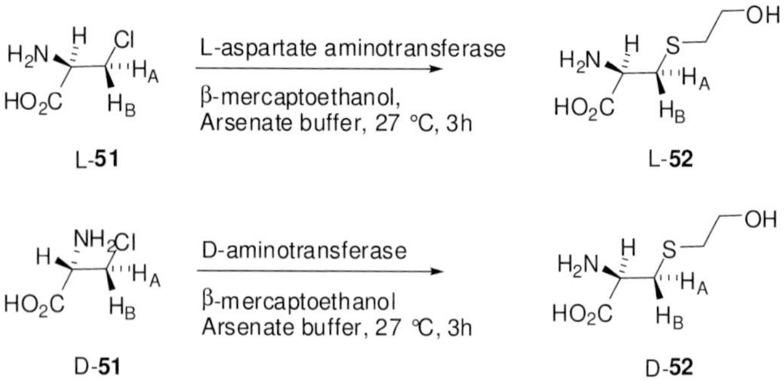

L-aspartate aminotransferase
β-mercaptoethanol,
Arsenate buffer, 27 °C, 3h

L-**51** L-**52**

D-aminotransferase
β-mercaptoethanol
Arsenate buffer, 27 °C, 3h

D-**51** D-**52**

**Scheme 2.22** Aminotransferase-catalyzed β-substitution reactions.

*Trigonopsis variabilis*
D-AAO
$NaBH_4$

R = n-Pr, Ph

**Scheme 2.23** Stereoinversion of β-substituted α-amino acids using a combination of D-AAO and sodium borohydride.

The same group has combined this approach with asymmetric transition metal-catalyzed reduction of dehydroamino acids to furnish a library of enantiomerically pure amino acids. Specifically, a range of β-methyl-β-aryl alanine (2*R*,3*R*) and (2*S*,3*S*)-**54** diastereoisomers were synthesized chemoenzymatically using L-threonine methyl ester as the ultimate starting material (Scheme 2.24). The (2*R*,3*S*) and (2*S*,3*R*) isomers were obtained via asymmetric hydrogenation of **53** followed by hydrolysis of the amide and ester protecting groups. Thereafter, the enantioselective AAOs, in combination with ammonia borane, were used to stereoinvert one of the asymmetric centers to obtain the (2*R*,3*R*) and (2*S*,3*S*) β-substituted alanines with yields ranging from 68 to 92% and greater than 98% e.e. [26].

Finally in this section, a novel approach has been developed for the enzyme-catalyzed synthesis of D-*tert*-leucine **55** from the corresponding racemate (Scheme

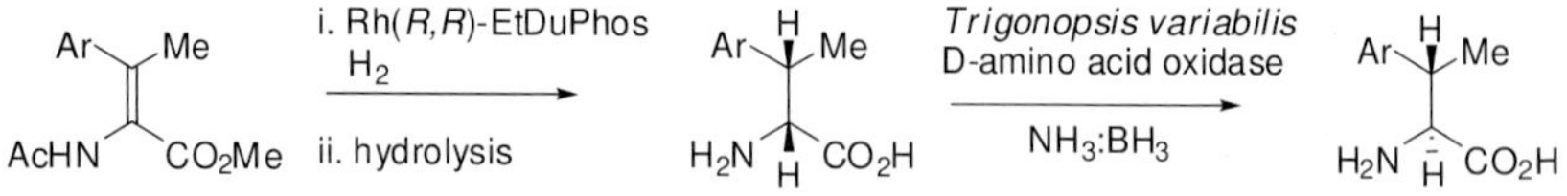

i. Rh(*R,R*)-EtDuPhos $H_2$
ii. hydrolysis

*Trigonopsis variabilis*
D-amino acid oxidase
$NH_3{:}BH_3$

(2*R*,3*S*)-**53** (2*S*, 3*S*)-**54**

**Scheme 2.24** Chemoenzymatic asymmetric hydrogenation and biocatalytic stereoinversion to generate β-substituted phenylalanines **54**.

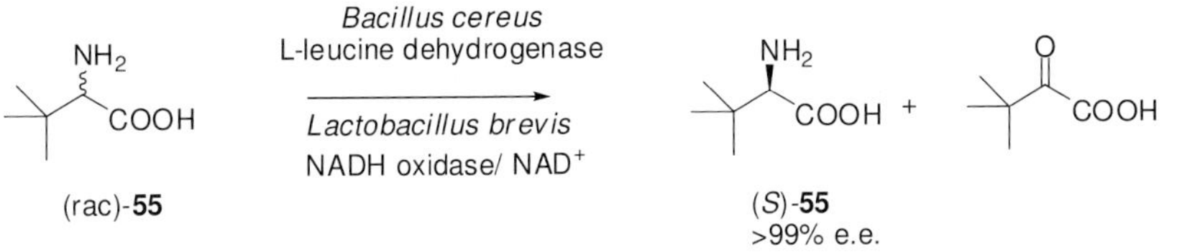

NH2 COOH (rac)-55 — Bacillus cereus L-leucine dehydrogenase; Lactobacillus brevis NADH oxidase/ NAD+ → NH2 COOH (S)-55 >99% e.e. + O COOH

**Scheme 2.25** Kinetic resolution of racemic *tert*-leucine **55** using leucine dehydrogenase.

2.25). The L-enantiomer was enantioselectively oxidized using the enzyme L-leucine dehydrogenase from *Bacillus cereus* in a reaction that involves concomitant reduction of $NAD^+$ to NADH. Reoxidation of NADH was achieved using NADH oxidase, hence ensuring cofactor recycling, resulting in a high enantiomeric excess (greater than 99%) of D-*tert*-leucine [27].

## 2.3 Amines

A number of different groups have recently investigated the DKR of racemic chiral amines using an enantioselective lipase (often CALB) in combination with a chemocatalyst which effects racemization of the unreactive amine enantiomer under the reaction conditions. A key issue with these type of DKR processes is finding conditions under which the bio- and chemocatalysts can function efficiently together. For example, the DKR of a range of amines **56a–k**, combining CALB with a ruthenium catalyst to promote racemization, has been reported (Scheme 2.26). This protocol allows unfunctionalized primary amines **56a–j** to be converted to the corresponding amides **57a–k** in good yields and high enantiomeric purities as shown. The addition of sodium carbonate increases the yield of the reaction

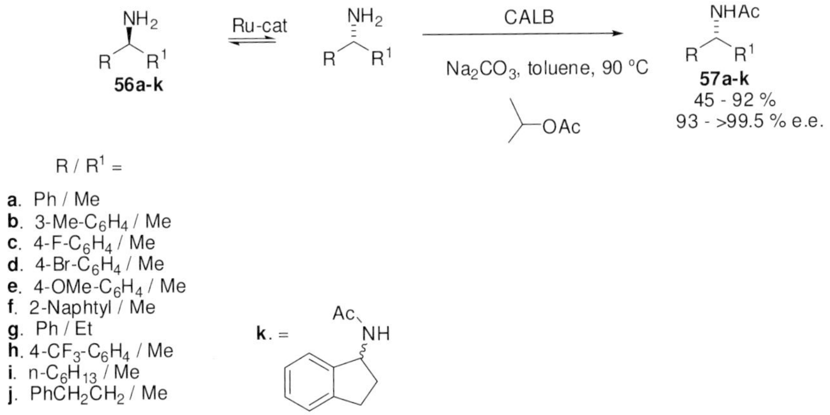

NH2 R R1 **56a-k** ⇌ (Ru-cat) NH2 R R1 → CALB, $Na_2CO_3$, toluene, 90 °C, OAc → NHAc R R1 **57a-k** 45 - 92 % 93 - >99.5 % e.e.

R / R1 =

**a**. Ph / Me
**b**. 3-Me-$C_6H_4$ / Me
**c**. 4-F-$C_6H_4$ / Me
**d**. 4-Br-$C_6H_4$ / Me
**e**. 4-OMe-$C_6H_4$ / Me
**f**. 2-Naphtyl / Me
**g**. Ph / Et
**h**. 4-$CF_3$-$C_6H_4$ / Me
**i**. n-$C_6H_{13}$ / Me
**j**. $PhCH_2CH_2$ / Me
**k**. = Ac NH (indane)

**Scheme 2.26** DKR of chiral amines using CALB and a ruthenium catalyst.

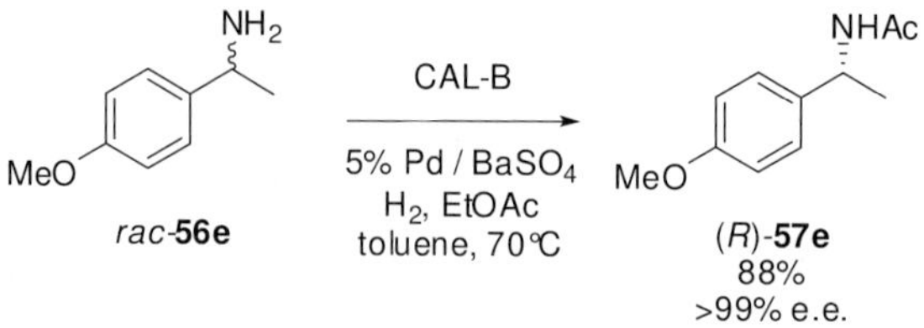

Scheme 2.27 DKR of (±)-**56c** using CALB and a palladium nanocatalyst.

and is believed to act by preventing traces of acid interfering with the activity of the metal catalyst [28].

Similarly, CALB has been used in combination with a palladium/alkaline earth metal-based racemization catalyst to effect a DKR on the benzylic amine **56e** (Scheme 2.27). The (*R*)-amide **57e** was obtained in very good yield and excellent optical purity. Several other substrates also underwent the reaction [29].

Kim *et al.* have developed a practical procedure for the DKR of primary amines illustrated by substrate **56c** (Scheme 2.28). They employed a supported palladium nanocatalyst as the racemization catalyst and commercially available CALB as the enantioselective catalyst for acylation of the amine using ethyl methoxyacetate as the acyl donor. High yields and enantiomeric excesses were achieved [30].

Gastaldi *et al.* discovered that *in situ* racemization of a chiral amine **59** was mediated by the addition of thiyl radicals (Scheme 2.29). Combination with CALB

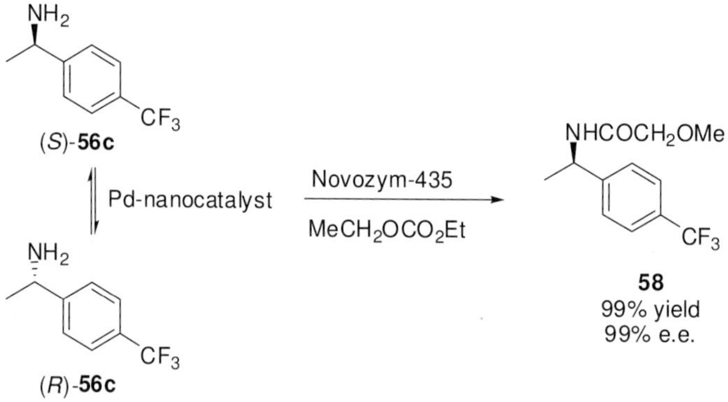

Scheme 2.28 DKR of racemic **56** using CALB and a palladium racemization catalyst.

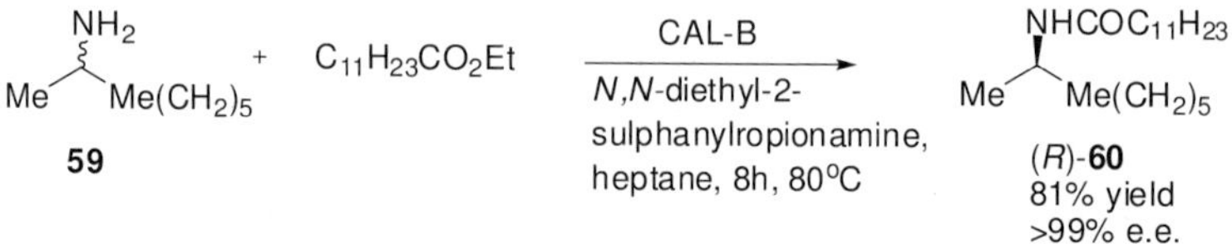

Scheme 2.29 Racemization of amine **59** via addition of thiyl radicals.

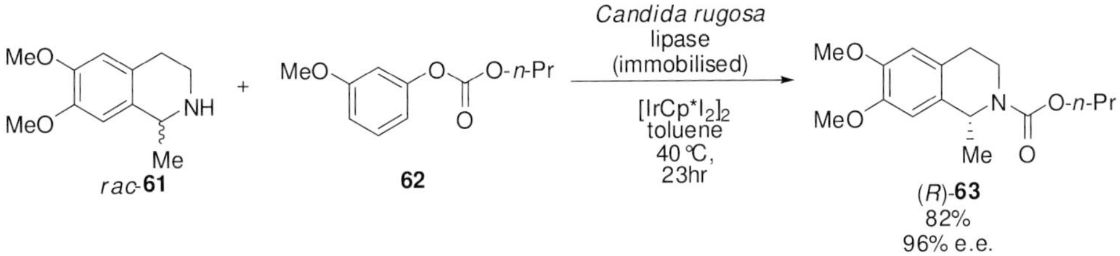

Candida rugosa lipase (immobilised)
$[IrCp^*I_2]_2$ toluene 40 °C, 23hr

rac-**61** + **62** → (R)-**63** 82% 96% e.e.

**Scheme 2.30** DKR of secondary amine **61** using a novel iridium catalyst.

enabled the DKR of non-benzylic amines **59** yielding amides (*R*)-**60** in good yield and high enantioselectivities [31].

Scientists at Huddersfield University in collaboration with Avecia have developed a DKR process involving the combination of immobilized *Candida rugosa* lipase and an iridium-based racemization catalyst (Scheme 2.30). By using carbonate **62** as the acyl donor, the racemic secondary amine **61** was converted to the corresponding carbamate (*R*)-**63** in high yield and enantiomeric excess [32].

Turner *et al.* have extended their method of chemoenzymatic deracemization, using an oxidase enzyme in combination with a chemical reducing agent, to encompass a broad range of chiral amines (Scheme 2.31). Implicit in this work was the need to identify an amine oxidase enzyme that possessed both broad substrate specificity and high enantioselectivity. Starting with the wild-type monoamine oxidase from *Aspergillus niger*, they were able to identify variants that possessed broader substrate specificity and hence could be used for the deracemization and stereoinversion of various chiral primary amines [33].

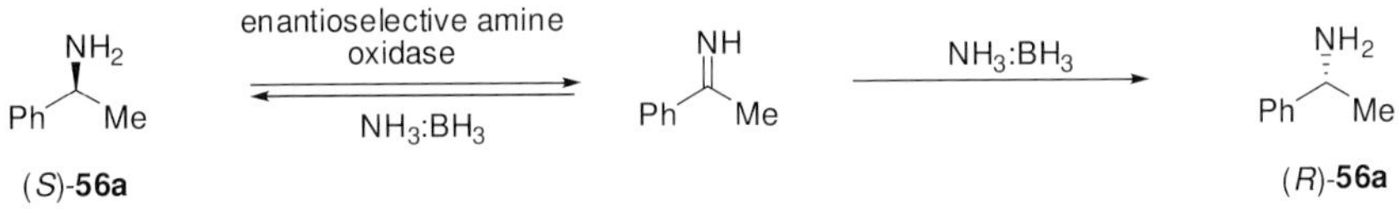

(S)-**56a** ⇄ (enantioselective amine oxidase / $NH_3:BH_3$) imine → ($NH_3:BH_3$) (R)-**56a**

**Scheme 2.31** Deracemization of amines via combined use of an enantioselective amine oxidase and ammonia borane.

In order to extend the approach to include deracemization of chiral secondary amines, this group carried out directed evolution on the monoamine oxidase (MAO) enzyme MAO-N (Scheme 2.32). A new variant was identified with improved catalytic properties towards a cyclic secondary amine **64**, the substrate used in the evolution experiments. This new variant had a single point mutation, Ile246Met, and was found to have improved catalytic properties towards a number of other cyclic secondary amines. The new variant was used in the deracemization of *rac*-**64** yielding (*R*)-**64** in high yield and enantiomeric excess [34].

Recently, Turner *et al.* have shown tertiary amines can also be used as substrates by using a further variant of MAO-N (MAO-N-D5) which had also been developed by directed evolution techniques (Scheme 2.33). For example, racemic *N*-methylpyrrolidine **65** was subjected to deracemization, via the intermediate iminium ion

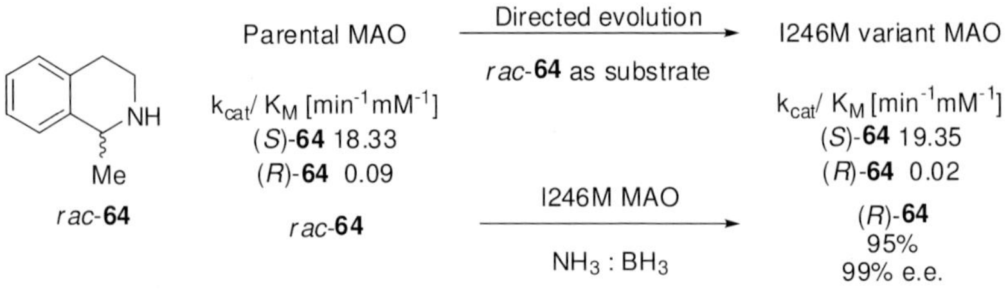

Parental MAO — Directed evolution, *rac*-**64** as substrate → I246M variant MAO

*rac*-**64**

Parental MAO: $k_{cat}/K_M$ [$min^{-1}mM^{-1}$] (*S*)-**64** 18.33; (*R*)-**64** 0.09

I246M variant MAO: $k_{cat}/K_M$ [$min^{-1}mM^{-1}$] (*S*)-**64** 19.35; (*R*)-**64** 0.02

*rac*-**64** → (I246M MAO, $NH_3:BH_3$) → (*R*)-**64**, 95%, 99% e.e.

**Scheme 2.32** Directed evolution to broaden the substrate specificity of MAO-N.

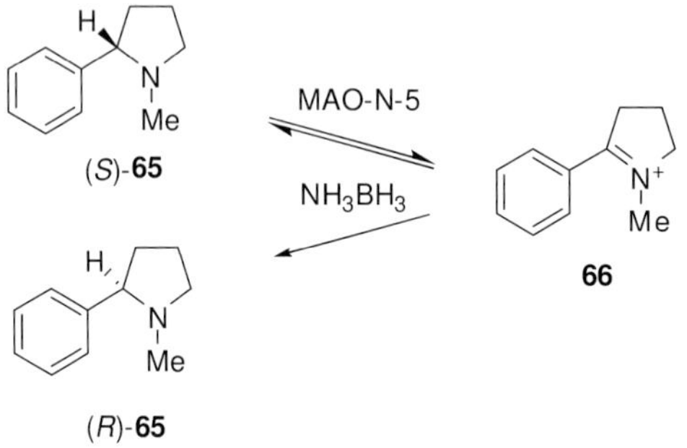

(*S*)-**65** ⇌ (MAO-N-5) **66**; **66** → ($NH_3BH_3$) (*R*)-**65**

**Scheme 2.33** Deracemization of *N*-methylpyrrolidine **65**.

**66**, yielding (*R*)-**65** in 75% isolated yield and 99% e.e. within 24 h. The possibility of using the approach for enantioselective intramolecular reductive amination reactions was also investigated [35].

A two-step approach, involving repeated use of the same enzyme, has been reported for the resolution of *rac*-1-phenylethylamine **56a** (Scheme 2.34). Penicillin acylase, from *Alcaligenes faecalis*, was initially used in aqueous medium with (*R*)-phenylglycine amide **67** as the acyl donor. Under these conditions, the enzyme catalyzed the enantioselective acylation of **56a** at pH 10–11. The product amide **68** was insoluble, and was collected and re-exposed to the enzyme at pH below 7.5. This resulted in the cleavage of the phenylglycinyl substituent. Excellent conversions, *E* values and enantiomeric excesses were achieved [36].

Researchers at the University of Graz, in collaboration with scientists from DSM, have developed an elegant and novel approach to the synthesis of β-amino alcohols using two different enzymes in one pot (Scheme 2.35). For example, a threonine aldolase-catalyzed reaction was initially used, under reversible conditions, to prepare L-**70** from glycine **69** and benzaldehyde **68**. L-**70** was then converted to (*R*)-**71** by an irreversible decarboxylation catalyzed by L-tyrosine decarboxylase. In a second example, D/L-*syn*-**70** was converted to (*R*)-**71** using the two enzymes shown combined with a D-threonine aldolase in greater than 99% e.e. and 67% yield [37, 38].

Kim *et al.* have recently reported [39] the use of an ω-transaminase from *Vibrio fluvialis* as shown in Scheme 2.36. In order to shift the equilibrium towards formation of the product amine they employed the use of excess L-alanine as the amino

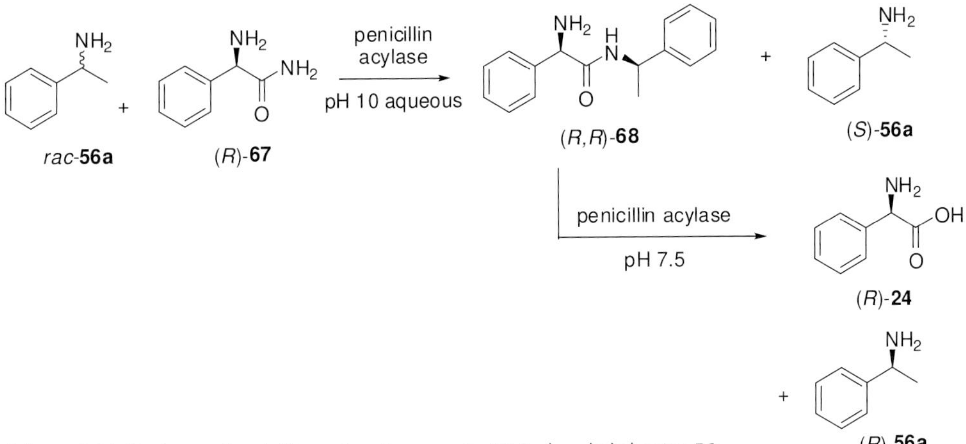

Scheme 2.34 Dual use of penicillin acylase to generate (*R*)-1-phenylethylamine **56a**.

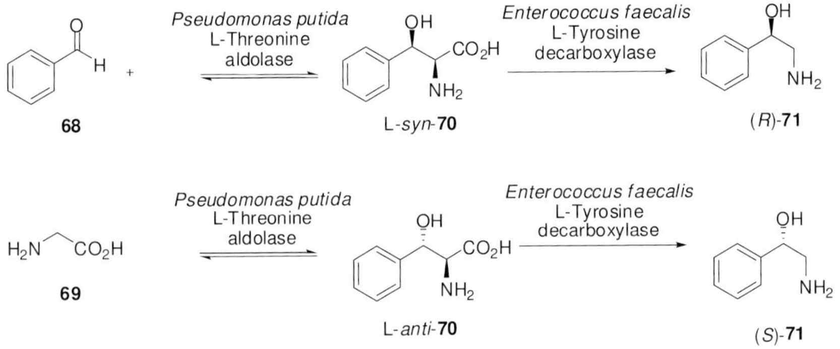

Scheme 2.35 Combined use of threonine aldolase and L-tyrosine decarboxylase.

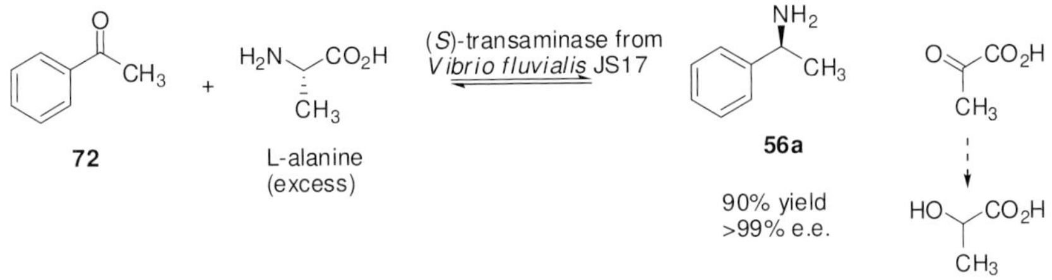

Scheme 2.36 Transaminase from *V. fluvialis*.

donor. However, they also found that the reaction was inhibited by the formation of pyruvate. This problem could be overcome by the addition of lactate dehydrogenase which converted the pyruvate to lactate. By using whole cells and a 1-fold excess of alanine they were able to obtain the product amine **56a** in around 90% yield in less than 24 h.

## References


1 Ager, D. (ed.) (2006) *Handbook of Chiral Chemicals*, 2nd edn, CRC Press/Taylor & Francis, Boca Raton, FL.

2 Groeger, H. and Drauz, K.-H. (2004) *Asymmetric Catalysis on Industrial Scale* (eds H.-U. Blaser and E. Schmidt), Wiley-VCH, Weinheim, pp. 131–47.

3 Turner, N.J. and Carr, R. (2007) *Biocatalysis in the Pharmaceutical and Biotechnology Industries* (ed. R.N. Patel), CRC Press, Boca Raton, FL, pp. 743–55.

4 Paál, T.A., Forró, E., Liljeblad, A., Kanerva, L.T. and Fülöp, F. (2007) *Tetrahedron: Asymmetry*, **18**, 1428.

5 Arosio, D., Caligiuri, A., D'Arrigo, P., Pedrocchi-Fantoni, G., Rossi, C., Saraceno, C., Servi, S. and Tessaro, D. (2007) *Advanced Synthesis and Catalysis*, **349**, 1345.

6 Liljeblad, A., Kiviniemi, A. and Kanerva, L.T. (2004) *Tetrahedron*, **60**, 671.

7 Asano, Y. and Yamaguchi, S. (2005) *Journal of the American Chemical Society*, **127**, 7696.

8 Cambiè, M., D'Arrigo, P., Fasoli, E., Servi, S., Tessaro, D., Canevotti, F. and Corona, L.D. (2003) *Tetrahedron: Asymmetry*, **14**, 3189.

9 Yokoyama, M., Kashiwagi, M., Iwasaki, M., Fuhshuku, K., Ohta, H. and Sugai, T. (2004) *Tetrahedron: Asymmetry*, **15**, 2817.

10 Wu, Z.-L. and Li, Z.-Y. (2003) *Tetrahedron: Asymmetry*, **14**, 2133.

11 Chaplin, J.A., Levin, M.D., Morgan, B., Farid, N., Li, J., Zhu, Z., McQuaid, J., Nicholson, L.W., Rand, C.A. and Burk, M.J. (2004) *Tetrahedron: Asymmetry*, **15**, 2793.

12 Hensel, M., Lutz-Wahl, S. and Fischer, I. (2002) *Tetrahedron Asymmetry*, **13**, 2629.

13 Wang, M.-X. and Lin, S.-J. (2002) *Journal of Organic Chemistry*, **67**, 6542.

14 Wang, M.-X. and Zhao, S.-M. (2002) *Tetrahedron Letters*, **43**, 6617.

15 Hanson, R.L., Goldberg, S.L., Brzozowski, D.B., Tully, T.P., Cazzulino, D., Parker, W.L., Lyngberg, O.K., Vu, T. C., Wong, M.K. and Patel, R.N. (2007) *Advanced Synthesis and Catalysis*, **349**, 1369.

16 Cainelli, G., Engel, P.C., Galletti, P., Giacomini, D., Gualandi, A. and Paradisi, F. (2005) *Organic and Biomolecular Chemistry*, **3**, 4316.

17 Muramatsu, H., Mihara, H., Kakutani, R., Yasuda, M., Ueda, M., Kurihara, T. and Esaki, N. (2004) *Tetrahedron: Asymmetry*, **15**, 2841.

18 Yasuda, M., Ueda, M., Muramatsu, H., Mihara, H. and Esaki, N. (2006) *Tetrahedron: Asymmetry*, **17**, 1775.

19 Gatto, G.J., Boyne, M.T., Kelleher, N.L. and Walsh, C.T. (2006) *Journal of the American Chemical Society*, **128**, 3838.

20 Swiderska, M.A. and Stewart, J.D. (2006) *Organic Letters*, **8**, 6131.

21 Reetz, M.T., Peyralans, J.J.-P., Maichele, A., Fu, Y. and Mywald, M. (2006) *Chemical Communications*, 4318.

22 Paizs, C., Katona, A. and Rétey, J. (2006) *European Journal of Organic Chemistry*, 1113.

23 Simons, C., Hanefeld, U., Maschmeyer, I.W.C.E., Arends, T. and Sheldon, R.A. (2006) *Advanced Synthesis and Catalysis*, **348**, 471.

24 Adams, B., Lowpetch, K., Thorndycroft, F., Whyte, S.M. and Young, D.W. (2005) *Organic and Biomolecular Chemistry*, **3**, 3357.

25 Enright, A., Alexandre, F.-R., Roff, G., Fotheringham, I.G., Dawson, M.J. and Turner, N.J. (2003) *Chemical Communications*, 2636.

26 Roff, G.J., Lloyd, R.C. and Turner, N.J. (2004) *Journal of the American Chemical Society*, **126**, 4098.

27 Hummel, W., Kuzu, M. and Geueke, B. (2003) *Organic Letters*, **5**, 3649.

28 Paetzold, J. and Bäckvall, J.E. (2005) *Journal of the American Chemical Society*, **127**, 17620.

29 Parvulescu, A., De Vos, D. and Jacobs, P. (2005) *Chemical Communications*, 5307.

30 Kim, M.-J., Kim, W.-H., Han, K., Choi, Y. and Park, J. (2007) *Organic Letters*, **9**, 1157.

31 Gastaldi, S., Escoubet, S., Vanthuyne, N., Gil, G. and Bertrand, M.P. (2007) *Organic Letters*, **9**, 837.

**32** (a) Stirling, M., Blacker, J. and Page, M.I. (2007) *Tetrahedron Letters*, **48**, 1247.
(b) Blacker, A.J., Stirling, M.J. and Page, M.I. (2007) *Organic Process Research and Development*, **11**, 642.

**33** Carr, R., Alexeeva, M., Enright, A., Eve, T.S.C., Dawson, M.J. and Turner, N.J. (2003) *Angewandte Chemie (International Edition)*, **42**, 4807.

**34** Carr, R., Alexeeva, M., Dawson, M.J., Gotor-Fernández, V., Humphrey, C.E. and Turner, N.J. (2005) *ChemBioChem*, **6**, 637.

**35** Dunsmore, C.J., Carr, R., Fleming, T. and Turner, N.J. (2006) *Journal of the American Chemical Society*, **128**, 2224.

**36** Guranda, D.T., Khimiuk, A.I., van Langen, L.M., van Rantwijk, F., Sheldon, R.A. and Švedas, V.K. (2004) *Tetrahedron: Asymmetry*, **15**, 2901.

**37** Steinreiber, J., Schürmann, M., Wolberg, M., van Assema, F., Reisinger, C., Fesko, K., Mink, D. and Griengl, H. (2007) *Angewandte Chemie (International Edition)*, **46**, 1624.

**38** Steinreiber, J., Schürmann, M., van Assema, F., Wolberg, M., Fesko, K., Reisinger, C., Mink, D. and Griengl, H. (2007) *Advanced Synthesis and Catalysis*, **349**, 1379.

**39** (a) Hwang, B.-Y., Cho, B.-K., Yun, H., Koteshwar, K. and Kim, B.-G. (2005) *Journal of Molecular Catalysis B: Enzymatic*, **37**, 47.
(b) Shin, J.-S. and Kim, B.-G. (1999) *Biotechnology and Bioengineering*, **65**, 206.

# 3
# Oxidizing Enzymes in Multi-Step Biotransformation Processes

*Stephanie G. Burton and Marilize le Roes-Hill*

## 3.1
## Oxidizing Enzymes in Biocatalysis

Oxidizing enzymes are those enzymes capable of catalyzing electron transfer or hydrogen transfer, resulting in the oxidation of the substrate. For the purposes of this chapter, this group includes oxidases, oxygenases and dehydrogenases, each of which is described briefly below. This chapter addresses issues which pertain specifically to oxidizing enzymes, including the widely varied characteristics of oxidizing enzymes, and the conditions and constraints of their application in multi-enzyme biocatalysis, and describes some examples of processes where this has been achieved. The field of redox enzyme biocatalysis is relatively young and underdeveloped, in comparison with other areas of biocatalysis such as the well-known and fully established hydrolytic systems, and relatively few multi-enzyme systems involving oxidation have been reported in the literature. Of those that have been reported, most have been conducted only at a small scale and have generally used well-known enzyme systems to prove concepts rather than to demonstrate the development of new biocatalysis systems. The characteristic that makes application of redox systems different from, and more challenging than, many other biocatalytic systems is that in the redox reactions, electron transfer must take place and therefore the redox potentials of the different enzymes in a multi-enzyme system need to be considered. Thus, for multiple enzymes to be directly coupled in multi-enzyme processes, their thermodynamic redox capacities and their kinetic properties must be understood.

## 3.2
## Classes of Oxidizing Enzymes

Oxidative enzymes catalyze a very wide range of reactions, via a number of different mechanisms, and their nomenclature can at times be unclear in indicating their function, but several reviews are available to provide clarity [1–4]. The

*Multi-Step Enzyme Catalysis: Biotransformations and Chemoenzymatic Synthesis*
Edited by Eduardo Garcia-Junceda

ISBN: 978-3-527-31921-3

following section describes some groups of oxidative enzymes relevant to biocatalytic systems.

**Oxygenases** catalyze the introduction of oxygen atoms into the reaction substrates, in some cases regioselectively. Monooxygenases require two electrons and two protons to reductively cleave oxygen ($O_2$) and generate a single molecule of water. The second oxygen atom is generally incorporated into the substrate resulting in an oxy-functionalized product. Dioxygenases, on the other hand, require four electrons and generate two water molecules (see below). Monooxygenases generally require a cofactor such as NADH or NADPH to provide reducing potential for the supply of electrons to the substrate:

$$\text{Substrate} + \text{Donor}-\text{H} + O_2 + H^+ \rightarrow \text{Substrate}-\text{O} + \text{Donor} + H_2O$$

**Cytochrome P450 enzymes** are hemoproteins that catalyze the incorporation of a single oxygen atom into a very wide range of substrates including drugs and xenobiotic aromatic compounds. Cytochrome P450-dependent systems act via the formation of an active heme–oxygen complex which then binds to generate a Fe(III)–substrate complex. This leads to reduction of the substrate by single electron reduction, followed by reaction with oxygen to give a Fe(II)–substrate–$O_2$ intermediate, addition of the second electron, and then release of water and oxygen addition to the substrate. The variation of selectivity among different cytochrome P450s provides a valuable source of different biocatalysts for oxidation reactions. Cytochrome P450 enzymes occur in almost all living cells and, as biocatalysts, are often cloned from eukaryotic sources as well as prokaryotes into practical expression systems. Since they are NAD(P)H dependent, application of P450s in biocatalytic reactions requires the presence and recycling of the cofactor, leading to their commonly being used in whole-cell biotransformation systems (see Section 3.7). The cytochrome P450s are known as mixed oxidases because they have both oxygenase and oxidase activity; their mechanism involves two one-electron transfer steps, resulting in one water molecule and one hydroxylated product molecule per catalytic cycle. However, they can also use peroxide in some circumstances and thereby generate oxygenated products [5].

The monooxygenase group of enzymes includes the non-P450 hydroxylases which catalyze the insertion of a hydroxyl group to replace a hydrogen atom at a saturated carbon [6–8] and the non-heme-dependent oxygenases such as the flavin–molybdenum–cobalt-dependent xanthine oxidase and aldehyde oxidase [9].

In flavin-dependent monooxygenases, a flavin–oxygen intermediate reacts with the substrate, also producing water in a second step, and requiring cofactors for regeneration of the flavin moiety. The unusual flavoprotein vanillyl-alcohol oxidase (EC 1.1.3.38), in which the flavin moiety is covalently bound, catalyzes the oxidation of *p*-substituted phenols as well as deamination, hydroxylation and dehydrogenation reactions [10].

**Oxidases** use molecular oxygen as their electron acceptor, while peroxidases use hydrogen peroxide. In both cases, reactive oxygen intermediates are produced which may then react further with the reducing substrates. These reactions can be non-specific or can produce several different products.

The **multi-copper oxidases**, such as the cofactor-independent copper oxidases tyrosinase and laccase, are monooxygenases in which multi-nuclear copper centers bind the dioxygen molecule so that it becomes polarized, resulting in its cleavage [1, 11]. The reducing substrate is then oxidized by the oxygen–copper complex. In the tyrosinasc reaction, a phenol is first hydroxylated and the resulting catechol is subsequently oxidized to give an *o*-quinone. In laccase-catalyzed reactions, four copper atoms are bound in three different redox sites and the primary product of the substrate conversion is a radical oxygen species which can later react further [1, 2].

**Dioxygenases** catalyze the incorporation of dioxygen into substrates. Thus, heme-dependent iron–sulfur plant dioxygenases and Rieske iron–sulfur non-heme dioxygenases catalyze reactions such as *cis*-dihydroxylation, aromatic ring cleavage and hydroperoxidation of varied substrates, including arenes and aromatic carboxylates [12]. Rieske iron–sulfur dioxygenases, in particular, have high redox potentials. Ring-cleaving diooxygenases, such as catechol dioxygenase, catalyze the incorporation of both atoms of oxygen as hydroxyl groups on adjacent carbons of aromatic rings, generating *cis*-dihydrodiols, which may be further oxidized to ring-opened products [4].

Copper-containing **amine oxidases** (non-blue copper proteins) catalyze the oxidative deamination of primary amines to the corresponding aldehydes with the release of ammonia and concomitant reduction of oxygen to hydrogen peroxide. They typically use a quinone redox cofactor [topaquinone (TPQ)], which is bound covalently in the active site, and are thought to form a Cu(I)–TPQ semi-quinone radical intermediate during the redox reaction [13].

**Peroxidases** (EC 1.11.1.7), which have ferric protoheme prosthetic groups, react non-selectively via free radical mechanisms, using hydrogen peroxide as the electron acceptor. A reactive Fe(IV)–O species and a radical heme intermediate are formed, and the intermediate then reacts with the reducing substrate to produce the oxidized product, regenerating the Fe(III) ion.

**Haloperoxidases** are peroxidases capable of halogenating substrates in the presence of halide and hydrogen peroxide [14] or other reactions such as sulfoxidation, epoxidation and aromatic hydroxylation. Here, the halide ion is initially bound to the active site which may incorporate heme or vanadium or be metal free. The halide ion is incorporated into the substrate after electron transfer [15].

**Dehydrogenases** catalyze the removal of two protons and two electrons from their substrates, generally using NAD or flavin cofactors, and are typically highly specific for their substrates. Alcohol dehydrogenases, for example, catalyze the production of alcohols and are well-recognized for their application in racemate resolution. 1,2-Diol dehydrogenases catalyze the conversion of unsaturated 1,2-diols to 1,2-dihydroxy aromatics and are involved in the metabolic degradation of aromatic compounds where dioxygenation is followed by dehydrogenation [16].

## 3.3 Mechanisms of Biological Oxidation and Implications for Multi-Enzyme Biocatalysis

Enzymatic oxidation reactions involve electron transfer from the substrate to an acceptor which may be bound in the enzyme itself or may be a small molecule electron acceptor acting as a mediator. The mechanisms of biological oxidation processes are very varied and invariably complex. This complexity arises from the cyclical nature of the metabolic processes involved, and the requirement that electron acceptors be regenerated and electron transfer chains be maintained. With a few exceptions, enzyme-catalyzed redox reactions require the presence of coenzymes, or cofactors, to act as an intermediate electron carrier. *In vivo*, cells utilize cofactors as electron carriers (such as the nicotinamide nucleotides or flavin prosthetic groups) in redox systems which involve pathways where a succession of enzymes operate in sequence. Thus, the pyridine nucleotide coenzymes such as $NAD^+$ and $NADP^+$, and their reduced forms NADH and NADPH, respectively, act by binding non-covalently at enzyme active sites, accepting electrons and hydrogen atoms in a reversible manner. The redox potential for $NAD^+$ is −0.32 V, which allows it to catalyze the oxidation of alcohols, primary and secondary amines, aldehydes, and alkyl groups. Reduced NADH must then be re-oxidized via the respiratory metabolic chain *in vivo* or by an external regeneration system in isolated enzyme reactions. Other cofactors such as flavin mononucleotides (FMN) and FAD act via similar mechanisms [17].

In cases where the electron-accepting cofactor is covalently bound, for example vanillyl oxidase [10] and the FAD-containing cytokinin dehydrogenase [18], electron transfer can occur directly from the substrate molecule to the cofactor and the cofactor must then be re-oxidized by subsequent reaction events at the enzyme active site. Alternatively, enzymes may bear metal ions in their active sites as redox-active prosthetic groups capable of activating oxygen species. Here, the electron acceptor is usually dioxygen or another oxygen species and the oxidation reaction involves electron transfer to a metal ion which then becomes reduced by other cofactors in the course of the catalytic cycle. In each case, the regeneration of oxidizing power is necessary for continued catalytic turnover. The condition required for a biocatalytic oxidative system to function would include adequate re-supply of the electron acceptor and concomitant efficient electron transfer from the electron acceptor to a 'downstream' electron acceptor, typically oxygen or hydrogen peroxide. Various approaches to cofactor recycle in biocatalytic systems have been investigated and some are discussed in Section 3.5.2.

In order to use oxidizing enzymes in biocatalysis, it is necessary to understand the redox potential, $E^{0\prime}$, of the enzyme-electron carrier systems in question. [The redox potential is a measure of the capacity of the enzyme system to catalyze a given oxidation reaction and itself be reduced, as compared with a standard redox reaction (usually $H^+/H_2$) under standard conditions.]. This gives an indication of whether a given enzyme-catalyzed reaction is thermodynamically feasible. In the present context, oxygen, which is the strongest oxidizing agent available, has a redox potential of 0.82 V in the four-electron reduction to water (see

**Table 3.1** Redox potentials of some common biological redox systems (against the standard hydrogen electrode at pH 7) [3].

| Redox system | Cofactor/metal ion | Redox potential (V) |
|---|---|---|
| Ferredoxins, iron–sulfur proteins | ferredoxin$_{ox}$ + 1e$^-$ → ferrodoxin$_{red}$ | −0.36 |
| | acetate → acetaldehyde | −0.06 |
| | Fe(III) → Fe(II) (in cytochrome *b*) | −0.10 |
| Molybdenum proteins (e.g. aldehyde oxidase, xanthine oxidase) | Mo(VI) → Mo(V) | −0.36 |
| | Mo(V) → Mo (IV) | −0.36 |
| Flavoproteins | NAD(P)$^+$ + H$^+$ + 2e$^-$ → NAD(P)H | −0.32 |
| | FAD + 2H$^+$ + 2e$^-$ → FADH$_2$ | −0.18 |
| | ascorbate$_{ox}$ + 2H$^+$ + 2e$^-$ → ascorbate$_{red}$ | +0.06 |
| | acetyl-CoA + 2H$^+$ + 2e$^-$ → acetyl-CoA | +0.19 |
| Oxidases (e.g. ascorbate oxidase) | $O_2$ + 2 $H_2O$ + 4e$^-$ → 2 $H_2O_2$ | +0.3 |
| Multi-copper oxidases (e.g. laccase) | Cu(II) → Cu(I) | +0.4 |
| Oxygenases | $O_2$ + 4 H$^+$ + 4e$^-$ → 2 $H_2O$ | +0.82 |

Table 3.1). Thus, for a redox reaction to be possible, the difference between the redox potential of the enzyme–cofactor system and that of the substrate must be above zero [3]. The catalytic role of the enzyme protein structure in a redox reaction is often to alter the electronic environment of the cofactor, thereby changing its redox potential and hence making the reaction more thermodynamically feasible. (For further in-depth discussion the reader is referred to the excellent text of Bugg [3].)

## 3.4 Multi-Step Biotransformation Processes Involving Oxidation

The role of oxidative enzymes in existing multi-step processes can be considered in various different contexts, each of which are discussed below. These include:

- Natural multi-step processes, where oxidation steps are involved in metabolic pathways.

- Applications of multi-step reaction processes, where oxidases are used in established industrial or laboratory processes including biocataytic processes.
- Chemoenzymatic processes, where oxidases are used in single steps within processes for production of synthetic compounds, including production of chiral products.
- Combinatorial biocatalysis, where redox enzymes are used in mulit-component systems for new molecule discovery.

**Natural multi-step biotransformation processes involving oxidation** include pathways such as polyketide synthesis and antibiotic production by microorganisms, lignin biosynthesis in plants and lignin degradation by microorganisms. The synthesis of polyketide antibiotics and non-ribosomal peptide antibiotics include postproduction modifications involving cytochrome P450 monooxygenases which catalyze the oxidative cross-linking of the aryl side-chains and epoxidation reactions which lead to the formation of pyran and furan ether ring systems in the production of polyether antibiotics [19–24]. Other oxidative enzymes involved in the production or postsynthesis modification of these aromatic antibiotics include L-amino acid oxidase [25], dehydrogenases [26], $NAD(P)^+$-dependent dehydrogenase [27] and phenoxazinone synthase [28]. The production of glycopeptide antibiotics involves a multi-enzyme process in which the enzyme chloroperoxidase (EC 1.11.1.10) plays an important role [29]. Clearly, all of these redox enzymes are potential candidates for future biotransformations, but many will need to be cloned and expressed in robust expression systems in order for the process to be developed to commercial scale.

**Industrial and applied multi-step reaction processes**, where oxidizing enzymes are components of industrial processes, include in the pulp and paper industry, in the food and beverage industry, for bioremediation, in biosensors and, more recently, in biofuel cells (discussed in more detail in Section 3.8).

The white rot fungi are well known for their ability to produce lignin-degrading enzymes, which include tyrosinases, laccases and peroxidases. Kirk and Chang [30] developed laccases for the pulp and paper industry where lignin-degrading enzymes are used not only to remove lignin, but also to degrade lignin-related pollutants from soil. This is attributable to the sequential action of laccases and peroxidases [31, 32]. Labat *et al.* [33] described the use of the ligninolytic enzymes, laccase and manganese peroxidase, to determine their effect on the quality of bread dough – the enzymes were found to act on different components within the dough, resulting in better dough consistency and mixing behavior. In a review on oxidative enzymes, Xu [31] reported the use of glucose oxidase and catalase in the improvement of freshness/preservation of shrimp and fish, while fruit juice and fermented alcoholic beverages can be treated with laccases and other oxidases to remove or modify phenolic compounds, color, flavor, taste or stability – after the treatment, unwanted phenolics are removed by silicate fining or filtration.

Oxidizing enzymes have also been used as key catalysts in multi-step reactions for the production of antibacterial and antiviral agents. Horseradish peroxidase and chloroperoxidase have been used in the production of the macrocyclic glycopeptide antibiotic, vancomycin [34, 35]. Peroxidases can be used to catalyze the successive C—O coupling of amino acid derivatives in the production of the diaryl ether backbone of the vancomycin molecule, whereas chloroperoxidases play a role in the chlorination of vancomycin and vancomycin derivatives [35]. A combination of α-galactosidase and chloroperoxidase in the production of bergenin analogs is an example of iterative enzymatic derivatization: the bergenin product was only produced if the α-galactosidase reaction was allowed to take place prior to the chloroperoxidase reaction and the reaction would not proceed if the enzymes are used in the reverse order [36].

In the field of bioremediation, oxidoreductases are considered to be excellent biocatalysts for environmentally friendly processes. Laccases and peroxidases are widely used to treat effluents from pulp/cotton mills, food/fruit processing plants and breweries [1, 2, 37]. Laccases, peroxidases and other oxygenases are also being studied for their ability to degrade hazardous coal substances, especially the sulfur-containing components, and in the treatment of industrial waste and contaminated soil and water in the transformation of xenobiotics, polycyclic aromatic hydrocarbons and other pollutants (biodetoxification and biodecontamination) [31]. Alcade *et al.* [38] have listed the enzymes, which to date, have been extensively studied and used in bioremediation; these include the bacterial mono- or dioxygenases, dehalogenases, cytochrome P450 monooxygenases and the enzymes involved in lignin degradation (laccases, lignin and manganese-dependent peroxidases). Studies are ongoing where 'cocktails' of these enzymes are used in the treatment of waste water. The use of these microbial enzymes have certain advantages: no toxic side-products are produced, enzyme activity is easier to control than whole-cell cultures, and recombinant DNA technology can be used to create enzymes that are more stable and more active under environmental conditions [38].

Another example of a industrial multi-step process is the combination of enzymes and chemicals in the making of denim fabric. Laccase, with peroxide, is used in the bleaching of cotton. Thereafter, the enzyme catalase is added to degrade the remaining peroxide once bleaching has been completed. Peroxidases are also used in the same process operation in the removal of any excess dye that might accumulate during the dyeing of the denim fabric [39].

In the field of novel carbohydrate production, oxidative enzymes have been used in multi-step processes [37, 40]. For example, galactose oxidase has been used in the oxidation of galactose of galactosamine, which is subsequently polymerized using a chemical reductive amination. In this system, galactose was incorporated into a polymer through the action of galactose oxidase. A chemical reductive step resulted in the production of sugar-containing polymers. These polymers may have potential in the production of hydrogels or water-soluble emulsifiers and open the door for other oxidative reactions coupled with chemical reactions [41].

**Chemoenzymatic processes** involving oxidizing enzymes have been reported particularly for specific chemical syntheses. For example, industrially important amino acids can be deracemized by exploiting the enantioselectivity of amino acid oxidases; a commercial process has recently been developed in which efficient biocatalysts have been developed for deracemization of a wider range of amino acids [42]. Here, an amino acid oxidase biocatalyst converts the racemic amino acid to the desired enantiopure product and non-selective chemical reduction is then used to stereo-invert the remaining isomer and hence to facilitate almost 100% conversion.

**Combinatorial biocatalysis** is a further approach to be considered in multi-step biocatalysis. This involves the generation of novel biologically active compounds by introduction of functional groups or the modification of existing functional groups, using enzymes capable of transforming non-natural substrates. Oxidoreductases are recognized participants in biosynthetic pathways leading to novel metabolites and their application in combinatorial processes has been reported, for example in the combination of polyketide synthase pathways with that of non-ribosomal peptide synthetases, which includes the combination of oxidoreductases, monooxygenases, amino transferases and *O*-methyltransferases [43].

## 3.5 Design and Development of New Multi-Enzyme Oxidizing Processes

### 3.5.1 Coupling Redox Enzymes

Bioprocesses incorporating more than one redox enzyme in an oxidative reaction system might involve, in the simplest case, two oxidizing enzymes coupled so that they act sequentially to effect two oxidation steps. A key issue in the development of such oxidative biocatalytic systems would be the determination of the values, for each enzyme involved, of the redox potentials. These can be determined by potentiometric titration using redox mediators (such as NADH) and techniques such as cyclic voltammetry or electrophoresis [44]. Knowledge of the redox potentials would facilitate the design and engineering of a process in which the two

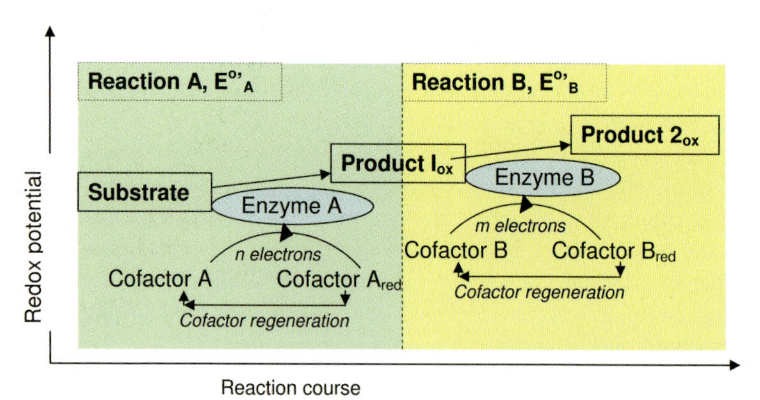


**Figure 3.1** Coupling of two oxidizing enzymes in a process, showing two reaction systems with different redox potentials.

enzymes were present at levels and under conditions such that their redox potentials and relative activities were accommodated to provide optimal product yields and efficient cofactor recycle (see Figure 3.1). Thus, an enzyme with a higher redox potential can be used to oxidize a second reaction in which the product of a previous reaction was produced by an enzyme with a lower redox potential. The necessary processes for regeneration of the cofactors will also need to be considered, since each cofactor system must be recycled at a sufficiently high rate. Again, redox potentials can be used to assist in design of the regeneration system (see Section 3.5.2).

An additional condition may be imposed, even when a cofactor-independent enzyme is used, if a mediator molecule is involved in the electron transfer process, as is often the case with oxidases. Laccases, for example, may employ small-molecule diffusible mediator compounds in their redox cycle to shuttle electrons between the redox center of the enzyme and the substrate or electrode (Scheme 3.1) [1, 2]. Similarly, certain dehydrogenases utilize pyrroloquinoline quinone. In biocatalytic systems, mediators based on metal complexes are often used.

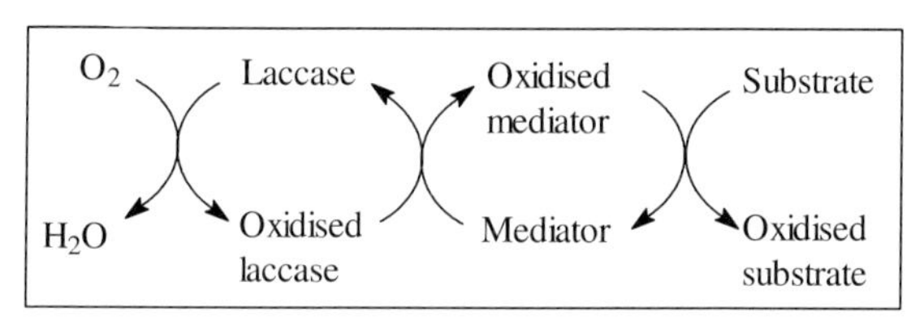


**Scheme 3.1** Illustration of the role of a mediator in a redox reaction cycle, using laccase as the example.

A further requirement for the development of a multi-enzyme oxidizing process would be the determination of the kinetic parameters of the enzymes and hence development of a model of the intended reaction system in terms of the relative productivities of the enzymes with respect to substrate conversion rates as well as electron transfer stoichiometry.

A detailed account of the relationship between redox potentials and enzyme kinetic parameters is given by Ikeda and Kano [44]. For example, for a mediated substrate oxidation reaction, the rate $\nu_E$ of the enzyme-catalyzed reaction can be measured and the kinetic parameters determined from:

$$\nu_E = k_{cat}[E]/(1 + K_M/[M_{ox}])$$

where $[M_{ox}]$ is the concentration of the oxidized form of the mediator, generated as a result of the oxidation reaction:

$$M_{red} \leftrightarrow M_{ox} + E_{ox} + n\,e^-$$

occurring at the electrode.

Taking this enzyme-electrochemical approach further, $k_{cat}$ and $K_M$ values can be obtained for redox enzymes in electrochemical cells and enzyme-mediator reactions [44, 45]. The approach can also be extended to whole-cell biocatalytic systems and kinetic parameters for the cells ($K_{cell.S}$, $K_{cell.M}$ and $k_{cell.cat}$) can be determined. This allows the investigation of the catalytic ability of the whole cells as oxidative biocatalysts and hence can lead to the development of models for oxidative multi-enzyme whole-cell systems. An approach to achieving this has been reported in which cytochrome *c* oxidase was immobilized on a metal electrode surface in an active form, so that the four consecutive electron transfer steps could be controlled and characterized electronically [46]. This represents significant progress towards biomimetic coupling of oxidases in membrane-bound complexes.

One approach to the development of a biocatalytic system coupling two oxidase enzymes which do not require a cofactor, used in our own laboratory, has involved design of a two-reactor process in which each of the two enzymes are immobilized separately, and the coupling is effected by designing and using two reactors in series. In order to successfully couple the enzymes, empirical experimental measurements of substrate conversion rates were used to determine the enzyme kinetic parameters of the biocatalysts and a model-based experimental design approach [47] was used to develop a model of the two-enzyme process. Thus, polyphenol oxidase (tyrosinase) and laccase were each immobilized and applied in the sequential process involving (i) the hydroxylation of tyrosol to hydroxytyrosol catalyzed by polyphenol oxidase and (ii) the subsequent oxidation of hydroxytyrosol to produce a dimeric product (Figure 3.2). In order to achieve this, the initial reaction and production rates of tyrosinase and laccase as immobilized biocatalysts were measured, operating in separate stirred-tank bioreactors. These data were then used to predict the theoretical productivity of the biocatalysts, both individually and then combined, in continuously operated fixed-bed reactors [48]. The two biocatalysts were then combined in an experimental continuously operated bioreactor system (Figure 3.3) and the biocatalytic production of the final product was

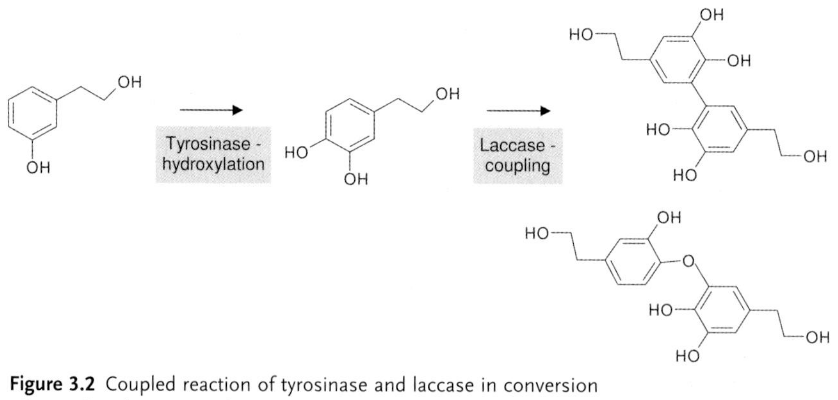

Tyrosinase - hydroxylation

Laccase - coupling

**Figure 3.2** Coupled reaction of tyrosinase and laccase in conversion of tyrosol to dimeric products [48].

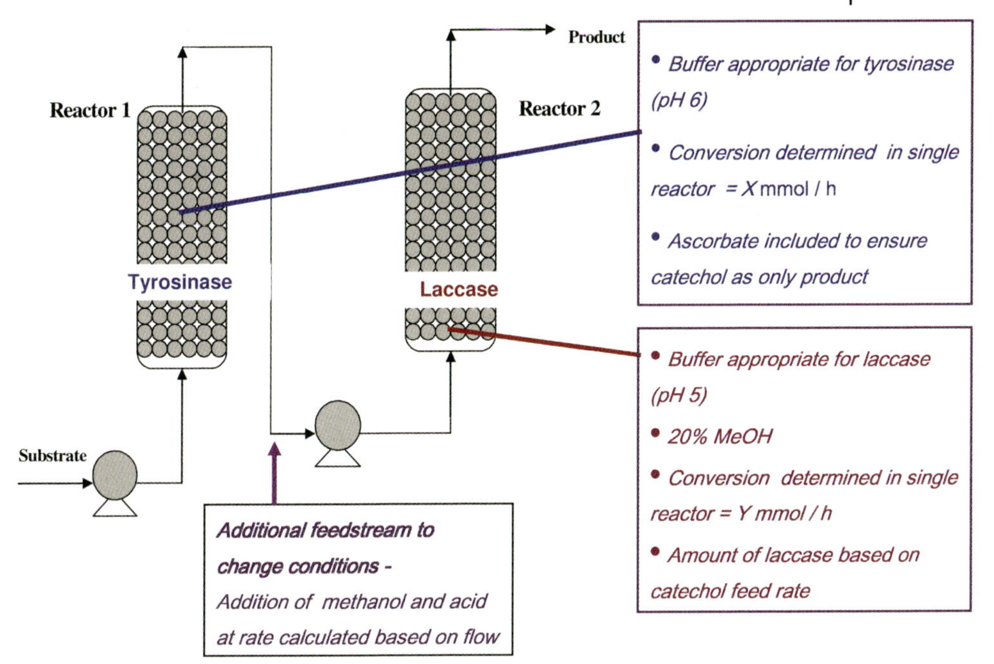


**Figure 3.3** Conceptual process model for application of a coupled tyrosinase–laccase reaction converting tyrosol. Immobilized enzymes are first characterized with respect to substrate conversion rates, using tyrosol and hydroxytyrosol as substrates for tyrosinase and laccase, respectively. One hundred percent conversion can be achieved in Reactor 1 by use of sufficient tyrosinase biocatalyst. Inclusion of equimolar ascorbate in the Reactor 1 feedstream ensures that all product entering Reactor 2 is hydroxytyrosol. The amount of laccase required and scale of Reactor 2 is determined by the rate of feed of hydroxytyrosol [49]. Methanol is fed at a flow rate sufficient to achieve 30% methanol : buffer in Reactor 2.

measured. In this study, to facilitate modeling of the dual process, it was assumed that the reactor volume and the volumetric flow rate, temperature, pH and yields were constant; and that the biocatalysts retained physical integrity and activity during the processing period. However, it was known that the for second reaction, in which laccase was used, the addition of an organic solvent was necessary to control polymerization of the final product [49] and thus methanol was required to be added to the system as an additional feed stream, introduced between the reactor units.

### 3.5.2
### Cofactor Recycle in Multi-Step Oxidizing Biocatalytic Systems

Technologies for recycle of cofactors in single-electron biocatalytic processes have been extensively studied and many reviews have been published (e.g. [50] and references therein). Clearly, if several redox enzymes are incorporated into

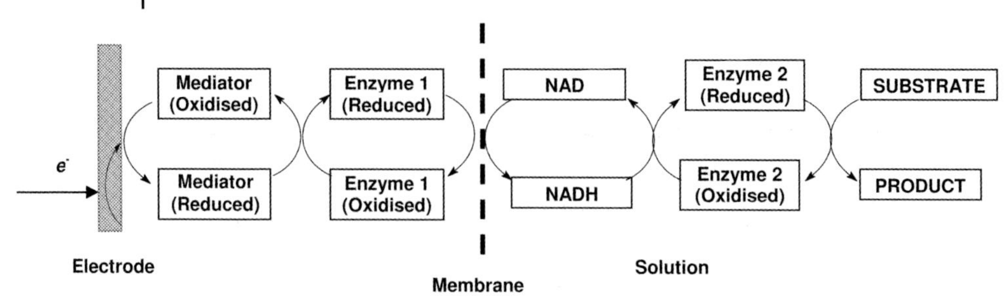


**Scheme 3.2** Cofactor regeneration using copolymerized mediator and lipoamide dehydrogenase at an electrode surface [51].

one system, the cofactor regeneration requirements of each must be accommodated. A membrane separation system was used to achieve cofactor regeneration in a dual-enzyme system where each enzyme required a mediator, and an enzyme electrode was used as the electron supply (Scheme 3.2) [51].

New supramolecular compounds mimicking biological cofactors have been proposed which have high redox potentials and, further, can be linked covalently onto supports in biotransformation systems [52]. Bioelectrochemical methods available for regeneration of nicotinamide- and flavin-dependent systems are comprehensively reviewed by Kohlman *et al.* [53].

Recent modeling approaches have shown that physical separation of the enzymes is not a prerequisite for successful cofactor recycle. Hogan and Woodley [54] demonstrated that kinetic data can be sued in conjunction with design principles to develop models for prediction of process requirements and hence to optimize the utilization of the cofactor, in this case NADPH required for cyclohexanone monooxygenase and regenerated using alcohol dehydrogenase. In another modeling approach, charged membrane bioreactors were proposed in which NADPH-dependent aldose reductase and glucose dehydrogenase were used for sorbitol synthesis from glucose. Here, the optimal cofactor and substrate concentrations were determined, and the modeling resulted in the conclusion that a continuously operated system would be most suitable in terms of productivity and economic considerations [55].

## 3.6 Examples of Multi-Enzyme Biotransformation Processes Involving Oxidizing Enzymes

A key consideration in development of all multi-step bioprocesses is the type of bioreactor; it may be necessary to accommodate a range of conditions including compartmentalization of the enzymes, cofactor recycle, adequate oxygen supply, variable temperature and pH requirements, and differential substrate feed rates. Examples described below include a range of different reactors, of which membrane bioreactors are clearly often particularly useful.

### 3.6.1
### Coupling of Oxidases with Non-Redox Enzymes

In cases where an oxidizing enzyme is coupled with a second, non-redox enzyme, obviously, only the oxidizing component needs to be considered in terms of electron transfer cofactor regeneration.

In an elegant and complex system, galactose oxidase was used for C6 oxidation of *p*-nitrophenyl-β-*N*-acetyl-D-galactosamine and the product *p*-nitrophenyl 2-acetamido-2-deoxy-β-D-galacto-hexodialdo-1,5-pyranoside was then converted, in a second enzymatic step, using a β-*N*-acetylhexosaminidase and further oxidation, to the biologically active disaccharide product 2-acetamido-2-deoxy-β-D-galactopyranosyluronic acid-(1–4)-2-acetamido-2-deoxy-D-glucopyranose [56]. The system was operated in a batch reactor, requiring optimization of oxygen supply and consideration for the preservation of the galactose oxidase activity under the high oxygen conditions required. Further, catalase was incorporated to remove the hydrogen peroxide produced by the galactose oxidase step. The process was operated at a scale of 100 mg/100 ml, and final yields were approximately 90% for the oxidase step and 40% for the final product. This report demonstrates the complexity and precision required to successfully use consecutive enzymatic reactions.

A liquid membrane bioreactor was developed as a means of encapsulation for a multi-enzyme system incorporating an oxidation and carbohydrate cleavage, demonstrated using α-glucosidase and glucose oxidase in the conversion of maltose to gluconic acid:

$$n\ \text{Maltose} \xrightarrow{\text{Glucosidase}} 2n\ \text{Glucose} \xrightarrow{\text{Glucose oxidase}} 2n\ \text{Gluconic acid}$$

Clearly, in such systems, the nature and stability of the liquid membranes is a critical issue [57].

### 3.6.2
### Biocatalytic Systems Involving Coupled Oxidizing Enzymes

Biocatalysis involving the concerted activity of two or more oxidizing enzymes requires consideration of suitable conditions for each enzyme, as described above. Approaches to achieving this generally include co-immobilization of the enzymes involved or development of combined or sequential bioreactors. Nanostructured multi-enzyme systems can be modeled on biological pathways, ensuring the close association of cofactors with the biocatalysts by physical tethering and/or co-encapsulation. Lactate dehydrogenase and L-glutamate dehydrogenase were used to demonstrate such a system in nanoparticles which were then encapsulated with NADH, as microcapsules [58].

Using an approach involving a tubular membrane reactor in which the biocatalysts were retained, Nasufi *et al.* [59] reported the production of ascorbic acid by a

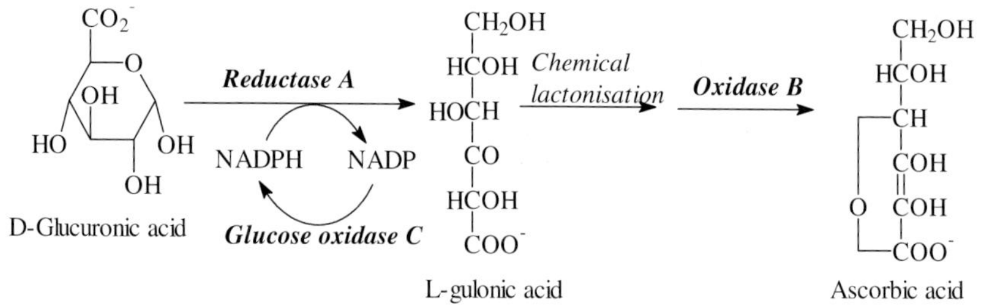

D-Glucuronic acid — Reductase A (NADPH → NADP; Glucose oxidase C) → L-gulonic acid — Chemical lactonisation → Oxidase B → Ascorbic acid

**Scheme 3.3** Chemoenzymatic conversion of D-glucuronic acid to L-ascorbic acid [59]. NADPH-dependent uronate reductase A from *Saccharomyces cerevisiae*, cloned into *E. coli* converted D-glucuronic acid at 90% yield. Chemical lactonization gave intermediate L-gulono 1,4-lactone, and this was converted by L-gulono 1,4-lactone oxidase or pyranose 2-oxidase B from *Trametes multicolor.* NDPH was regenerated by glucose oxidase C in an ultrafiltration membrane reactor.

three-step process in which D-glucuronic acid was first reduced to L-gulonic acid by NADPH-dependent uronate reductase. This intermediate was lactonized and converted using L-gulono 1,4-lactone oxidase or a cytosolic pyranose 2-oxidase, to produce 2-keto-L-gulono 1,4-lactone which rearranged to ascorbic acid under acid conditions. The NADPH used in the first step was regenerated by NAD(P)-dependent glucose dehydrogenase, using an ultrafiltration membrane reactor (Scheme 3.3). In this case, the three enzymes were isolated from three different microbial sources and the process was operated in separate steps; clearly there is potential for development of this system to a combined, continuous process.

An alternative method for coupling enzymes is to employ a mixed enzyme bioreactor; Pezzotti and Therisod [60] reported a bienzymatic system in which alcohol oxidase and peroxidase were coupled to effect the enantioselective oxidation of the sulfide thioanisole (Scheme 3.4). Here, the peroxidase from *Coprinus cinereus* was mixed with a crude extract of *Pichia pastoriz* alcohol oxidase and the two enzyme mixture was successfully used to convert gram quantities of thioanisole enantioselectively to *S*-methyl-phenyl-sulfoxide with an enantiomeric excess of 75%.

In an early report to a process using three oxidoreductases, namely hydrogenase (EC1.12.2.1), lipoamide dehydrogenase (EC 1.6.4.3) and 20β-hydroxysteroid dehydrogenase (EC1.1.1.53), a reverse micelle system was used to facilitate stereo- and site-specific reduction of apolar ketosteroids, assisted by the *in situ* NADH-regenerating enzyme system [61].

The cofactor regeneration process is not necessarily separated from the biocatalytic system; Morgan *et al.* [62] described the enzymatic oxidation of hydrocarbons using chloroperoxidase, chosen because of its ability to catalyze the same type of reactions catalyzed by P450 monooxygenases (i.e. halogenation, epoxidation, sulfoxidation and the oxidation of alcohols to aldehydes). In this system, to ensure the constant generation of hydrogen peroxide, the peroxidase reaction was conducted in conjunction with xanthine oxidase which produces hydrogen peroxide.

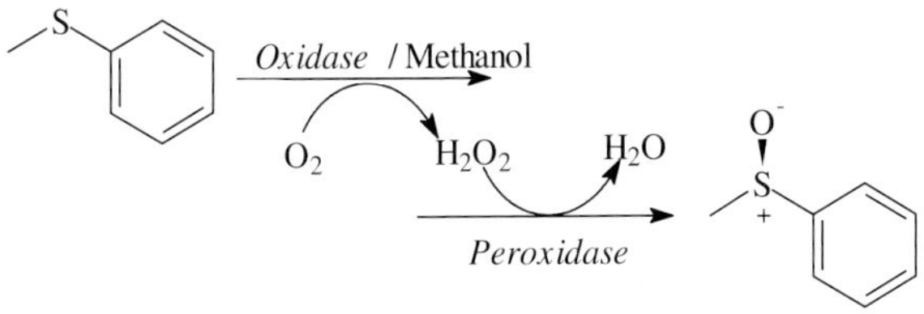

*Oxidase* / Methanol
$O_2$ $H_2O_2$ $H_2O$
*Peroxidase*

**Scheme 3.4** Enantioselective oxidation of anisole using alcohol oxidase and peroxidase [60]. *S*-methyl-phenyl sulfoxide was obtained in 72% yield and at 75% enantiomeric excess.

## 3.7 Multi-Enzyme Systems in Whole-Cell Biotransformations and Expression of Redox Systems in Recombinant Hosts

Whole-cell systems are widely recognized as being particularly suitable for redox biocatalysis because of the inherent capacity for cofactor recycle. Of concern in such whole-cell processes is the necessity for the substrates to enter the cells; while fungal cells are generally permeable to substrate molecules, bacterial cells may be less so. The solubility of the substrate in the medium is also an issue to be addressed in the design of the process, as is product recovery. The key benefit is, obviously, the *in vivo* cofactor regeneration. Many biotransformations have been reported in which both natural and xenobiotic compounds can be converted by redox enzymes such as the P450 oxidases, and some processes based on these reactions are now being operated at industrial scale [4].

Cloning and heterologous expression of oxidizing enzymes are receiving increasing attention as the demand for application of these enzymes increases. In the whole-cell application of recombinant oxidizing enzymes in their heterologous hosts, it is, of course, necessary to ensure that cofactor regeneration systems are sufficiently active and effective. This may be achieved by use of growing cells supplied adequately with nutrients and/or by cloning the cofactor regeneration system appropriate for the oxidation system into the expression host.

A whole-cell biocatalyst was developed carrying naphthalene dioxygenase and dihydrodiol dehydrogenase genes inducible in aerobic and anaerobic conditions. The naphthalene dioxygenase and dihydrodiol dehydrogenase genes were cloned into *Pseudomonas fluorescens* N3, allowing for the expression of both oxidative enzymes within one host [63]. More recently, naphthalene dihydrodiol dehydrogenase was successfully cloned and expressed in *Escherichia coli*, and shown to catalyze both the forward reaction of 1,2-dihydrodiol dehydrogenation and the reverse reaction of aromatic hydrogenation, demonstrating the significant redox power of the enzyme [64].

In a biotransformation system designed to mediate the *o*-dealkylation of 7-ethoxycumarin to 7-hydroxycoumarin, the genes for the P450 from *Streptomyces peucetius* and putidaredoxin reductase (CamA) and putidaredoxin (CamB) from

*Pseudomonas putida* were coexpressed in an *E. coli* strain. CamA is an NAD-dependent flavin-containing protein which requires CamB (an iron sulfur protein) as an electron shuttle [65].

A significant step in progress was reported by Prachayaittiul *et al.* [66] who successfully cloned both the genes for two hydrogenases into one plasmid in a continuous reading frame, resulting in the expression of a chimeric protein with the ability to catalyze both galactose and lactate dehydrogenase reactions. The chimeric enzyme was also shown to recycle NAD at rates such that no externally added NADH was required. Thus, effective intracellular cofactor recycling has been achieved in a heterologous host by metabolic engineering approaches.

In an alternative approach, a whole-cell system using two mutants of *E. coli*, expressing sequential metabolic genes for toluene/biphenyl dioxygenase and dihydridiol dehydrogenase was reported to produce monocyclic arene-dihydrodiols and vicinal diols for application in pharmaceutical and agricultural chemical products [67]. Here, *E. coli* mutants expressing each of two genes (i.e. the gene encoding a hybrid toluene/biphenyl dioxygenase and that encoding a dihydrodiol oxygenase respectively) were cocultured and used, as growing cells in a mixed batch reactor, to biotransform a range of aromatic substrates at a scale of 10 mg.

The production of oxidase enzymes in native strains is often not constitutive, since these enzymes are only required under specific conditions, such as nutrient stress. This imposes limitations on their application in continuous bioprocesses and one way of dealing with this issue is to develop improved expression systems. Laccase from *Trametes* sp. C30, which in its native strain is inducible and only weakly expressed, has been cloned and expressed in yeast to provide a functional recombinant enzyme with the same high activity and low redox potential as the native enzyme [68].

In a recent investigation to develop novel cytochrome P450 biocatalysts, DNA shuffling was used to produce chimeric cytochrome P450s mutants with enhanced biocatalytic activities, which were then co-expressed with NADPH-cytochrome reductase in *E. coli* to form an efficient system, in this case demonstrated to be effective for indole oxidation [69].

## 3.8
## Other Applications of Multi-Enzyme Oxidizing Systems

Bioelectrocatalysis involves the coupling of redox enzymes with electrochemical reactions [44]. Thus, oxidizing enzymes can be incorporated into redox systems applied in bioreactors, biosensors and biofuel cells. While biosensors and enzyme electrodes are not synthetic systems, they are, essentially, biocatalytic in nature (Scheme 3.5) and are therefore worthy of mention here. Oxidases are frequently used as the biological agent in biosensors, in combinations designed to detect specific target molecules. Enzyme electrodes are possibly one of the more common applications of oxidase biocatalysts. Enzymes such as glucose oxidase or cholesterol oxidase can be combined with a peroxidase such as horseradish peroxidase.

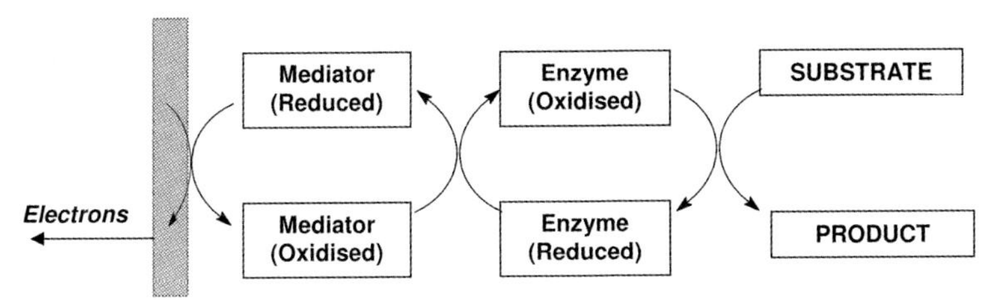


**Scheme 3.5** Electron transfer between the substrate and enzyme/mediator system in biosensors.

Thus, the oxidase–substrate reaction yields oxidized substrate and inactive reduced enzyme; the enzyme is returned to its active state by electron transfer, producing hydrogen peroxide. The peroxidase converts the peroxide, generating an electrochemical signal in the electrode.

In an early application, an enzyme electrode system was reported for the determination of creatinine and creatine, using a combination of creatinine amidohydrolase, creatine amidinohydrolase and sarcosine oxidase, co-immobilized on an asymmetric cellulose acetate membrane. Thus, the hydrogen peroxide produced was detected to give a quantitative measure of creatine and creatinine in biological fluids [70].

In a system devised to oxidize a dye, glucose oxidase and horseradish peroxidase were coassembled by layer-by-layer alternative adsorption to construct multi-enzyme films. The peroxidase was adsorbed to poly(styrene sulfonate) while glucose oxidase was adsorbed to poly(ethylene imine), allowing for sequential redox reactions to take place resulting in the subsequent oxidation of the textile dye DA67 [71].

In another example, a sucrose biosensor was developed, incorporating glucose oxidase, mutarotase and invertase, combined in optimal proportions, and cross-linked with glutaraldehyde and bovine albumin serum, and immobilized on a Prussian Blue-modified glassy carbon electrode [72]. An enzyme biosensor for monitoring phenolics and pesticides was developed using coimmobilized oxidases (i.e. tyrosinase, peroxidase or hydrogenase) with cholinesterase. In this case, the multi-enzyme system was used in a screen-printed array capable of detecting nanomolar concentrations of target pollutants [73]. Similarly, a multi-enzyme electrode for glycerol detection in wine was developed, demonstrating the successful application of combined glycerol dehydrogenase and diaphorase (dihydrolipoamide dehydrogenase) [74]. In this case, the enzymes were linked via tetrathiafulvalene linkers to the electrodes, along with NADH as cofactor.

More recently, nanotechnology has facilitated progress in miniaturizing redox enzyme electrodes and extending their application. In order to achieve contact between the active site of the redox enzyme where electron transfer takes place, usually buried within the protein structure, and the electrode electrical contact, cofactor-functionalized nanomaterials have been developed [75]. Diffusible cofactors such as FAD can be used as the relay system for carrying electrons to electrical wiring, or alternatively an apoenzyme can be generated which is then tethered to the electrical system though binding to a cofactor monolayer which is bound to the electrode [75].

An ingenious new development in the co-application of redox enzymes is exemplified in the concept of biofuel cells. A prototype implantable biofuel cell was recently demonstrated in which the specificity of laccase and glucose oxidase are used to ensure the safe application of the biofuel cell under physiological conditions. The system involved immobilization of glucose oxidase at the anode and laccase at the cathode to reduce dioxygen. Thus, water-soluble fuel substrates (i.e. glucose and oxygen) can be used to generate electrical voltage in the order of 1 V, giving non-toxic products (i.e. gluconolactone and water, respectively). Osmium redox polymer mediators were incorporated to facilitate the electron transfer to the electrodes [76].

## 3.9 Conclusions

The potential value of oxidizing enzymes in multi-step biotransformation processes is very clear. The limitations to their wider application are essentially 2-fold. First, there is a need to produce large amounts of the enzymes; much recent progress in heterologous expression of eukaryotic proteins and in genetic modification to enhance the stability and activity of redox enzymes is currently providing solutions to this issue. Second, we need effective design of efficient (and large-scale) bioreactors in which the thermodynamics and kinetics of the composite redox pathways are adequately addressed, and this is the current challenge for biochemical engineers and industrial chemists seeking to employ oxidizing biotransformations.

## References


1 Burton, S.G. (2003) *Trends in Biotechnology*, **21**, 543–9.

2 Burton, S.G. (2003) *Current Opinion in Biotechnology*, **7**, 1317–31.

3 Bugg, T. (1997) *An Introduction to Enzyme and CoEnzyme Chemistry*, Blackwell, Oxford.

4 Holland, H.L. (1992) *Organic Synthesis with Oxidative Enzymes*, Wiley-VCH, Weinheim.

5 Sligar, S.G., Makris, T.M. and Denisov, I.G. (2005) *Biochemical and Biophysical Research Communications*, **338**, 346–54.

6 Duetz, W.A., Van Beilen, J.B. and Witholt, B. (2001) *Current Opinion in Biotechnology*, **12**, 419–25.

7 Holland, H.L. (1999) *Current Opinion in Chemical Biology*, **3**, 22–7.

8 Holland, H.L. and Weber, H.K. (2000) *Current Opinion in Biotechnology*, **11**, 547–53.

9 Fettzner, S. (2000) *Die Naturwissenschaften*, **87**, 59–69.

10 Van den Heuvel, R.H.H., Fraaije, M.W., Mattevi, A., Laane, C. and van Berkel, W.J.H. (2001) *Journal of Molecular Catalysis B: Enzymatic*, **11**, 185–8.

11 McGuirl, M.A. and Dooley, D.M. (1999) *Current Opinion in Chemical Biology*, **3**, 138–44.

12 Wackett, L.P. (2002) *Enzyme and Microbial Technology*, **31**, 577–81.

**13** Mukherjee, C., Weyhermüller, T., Bothe, E. and Chaudhuri, P. (2007) *Comptes Rendus Chimie*, **10**, 313–25.

**14** Colonna, S., Gaggero, N., Richelmi, C. and Pasta, P. (1999) *Trends in Biotechnology*, **17**, 163–8.

**15** Littlechild, J. (1999) *Current Opinion in Chemical Biology*, **3**, 28–34.

**16** Sello, G., Bernasconi, S., Orsini, F., Mattavelli, P., Di Gennaro, P. and Bestetti, G. (2007) *Journal of Molecular Catalysis B: Enzymatic*, **52/53**, 67–73.

**17** Silverman, R.B. (2000) *The Organic Chemistry of Enzyme-Catalyzed Reactions*, Academic Press, New York.

**18** Malito, E., Coda, A., Bilyeu, K.D., Fraaije, M.W. and Mettevi, A. (2004) *Journal of Molecular Biology*, **341**, 1237–49.

**19** Walsh, C.T. (2004) *Science*, **303**, 1805–10.

**20** Sosio, M., Bianchi, A., Bossi, E. and Donadio, S. (2000) *Antonie van Leeuwenhoek*, **78**, 379–84.

**21** Recktenwald, J., Shawky, R., Puk, O., Pfennig, F., Keller, U., Wohlleben, W. and Pelzer, S. (2002) *Microbiology*, **148**, 1105–18.

**22** Weber, T., Welzel, K., Pelzer, S., Vente, A. and Wohlleben, W. (2003) *Journal of Biotechnology*, **106**, 221–32.

**23** Podust, L.M., Bach, H., Kim, Y., Lamb, D.C., Arase, M., Sherman, D.H., Kelly, S.L. and Waterman, M.R. (2004) *Protein Science*, **13**, 255–68.

**24** Urlacher, V.B. and Eiben, S. (2006) *Trends in Biotechnology*, **24**, 324–30.

**25** Nishizawa, T., Aldrich, C.C. and Sherman, D.H. (2005) *Journal of Bacteriology*, **187**, 2084–92.

**26** Yin, X. and Zabriskie, T.M. (2006) *Microbiology*, **152**, 2969–83.

**27** Pfeifer, V., Nicholson, G.J., Ries, J., Recktenwald, J., Schefer, A.B., Shawky, R.M., Schröder, J., Wohlleben, W. and Pelzer, S. (2001) *Journal of Biological Chemistry*, **276**, 38370–7.

**28** Szigyártó, I.C.S., Simándi, T.M., Simándi, L.I., Korecz, L. and Nagy, N. (2006) *Journal of Molecular Catalysis A: Chemical*, **251**, 270–6.

**29** Marshall, G.C. and Wright, G.D. (1996) *Biochemical and Biophysical Research Communications*, **219**, 580–3.

**30** Kirk, T.K. and Chang, H.-M. (1981) *Enzyme and Microbial Technology*, **3**, 189–96.

**31** Xu, F. (2005) *Industrial Biotechnology*, **1**, 38–49.

**32** Mester, T. and Tien, M. (2000) *International Biodeterioration & Biodegradation*, **46**, 51–9.

**33** Labat, E., Morel, M.H. and Rouau, X. (2001) *Food Hydrocolloids*, **15**, 47–52.

**34** Malnar, I. and Sih, C.J. (2000) *Tetrahedron Letters*, **41**, 1907–11.

**35** Malnar, I. and Sih, C.J. (2000) *Journal of Molecular Catalysis B: Enzymatic*, **10**, 545–9.

**36** Krstenansky, J.L. and Khmelnitsky, Y. (1999) *Bioorganic & Medicinal Chemistry Letters*, **7**, 2157–62.

**37** Riva, S. (2001) *Current Opinion in Biotechnology*, **5**, 106–11.

**38** Alcalde, M., Ferrer, M., Plou, F.J. and Ballesteros, A. (2006) *Trends in Biotechnology*, **24**, 281–7.

**39** Kirk, O., Borchert, T.V. and Fuglsang, C.C. (2002) *Current Opinion in Biotechnology*, **13**, 345–51.

**40** Riva, S. (2006) *Trends in Biotechnology*, **24**, 219–26.

**41** Liu, X.C. and Dordick, J.S. (1999) *Journal of the American Chemical Society*, **121**, 466–7.

**42** Fotheringham, I., Archer, I., Carr, R., Speight, R. and Turner, N.J. (2006) *Biochemical Society Transactions*, **34**, 287–90.

**43** Rich, J.O., Michels, P.C. and Khmelnitsky, Y.L. (2002) *Current Opinion in Chemical Biology*, **6**, 161–7.

**44** Ikeda, T. and Kano, K. (2001) *Journal of Bioscience and Bioengineering*, **92**, 9–18.

**45** Iswantini, D. (2000) *Biochemical Journal*, **350**, 917–23.

**46** Friedrich, M., Robertson, J., Walz, D., Knoll, W. and Naumann, R. (2008) *Biophysical Journal – BioFAST*, **94**, 3698-705.

**47** Berkholz, R., Guthke, R., Schmidt-Heck, W. and Röhlig, D. (2000) International Biotechnology Symposium, Berlin.

**48** Ncanana, S. (2007) PhD thesis, University of Cape Town. S.A.

**49** Ncanana, S. and Burton, S. (2007) *Journal of Molecular Catalysis B: Enzymatic*, **44**, 66–71.

**50** Rozzell, J.D. (1999) *Bioorganic & Medicinal Chemistry*, **7**, 2253–61.

**51** Leonida, M.D. (1996) *Bioorganic & Medicinal Chemistry Letters*, **6**, 1663–6.

**52** Rotello, V.M. (1999) *Current Opinion in Chemical Biology*, **3**, 747–51.

**53** Kohlman, C., Märkle, W. and Lütz, S. (2008) *Journal of Molecular Catalysis B: Enzymatic*, **51**, 57–72.

**54** Hogan, M.C. and Woodley, J.M. (2000) *Chemical Engineering Science*, **55**, 2001–8.

**55** Ikemi, M., Ishimatsu, Y. and Kise, S. (1990) *Biotechnology and Bioengineering*, **36**, 155–65.

**56** Fialová, P., Namdjou, D., Ettrich, R., Prikrylová, V., Rauvolfová, J., Krenek, K., Kuzma, M., Elling, L., Bezouška, K. and Kren, V. (2005) *Advanced Synthesis & Catalysis*, **347**, 997–1006.

**57** Pal, P., Datta, S. and Bhattacharya, P. (2002) *Separation and Purification Technology*, **27**, 145–54.

**58** Wang, P., Ma, G., Gao, F. and Liao, L. (2005) *China Particuology*, **3**, 304–9.

**59** Nasufi, L., Leitner, C., Zámocký, M., Haltrich, D., Kulbe, K.D. (2005) MECP Symposium, Graz.

**60** Pezzotti, F. and Therisod, M. (2007) *Tetrahedron: Asymmetry*, **18**, 701–4.

**61** Hillhorst, R., Laane, C. and Veeger, C. (1983) *FEBS Letters*, **159**, 225–8.

**62** Morgan, J.A., Lu, Z. and Clark, D.S. (2002) *Journal of Molecular Catalysis B: Enzymatic*, **18**, 147–54.

**63** Di Gennaro, P., Galli, E., Orsini, F., Pelizzoni, F., Sello, G. and Bestetti, G. (2000) *Research in Microbiology*, **151**, 383–91.

**64** Sello, G. (2008) *Journal of Molecular Catalysis B: Enzymatic*, 52–53, 67–73.

**65** Shresta, P., Oh, T.-J. and Sohng, J.K. (2008) *Biotechnology Letters*, **30**, 1101–1106.

**66** Prachayasittikul, V., Ljung, S., Isarankura-Na-Ayudhya, C. and Bülow, L. (2006) *International Journal of Biological Sciences*, **2**, 10–16.

**67** Shindo, K., Nakamura, R., Osawa, A., Kagami, O., Kanoh, K., Furukawa, K. and Misawa, N. (2005) *Journal of Molecular Catalysis B: Enzymatic*, **35**, 134–41.

**68** Klonowska, A., Gaudin, C., Asso, M., Fournel, A., Réglier, M. and Tron, T. (2005) *Enzyme and Microbial Technology*, **36**, 34–41.

**69** Rosic, N.N., Huang, W., Johnston, W.A., DeVoss, J.J. and Gillam, E.M.J. (2007) *Gene*, **395**, 40–8.

**70** Tsuchida, T. and Yoda, K. (1983) *Clinical Chemistry*, **29**, 51 5.

**71** Onda, M., Lvov, Y., Ariga, K. and Kunitake, T. (1996) *Biotechnology and Bioengineering*, **51**, 163–7.

**72** Haghighi, B., Varma, S., Alizadeh, S.F.M., Yigzaw, Y. and Gorton, L. (2004) *Talanta*, **64**, 3–12.

**73** Solná, R., Dock, E., Christenson, A., Winther-Nielsen, M., Carlsson, C., Emnéus, J., Ruzgas, T. and Skládal, P. (2005) *Analytica Chimica Acta*, **528**, 9–19.

**74** Gamella, M., Campuzano, S., Reviejo, A.J. and Pingarrón, J.M. (2008) *Analytica Chimica Acta*, **609**, 201–9.

**75** Willner, B., Katz, E. and Willner, I. (2006) *Current Opinion in Biotechnology*, **17**, 589–96.

**76** Barrière, F., Kavanagh, P. and Leech, D. (2006) *Electrochimica Acta*, **51**, 5187–92.

# 4
# Dihydroxyacetone Phosphate-Dependent Aldolases in the Core of Multi-Step Processes

*Laura Iturrate and Eduardo García-Junceda*

## 4.1
## Introduction

Organic synthesis can be considered, with only a small literary license, like a Meccano set that brings together small and simple pieces to build up more complicated structures (Figure 4.1).

In synthesis, the main screws that maintain the different pieces together are the C—C bonds. Therefore, C—C bond formation reactions can be considered the essence of organic synthesis [1].

The aldol addition reaction has long been recognized as one of the most useful tools that the synthetic chemist has for the construction of new C—C bonds [2]. Concomitant with the C—C bond-forming process is the formation of one or two new stereocenters, allowing us to approach a broad range of both natural and novel compounds.

Nature builds up carbohydrate, amino acids, α-hydroxy acids and other molecules by the use of the aldol reaction. The enzyme aldolases are among the most important biocatalysts for C—C bond formation in nature [3].

Aldolases are part of a large group of enzymes called lyases and are present in all organisms. They usually catalyze the reversible stereo-specific aldol addition of a donor ketone to an acceptor aldehyde. Mechanistically, two classes of aldolases can be recognized [4]: (i) type I aldolases form a Schiff-base intermediate between the donor substrate and a highly conserved lysine residue in the active site of the enzyme, and (ii) type II aldolases are dependent of a metal cation as cofactor, mainly $Zn^{2+}$, which acts as a Lewis acid in the activation of the donor substrate (Scheme 4.1).

The stereochemistry of the aldol reaction is highly predictable since it is generally controlled by the enzyme and does not depend on the structure or stereochemistry of the substrates. Aldolases generally show a very strict specificity for the donor substrate (the ketone), but tolerate a broad range of acceptor substrates (the aldehyde). Thus, they can be functionally classified on the base of the donor substrate accepted by the enzyme.

*Multi-Step Enzyme Catalysis: Biotransformations and Chemoenzymatic Synthesis*
Edited by Eduardo Garcia-Junceda

ISBN: 978-3-527-31921-3

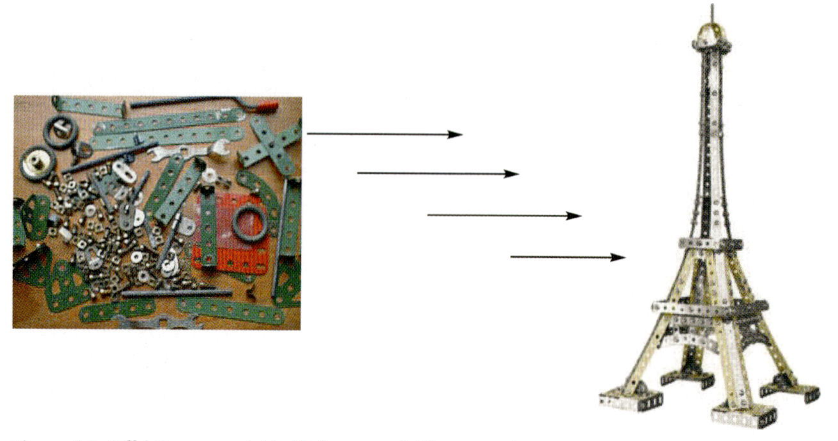

**Figure 4.1** Eiffel Tower model built from small Meccano pieces.

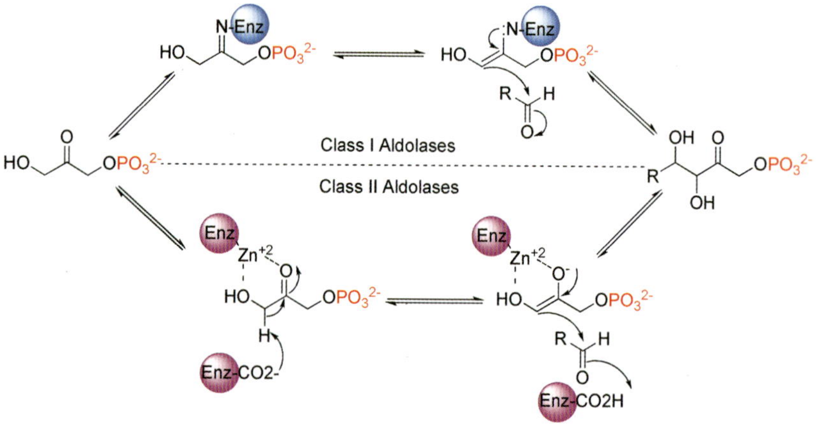

**Scheme 4.1** Simplified representation of the reaction mechanisms of class I and class II aldolases.

Aldolases have been extensively used in chemoenzymatic syntheses and many of these applications have been compiled in several superb reviews over the last 10 years [5].

The main group of aldolases from the biocatalytic point of view is, arguably, the one that uses dihydroxyacetone phosphate (DHAP) as donor. Here, we will concentrate on that applications in which DHAP-dependent aldolase are part of a multi-enzyme system or, alternatively, on those in which the aldolase-catalyzed reaction is key in a multi-step synthetic pathway.

## 4.2
## DHAP-Dependent Aldolases

DHAP-dependent aldolases produce 2-keto-3,4-dihydroxy adducts with high control of the configuration of the two newly formed stereogenic centers. However, while it can be assumed that the absolute configuration at C3 is independent on the acceptor used in the reaction, the configuration of the stereocenter generated from the addition to the aldehyde (C4 position) in some cases may depend on the structure and stereochemistry of the acceptor [6].

An advantage of these enzymes is that they are stereocomplementary, in that they can synthesize the four possible diastereoisomers of vicinal diols from achiral aldehyde acceptors and DHAP (Scheme 4.2). Although this statement is generally used and accepted, it is not completely true since tagatose-1,6-bisphosphate aldolase (TBPA) from *Escherichia coli*–the only TBPA that has been investigated in terms of its use in synthesis–does not seems to control the stereochemistry of the aldol reaction when aldehydes different from the natural substrate were used as acceptors [7]. However, this situation could be modified soon since it has been demonstrated that the stereochemical course of TBPA-catalyzed C—C bond formation may be modified by enzyme-directed evolution [8].

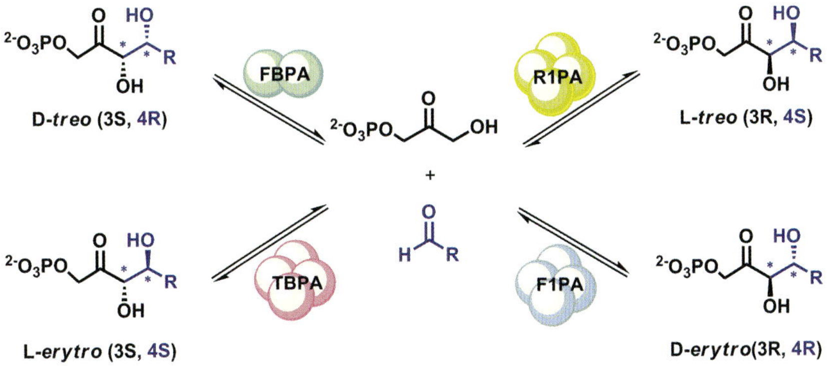

**Scheme 4.2** Complementary stereochemistry of DHAP-dependent aldolases.

### 4.2.1
### Problem of DHAP Dependence

The main drawback of the DHAP-dependent aldolases is their strict specificity for the donor substrate. Apart from the scope limitation that this fact represents, DHAP is expensive to be used stoichiometrically in high-scale synthesis, and labile at neutral and basic pH, and therefore its effective concentration decreases over time in enzymatic reaction media, hindering the overall yield of the aldol reaction. In addition, due to the presence of a phosphate group in both DHAP and the

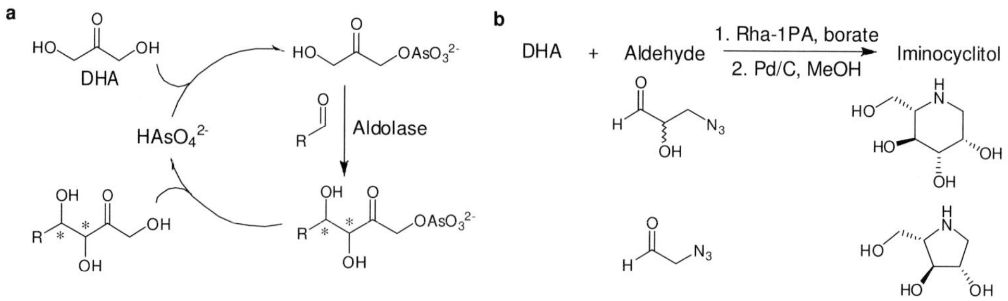

Scheme 4.3 (a) Use of arsenate ester of DHA as mimic of DHAP. (b) Two-step synthesis of iminocyclitols using borate ester of DHA as substrate of the Rha-1PA.

aldol adduct, purification of the reaction product by chromatographic steps is hampered.

Efforts to overcome the DHAP dependence of aldolases have involved the *in situ* formation of arsenate [9] or borate [10] complexes with dihydroxyacetone (DHA), which could mimic a phosphate ester (Scheme 4.3).

The synthetic applicability of arsenates is restricted by their toxicity that avoids the green aspect of the enzymatic processes. Wong *et al.* have shown that the use of inorganic borate buffer allows L-rhamnulose-1-phosphate aldolase (Rha-1PA) to accept DHA as substrate, although the $v_{max}$ of the reaction is about 50 times lower than with the natural substrate [10]. In spite of this fact, these authors have successfully used this approach for the one-step synthesis of L-fructose and L-rhamnulose, and for the facile two-step synthesis of several L-iminocyclitols.

Another approach to avoid the use of DHAP is the use of enzyme-directed evolution strategies to modify the substrate specificity of these aldolases. In a pioneering work in this field [11], Wong *et al.* have reported the development of an *in vivo* selection system for the directed evolution of Rha-1PA. Using this selection system, in a first round screening of an error-prone polymerase chain reaction-generated library, some colonies showed L-rhamnulose aldolase activity. Although preliminary, these results are quite promising and could be applied for directed evolution of other DHAP-dependent aldolases.

Regardless of the above, an efficient method of DHAP preparation is still essential. Several chemical routes of DHAP synthesis have been described in the literature. Of these routes, those starting from the DHA dimmer [12] or 1,3-dibromoacetone [13] are the most attractive since they provide a stable precursor of DHAP (Scheme 4.4). Their main drawback is the relatively low overall yield.

Here, we will focus on the enzymatic routes since enzymatic preparation of DHAP is usually coupled with the aldol addition catalyzed by the aldolase representing genuine multi-enzyme systems.

DHAP can be prepared by oxidation of L-glycerol-3-phosphate (L- G3P) catalyzed by glycerophosphate oxidase, coupled with hydrogen peroxide decomposition in the presence of catalase (Scheme 4.5) [14]. More recently, this synthetic route has

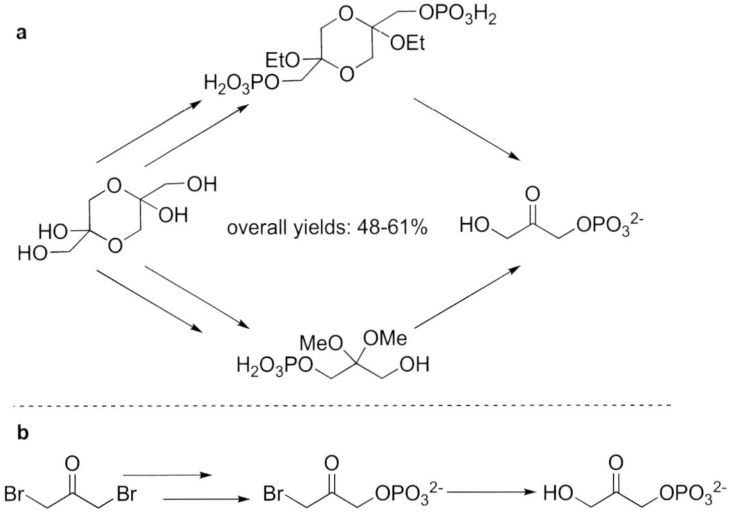

a

overall yields: 48-61%

b

**Scheme 4.4** Chemical routes to DHAP. (a) Routes from dihydroxyacetone dimer. The stable precursors are converted to DHAP by acid hydrolysis. (b) Route from 1,3-dibromoacetone. The stable precursor is converted to DHAP by treatment with NaOH.

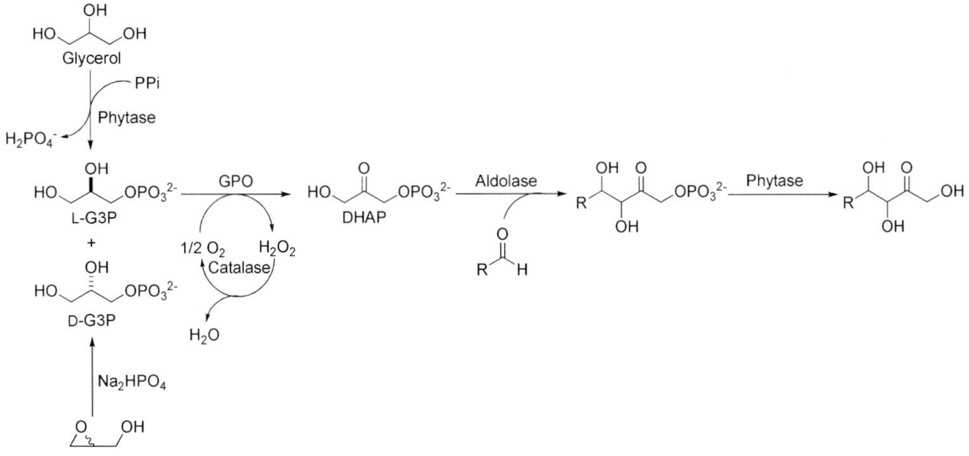

Glycerol

PPi

Phytase

$H_2PO_4^-$

L-G3P

+

D-G3P

$Na_2HPO_4$

*rac*-glycidol

GPO

1/2 $O_2$ $H_2O_2$

Catalase

$H_2O$

DHAP

Aldolase

Phytase

**Scheme 4.5** Enzymatic routes to DHAP based in the use of glycerophosphate oxidase (GPO) coupled with the aldolase-catalyzed reaction and with dephosphorylation of the aldol adduct.

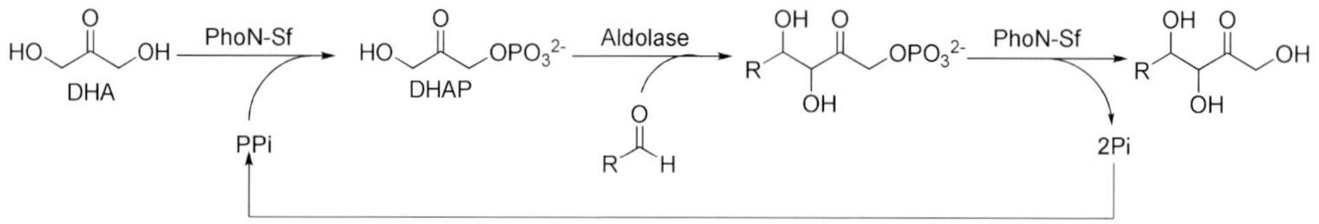

**Scheme 4.6** Cascade reaction for aldol adduct production with *in situ* formation of DHAP from DHA and $PP_i$, and recycling of the phosphate group.

been coupled to the *in situ* formation of D,L-G3P, either by phosphorylation of glycerol by the phosphatase phytase [15] or by regioselective opening of the *rac*-glycidol epoxide ring with phosphate [16].

Fessner and Sinerius, in their seminal work [14], showed that the depicted strategy was adaptable to the synthesis of DHAP analogs modified at the phosphate group and from two of them they could identify the aldol adduct as mixtures with remaining starting material.

In the multi-enzyme system described by Sheldon *et al.* [15], the key point is the use of the phosphatase phytase from *Aspergillus ficuum*, which is a cheap and readily available industrial enzyme. Phytase is active at acid pH and becomes inactive at neutral pH. Thus, the pH can be used to switch on/off the activities of the various enzymes, allowing us to carry out the four-enzyme cascade in one pot.

Wever *et al.* [17] have described another cascade reaction for the *in situ* generation of DHAP using the acid phosphatase from *Shigella flexneri* (PhoN-Sf) and $PP_i$ as phosphate donor (Scheme 4.6).

This procedure takes advantage of the reversibility of the phosphatase reaction, since it allows the recycling of the phosphate group. The phosphatase (i) catalyzes the phosphorylation of DHA and (ii) dephosphorylates the aldol adduct. This phosphate group is used to phosphorylate another DHA molecule when present in excess to generate more DHAP, which is used in aldol coupling. The main drawback of this method, as the own authors recognize, it is the large quantities of DHA and $P_i$ that are present during purification.

Kinase-catalyzed DHA phosphorylation, using ATP as phosphoryl donor, is another strategy to synthesize DHAP. This approach was first described in 1983 by Wong and Whitesides using the enzyme glycerol kinase [18]. Owing to the high cost of the ATP and the fact that the resulting ADP is usually a potent inhibitor of the kinases, synthesis requires to be coupled with an ATP regeneration system (Scheme 4.7) (for some reviews on ATP regeneration, see [19]). The most used ATP regeneration systems are those based on the use of phosphoenolpyruvate as the phosphate donor in a coupled reaction catalyzed by pyruvate kinase and those based on the use of acetylphosphate coupled with acetate kinase (AK). Although it has never been used in the phosphorylation of DHA, regeneration of ATP is possible by phosphorylation of ADP catalyzed by polyphosphate kinase using inorganic polyphosphate PPi as phosphoryl donor [20].

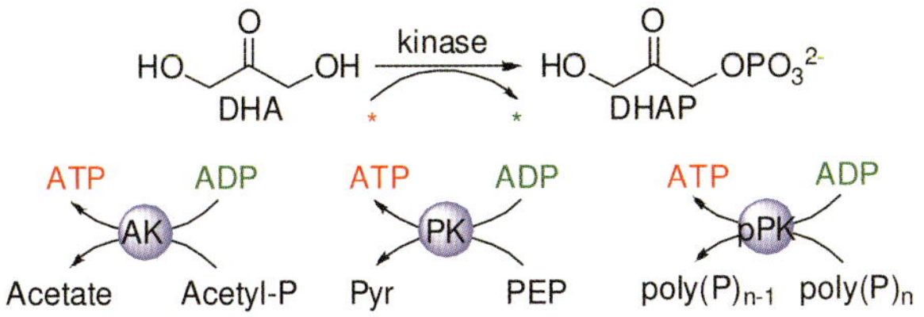

**Scheme 4.7** General scheme of the kinase-dependent phosphorylation of DHA with *in situ* regeneration of ATP.

Dihydroxyacetone kinases (DHAKs) have also been employed for DHA phosphorylation. DHAKs are a family of enzymes that catalyze the formation of a phosphoester bond between a hydroxy group of the DHA and a phosphate molecule to yield the glycolytic intermediate DHAP. According with the source of high-energy phosphate used, DHAKs can be classified in two groups: (i) the phosphotransferase-dependent group, which utilizes a phosphoprotein that belongs to the bacterial phosphoenolpyruvate:carbohydrate phosphotransferase system as phosphoryl donor, and (ii) the ATP-dependent group, which employs ATP as phosphoryl donor. Although DHAKs are widely distributed in the three biological kingdoms, only their roles in the catabolism of glycerol and in methanol assimilation in microorganisms have been well characterized (for a review on the function, structure and phylogeny of dihydroxyacetone kinases, see [21]). From the biocatalysis point of view, ATP-dependent DHAKs have been given considerable attention because of the feasibility of using them for the simple and efficient production of DHAP. Thus, Itoh *et al.* have shown that the DHAK isoenzyme I, present in the yeast *Schizosaccharomyces pombe* IFO 0354 strain, is a useful biocatalyst for the production of DHAP when coupled with an *in situ* regeneration system for ATP using AK [22]. In our research group, we have developed a straightforward multi-enzyme system for one-pot C–C bond formation [23]. This route integrates a recombinant ATP-dependent DHAK from *Citrobacter freundii* CECT 4626 for *in situ* DHAP formation, fuculose-1-phosphate aldolase (Fuc-1PA) for the aldolase-catalyzed reaction and the regeneration of ATP by AK (Scheme 4.8).

For the regeneration of ATP, we chose the system based in the use of acetyl phosphate as final phosphoryl donor because this affords several advantages: (i) acetyl phosphate is easily obtained by acylation of phosphoric acid with acetic anhydride in ethyl acetate [24], and (ii) the resulting sodium acetate is a non-toxic and an environmentally compatible compound. However, this regeneration system is quite sensitive to pH changes. Thus, a continuous adjustment of the pH to 7.5 is needed to maintain the proper operation of the system. Perhaps the main aspect of this approach is that the DHAP must be formed at the same rate as it is consumed by the aldolase. To avoid the accumulation of DHAP and minimize its non-enzymatic degradation, fine tuning of the aldolase/DHAK activities is needed. This adjustment must be experimentally optimized for some acceptors.

Although all the routes described above are quite attractive since they are one-pot routes to the phosphorylated aldol adduct, a considerable number of issues remain

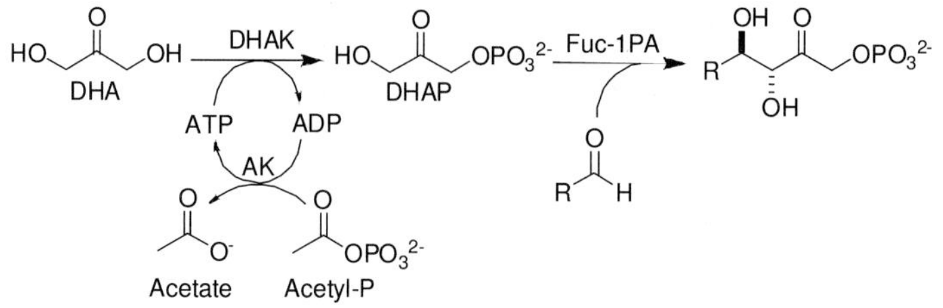

| R | $H_3C$ / HO | $CH_3$– | $CH_3CH_2$– | $CH_3CHCH_2CH_2$– | $H_3C$ / $H_3C$ | $H_3CO$ / $H_3CO$ | HO / OH |
|---|---|---|---|---|---|---|---|
| Aldol yield | 88.8 | 73.0 | 28.3 | 27.3 | 26.3 | 93.5 | 64.4 |

**Scheme 4.8** Multi-enzyme system for the facile one-pot C–C bond formation catalyzed by Fuc-1PA from readily available DHA and an aldehyde acceptor.

to be addressed and solved (for an in-depth discussion of the chemical and enzymatic routes to DHAP, see [25]).

## 4.2.2
## DHAP-Dependent Aldolases in the Core of Aza Sugar Synthesis

Apart from these drawbacks and limitations, DHAP-dependent aldolases have been extensively used in synthesis. One of the main applications of DHAP-dependent aldolases has been in the synthesis of iminocyclitols. Sugar mimics, in which the ring oxygen atom has been replaced by nitrogen (aza sugars), inhibit glycosidases and glycosyltransferases by mimicking the transition state of the enzymatic reaction (Scheme 4.9).

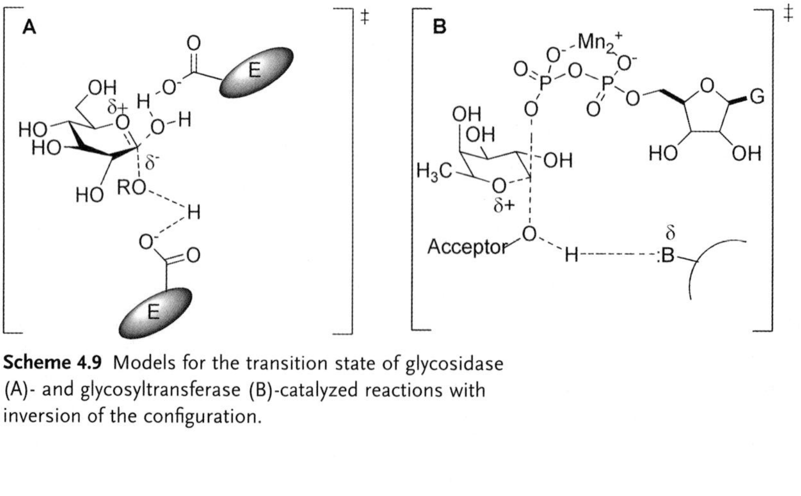

**Scheme 4.9** Models for the transition state of glycosidase (A)- and glycosyltransferase (B)-catalyzed reactions with inversion of the configuration.

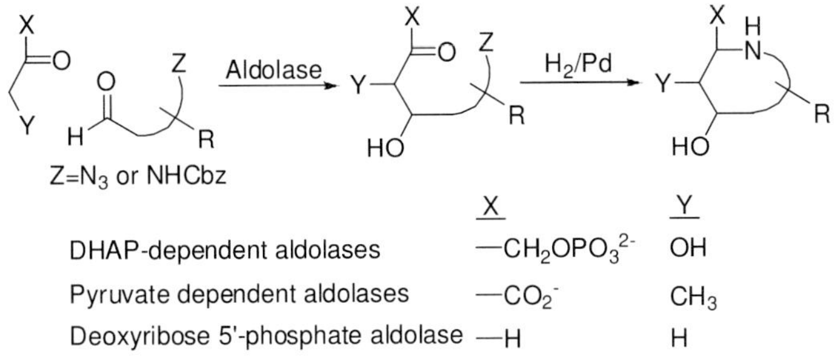

Aldolase $H_2/Pd$

$Z = N_3$ or NHCbz

| | X | Y |
|---|---|---|
| DHAP-dependent aldolases | $—CH_2OPO_3^{2-}$ | OH |
| Pyruvate dependent aldolases | $—CO_2^-$ | $CH_3$ |
| Deoxyribose 5'-phosphate aldolase | —H | H |

**Scheme 4.10** General strategy for the chemoenzymatic synthesis of iminocyclitols based in the use of aldolases and palladium-mediated reductive amination.

The realization that imino sugar glycosidase inhibitors might have enormous therapeutic potential in many diseases by altering the glycosylation or catabolism of glycoproteins has led to a tremendous interest in and demand for these compounds [26]. The aldolase-based approach for the synthesis of iminocyclitols is valid for all the aldolases, regardless of the donor used, and consists of two main steps: (i) the aldolase-catalyzed aldol reaction between a nitrogen-containing aldehyde and the corresponding ketone donor, and (ii) the reductive amination catalyzed by Pd/C (Scheme 4.10) [27].

Azido aldehydes and α-protected amino aldehydes have been used to incorporate the nitrogen in the aldolase-catalyzed reaction (for an extensive review on aldolase-mediated synthesis of iminocyclitols, see [28] and references therein). The sterically unhindered azido and *N*-formylamino aldehydes display a marked kinetic advantage over analogs having larger and/or poor water-soluble N-protecting groups [29].

However, the use of the well-known benzyloxycarbonyl (Cbz) or *tert*-butyloxycarbonyl N-protecting groups for aminoaldehydes is interesting, mainly because they provide a complement for orthogonal protection schemes very useful when the aldol adducts are to be used as chiral building blocks. To avoid the poor water solubility of these N-protected aminoaldehydes, Clapés *et al.* have recently developed new reaction systems based on colloidal dispersions, namely highly concentrated water-in-oil (gel) emulsions, which overcome most of the disadvantages of the aqueous/cosolvent mixtures such as inactivation of the aldolase and incomplete aldehyde solubilization in the medium [6b, 30].

When DHAP-dependent aldolases are used as catalyst of the aldol reaction, a phosphorylated azido or amino polyhydroxyketone is obtained. The phosphate may be cleaved enzymatically or reductively cleaved under the hydrogenation conditions of the next step in which the azide is reduced to the amine. Intramolecular imine formation occurs spontaneously when the azide is reduced. The intramolecular reductive amination is the second key step of the aldolase-mediated synthesis of iminocyclitols. In general, delivery of hydrogen onto five- and six-membered ring imines occurs from the face opposite to the C4 hydroxyl group,

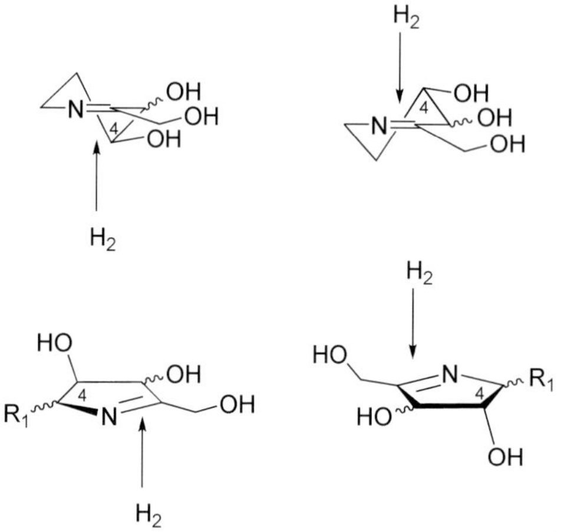

**Scheme 4.11** Diastereoselectivity of the reductive amination. Hydrogenation takes place from the face opposite to the C4.

regardless of the relative stereochemistry of the other substituents, or from the face that affords the least torsional strain in the product (Scheme 4.11) [31]. Therefore, the stereochemistry observed at C2 is controlled exclusively by the configuration at C4 [6b].

Several seven-membered iminocyclitols are potent inhibitors of glycosidases, and some exhibit even higher inhibition potencies than the five- and six-membered counterparts [32]. This observation can be explained because the seven-membered ring species are conformationally more flexible than the corresponding five- and six-membered species, and for this reason may adopt a quasi-flattened conformation with minimum energetic demand, which could lead to a favorable binding in the enzyme active site. The enzymatic syntheses of seven-membered iminocyclitols involve the combined use of aldolases and isomerases. The isomerase displaces the carbonyl group one carbon atom and subsequently the ring is expanded to seven members. Reaction of (±)-3-azido-2-hydroxypropanaldehyde with DHAP catalyzed by an aldolase followed by treatment with acid phosphatase (APase) and an isomerase gave 6-azido-6-deoxyaldopyranoses, which upon reductive amination afforded the corresponding 3,4,5,6-tetrahydroxyazepane [33]. Isomerization of the ketose to aldose is only partial although the equilibrium favors the aldose product (Scheme 4.12).

Lemaire *et al.* have developed a efficient fructose-1,6-bisphosphate aldolase (FBPA)-mediated synthesis of aminocyclitol analogs of valiolamine [34]. This one-pot route involves the formation of two C—C bonds where four stereocenters are created. The first C—C bond formation reaction is catalyzed by the aldolase, coupling DHAP to nitrobutyraldehydes; the other one is the result of a highly stereoselective intramolecular Henry reaction occurring on the intermediate nitroketone under acidic conditions during the aldolase-catalyzed reaction and phytase-catalyzed phosphate hydrolysis coupled step (Scheme 4.13).

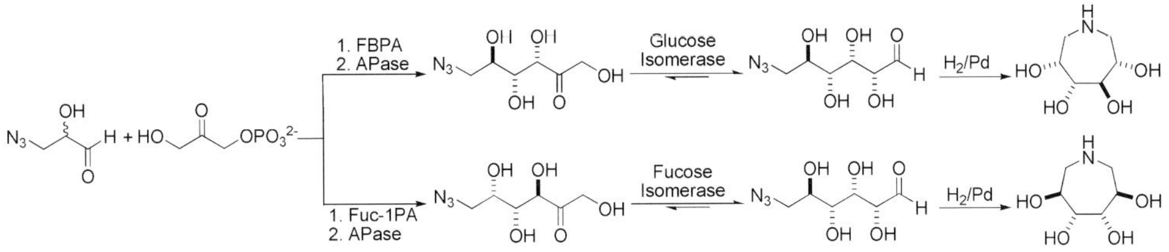

Scheme 4.12 Synthesis of polyhydroxylated azepanes based in the combined use of aldolases and isomerases.

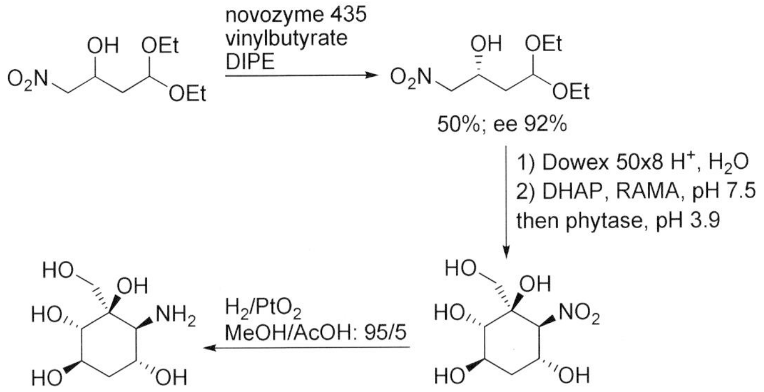

Scheme 4.13 Multi-enzyme route for the synthesis of aminocyclitol analogs of valiolamine.

The nitroalcohol precursor was resolved by the lipase-catalyzed acylation of the hydroxy group. The nitrobutyraldehyde was obtained by acid-mediated hydrolysis of the nitrodiethylacetal and used directly in the aldolization reaction after pH adjustment to 7.5.

### 4.2.3
### Combined Use of Aldolases and Isomerases for the Synthesis of Natural and Unnatural Sugars

Rare or unnatural monosaccharides have many useful applications as non-nutritive sweeteners, glycosidase inhibitors and so on. For example, L-glucose and L-fructose are known to be low-calorie sweeteners. In addition, rare or unnatural monosaccharides are potentially useful as chiral building blocks for the synthesis of biologically active compounds. Therefore, these compounds have been important targets for the development of enzymatic synthesis based in the use of DHAP-dependent aldolases alone or in combination with isomerases. Fessner *et al.* showed that rare ketose-1-phosphates could be reached not only by aldol addition catalyzed by DHAP-dependent aldolases, but by enzymatic isomerization/phosphorylation of aldoses [35]. Thus, for example, L-fructose can be prepared

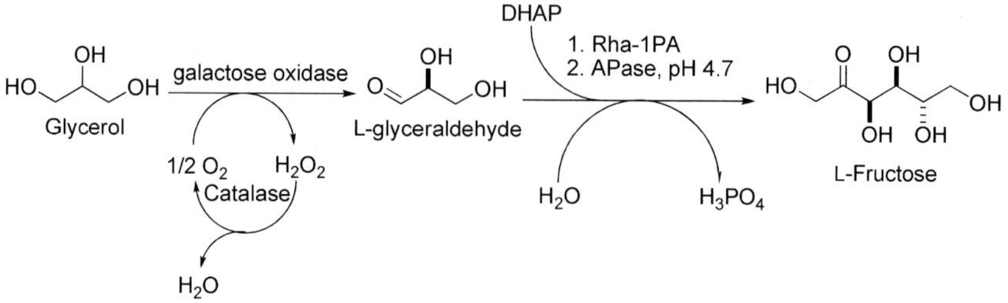

**Scheme 4.14** L-Fructose synthesis starting from glycerol and DHAP using a multi-enzyme system with galactose oxidase, Rha1PA, catalase and APase.

by L-mannose isomerization catalyzed by rhamnulose isomerase or by the aldol reaction of L-glyceraldehyde with DHAP in the presence of Rha-1PA followed by removal of the phosphate group catalyzed by APase [35a, 36]. The aldol addition route suffers from two areas of limitation: (i) L-glyceraldehyde is not commercially available, its preparation on the gram scale is inefficient and the use of toxic heavy metals limits its applications, and (ii) the intermediate L-fructose-1-phosphate was isolated as a barium salt, and additional steps were required to remove and recycle the toxic metal salts. To avoid these limitations, Wong *et al.* coupled the *in situ* stereoselective oxidation of glycerol by galactose oxidase with the aldolase-catalyzed reaction (Scheme 4.14) [37].

Unfortunately, this multi-enzyme system cannot be performed in one pot, probably because DHAP is oxidized by galactose oxidase faster than the glycerol. In fact, the $v_{max}/K_M$ of the galactose oxidase for DHAP is $0.048\,mM^{-1}\,min^{-1}$, whereas for glycerol it is $0.38\,M^{-1}\,min^{-1}$.

The synthesis of the more prevalent aldose sugars can be achieved by enzymatic isomerization of aldolase-derived ketose sugars. This strategy was first described by Wong *et al.* for the synthesis of several hexoaldose derivatives using fructose bisphosphate aldolase and glucose isomerase as catalysts [38]. Combining different aldolases and isomerases, and following the same strategy, other aldoses have been prepared. Thus, L-glucose has been synthesized from L-glyceraldehyde with Rha-1PA, APase and fucose isomerase (FucI) [36], and L-fucose has been enzymatically synthesized from DHAP and D,L-lactaldehyde catalyzed by L-Fuc-1PA, followed by reaction with APase and FucI (Scheme 4.15) [39].

However, the applicability of this strategy is limited by the substrate specificity of the isomerases so that only a fraction of the ketoses that can be obtained from the aldose-catalyzed reaction can be enzymatically isomerized to the corresponding aldose. Moreover, the isomerization reaction is reversible and, as a ketone is generally more stable than an isomeric aldehyde, the equilibrium produces substantial aldose isomer only if the aldose sugar can exist in a very stable aldopyranose form [38b].

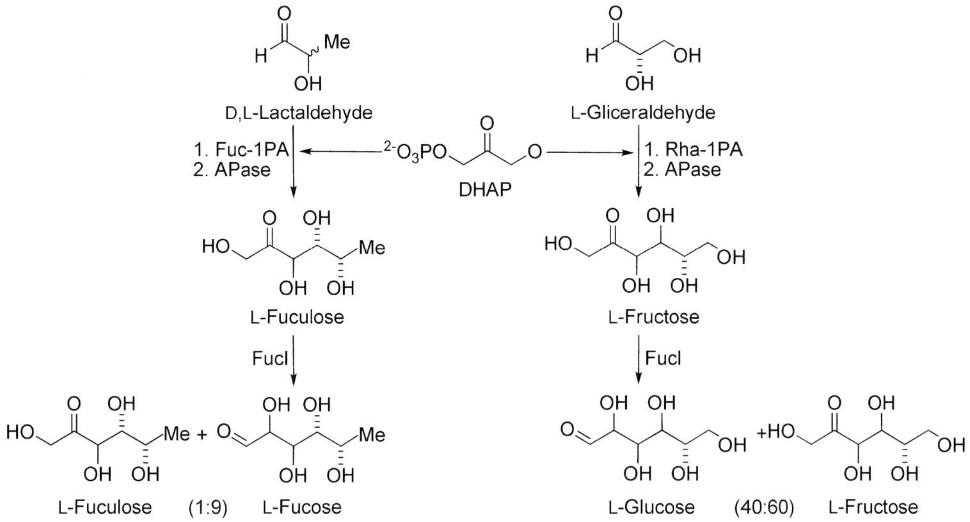

Scheme 4.15 Enzymatic synthesis of L-glucose and L-fucose by aldolization/isomerization.

### 4.2.4 DHAP-Dependent Aldolases in the Synthesis of Natural Products

DHAP-dependent aldolases have also been used as key step in the synthesis of several complex natural products starting from achiral precursors. Thus, the sex pheromone (+)-*exo*-brevicomin can be synthesized in a multi-step route starting with the stereospecific aldol addition between DHAP and 5-oxohexanal or its 5-dithiane-protected analog catalyzed by FBPA from rabbit muscle ('RAMA') as the key step by which the absolute configuration of the target is established (Scheme 4.16) [40].

To complete the synthesis of brevicomin, the side-chain had to be deoxygenated. This could be achieved by a reduction of the keto group, elimination of the resulting diol to the olefin and reduction of the double bond.

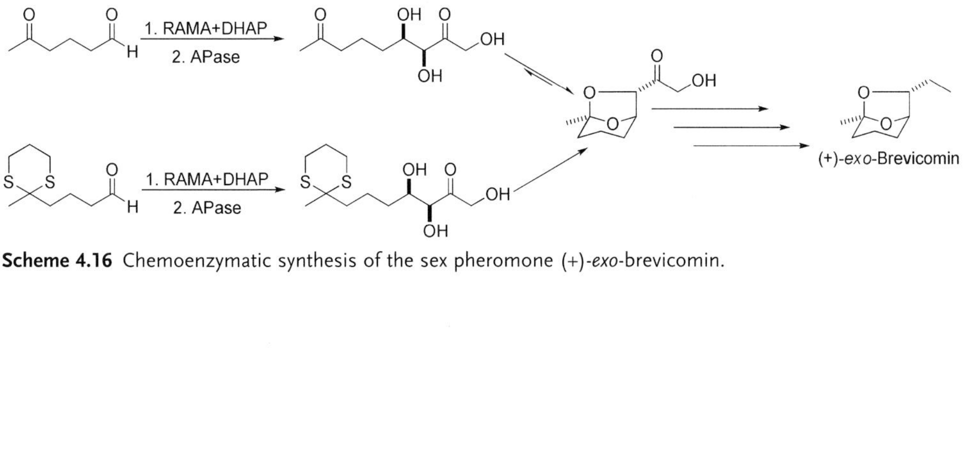

Scheme 4.16 Chemoenzymatic synthesis of the sex pheromone (+)-*exo*-brevicomin.

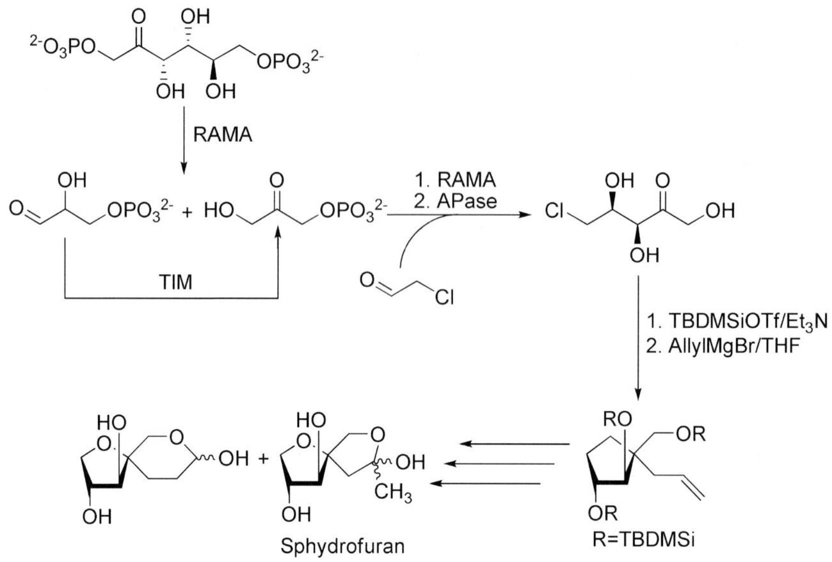

**Scheme 4.17** Chemoenzymatic synthesis of sphydrofuran.

Sphydrofuran is a structurally unique secondary metabolite produced by a variety of *Streptomycetes* strains. The first total synthesis of this compound was based in the use of RAMA to catalyze the aldol addition of chloroacetaldehyde with DHAP that provides two of the three chiral centers of the target molecule. The third quaternary center was introduced via a highly diastereoselective Grignard addition of allylmagnesium bromide (Scheme 4.17) [41].

In this case, DHAP was generated *in situ* from fructose-1,6-bisphosphate using RAMA and triosephosphate isomerase (TIM). To introduce the carbonyl functionality present in the target compound via a Wacker reaction was necessary to change the *tert*-butyldimethylsilyl protecting groups to acetyl groups. Sphydrofuran and its aldehyde analog were easily separated by column chromatography.

Fessner *et al.* have reported an elegant strategy for the stereospecific synthesis of novel pancratistatin analogs [42]. The pancratistatin alkaloid and its closely related natural congeners, including notably *trans*-dihydrolycoricidine and the anhydro and deoxy derivatives narciclasine and lycoricidine (Figure 4.2), have attracted considerable attention due to their biological activities [43].

The major difficulty in the synthesis of these molecules is the generation of the contiguous four to six chiral centers. The synthetic target in the strategy developed by Fessner *et al.* was a pancratistatin analog in which the cyclitol C is substituted by a pyranose ring. The strategy combines the enzymatic dihydroxylation and aldolization to create four contiguous stereocenters (Scheme 4.18).

The naphtho[2,3-*d*]-1,3-dioxole was oxidized using recombinant *E. coli* whole cells overexpressing the gene for the naphthalene dioxygenase from *Pseudomonas putida* G7. The linear carbohydrate fragment was enzymatically formed after ozonization of the diol, by chain extension with a dihydroxyacetone fragment in an

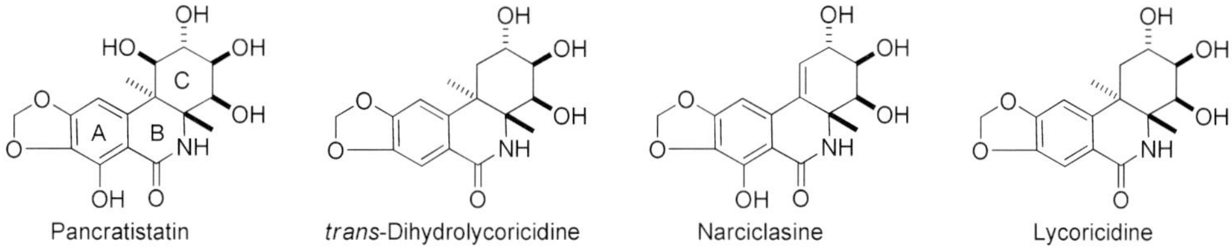

Pancratistatin trans-Dihydrolycoricidine Narciclasine Lycoricidine

**Figure 4.2** Structure of the alkaloid pancratistatin and analogs.

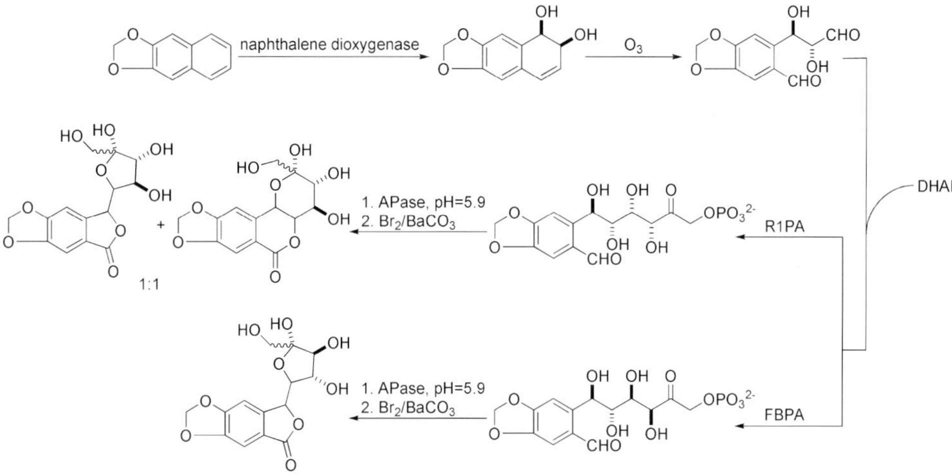

naphthalene dioxygenase $O_3$ DHAP R1PA FBPA 1. APase, pH=5.9 2. $Br_2/BaCO_3$ 1:1

**Scheme 4.18** Multi-step process for the chemoenzymatic synthesis of pancratistatin analogs.

aldolase-catalyzed reaction, which simultaneous provides the desired absolute configuration. This approach takes advantage of the different reactivity of the aromatic carbaldehydes and hydroxyaldehydes with the aldolases to regiospecifically modify only the former aldehyde [44]. After enzymatic dephosphorylation, stable lactones were furnished by mild oxidation. In the reaction catalyzed by the Rha-1PA, both the desired pyranoid isomer and an equal fraction of the furanoid isomer were formed. However, in the reaction catalyzed by the FBPA, only the undesired furanoid product was found.

Shimagaki *et al.* reported the synthesis of the C11–C16 fragment of the pentamycin based in the stereoselective C—C bond formation reaction catalyzed by FBPA [45]. Pentamycin is a polyene macrolide antibiotic, whose configurations at C15 and C14 would correspond to those of the C3 and C4 positions of an aldol constructed from addition of DHAP–derived from FBP by use of FBPA and TIM–to the corresponding aldehyde catalyzed by FBPA (Scheme 4.19).

The reaction catalyzed by FBPA followed of dephosphorylation by APase, gave a mixture (3.7 : 1) at the anomeric position of 7-*O*-benzyl-6-deoxy-D-xyloheptulose (Scheme 4.20).

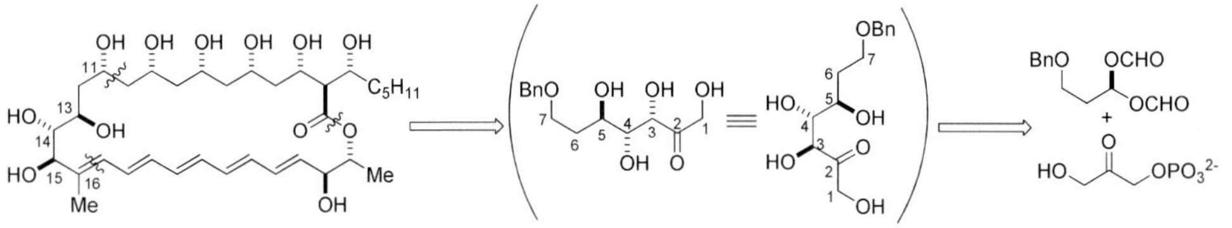

**Scheme 4.19** Correspondence between the configurations at C15 and C14 of pentamycin and those of the C3 and C4 positions of an aldol constructed by FBPA.

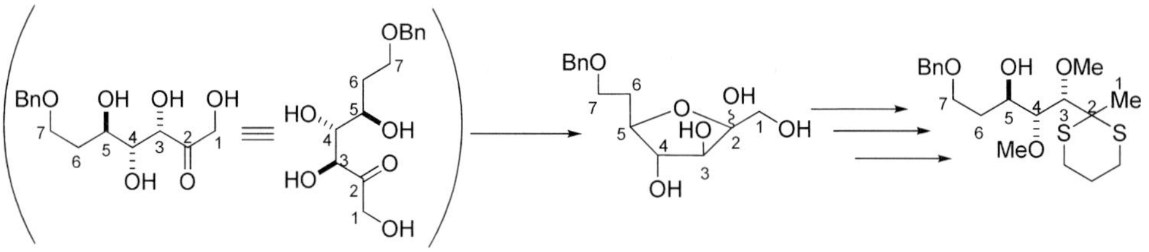

**Scheme 4.20** Chemical transformation of 7-*O*-benzyl-6-deoxy-D-xyloheptulose in the precursor of the C11–C16 fragment of pentamycin.

Further chemical transformation of the xyloheptulose derivative afforded a direct precursor of the C11–C16 fragment of pentamycin.

DHAP-dependent aldolases have also been used in the synthesis of the C3–C9 fragment of aspicilin [46] and of the C12–C20 fragment of amphotericin [47].

Other interesting natural product in which a key synthesis step is the aldol reaction catalyzed by a DHAP-dependent aldolase is syringolide 2. Syringolides are *C*-glycosidic microbial elicitors that have been isolated from the bacterial plant pathogen *Pseudomonas syringae* pv. *Tomato* [48]. These compounds triggers a hypersensitive response in resistant cultivars of soybeans that involves a rapid, localized cell death and accumulation of phytoalexins around the infection site. Chênevert and Dasser [49] synthesized this compound in five steps through a FBPA-catalyzed reaction (Scheme 4.21).

## 4.3
## Fructose-6-Phosphate Aldolase: An Alternative to DHAP-Dependent Aldolases?

Although fructose-6-phosphate aldolase (FSA) does not belong to the DHAP-dependent aldolases group, it deserves to be mentioned in this chapter as it can be considered as an alternative to those enzymes, or at least, an alternative to FBPA. FSA was described for the first time by Schürmann and Sprenger in *E. coli* K-12 strain MG1655 [50]. The enzyme is a class I aldolase with a homodecameric

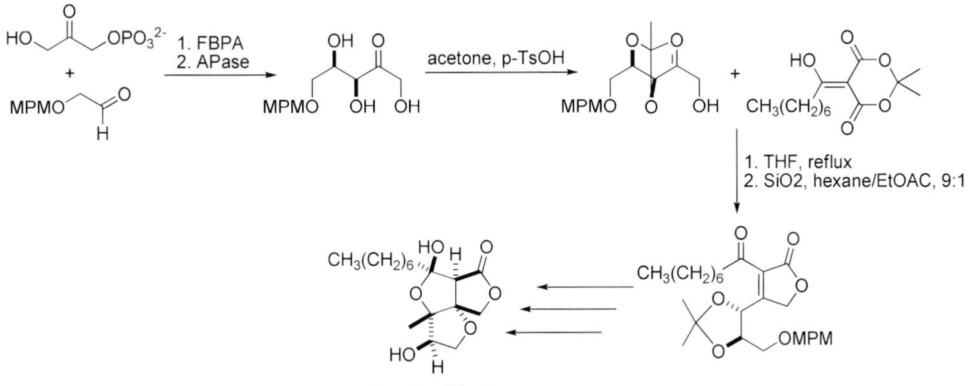

**Scheme 4.21** Chemoenzymatic synthesis of syringolide 2.

structure. It is highly thermostable with a half-life that ranges between 200 h at 55 °C and 16 h at 75 °C, and also presents a broad range of activity in buffers from pH 6.0 to 12.0. The enzyme is enantiospecific forming (3*S*,4*R*)-diols like FBPA. The substrate specificity appeared to be narrow with fructose-6-phosphate being the only substrate for aldol cleavage, but broader in the aldol-forming activity. Thus, as well as DHA the FSA from *E. coli* can use hydroxyacetone [51] and 1-hydroxy-2-butanone [52] as donor, and several hydroxyaldehydes and their respective phosphorylated forms as acceptors [50, 51]. However, DHAP is not accepted as a donor. The flexibility shown by FSA with respect to the donor is quite unusual between aldolases and it is another interesting characteristic of this enzyme. Thus, several rare 1-deoxy sugars and derivatives have been synthesized using hydroxyacetone as donor [51].

The use of this enzyme in multi-step synthesis is relatively recent. Clapés *et al.* have reported the first example of FSA-mediated synthesis of iminocyclitols [53]. The synthetic strategy is similar to the one previously described for DHAP-dependent aldolases without the need for the dephosphorylation step. Aldolic reaction of DHA with *N*-Cbz-3-aminopropanal catalyzed by FSA followed by selective catalytic reductive amination furnishes the naturally occurring imino-sugar D-fagomine (Scheme 4.22).

*N*-Alkylated derivatives could be obtained by catalytic reductive amination from a mixture of D-fagomine, or its linear precursor, and the corresponding aldehyde.

Using a similar strategy, Wong *et al.* have also reported the synthesis of several iminocyclitols [52]. These authors use 2-azido aldehydes as acceptor to prepare five-membered iminocyclitols and *N*-Cbz-3-amino aldehydes to prepare six-membered iminocyclitols. Taking advantage of the donor tolerance of FSA, they also employ hydroxyacetone and 1-hydroxy-2-butanone to obtain several known and novel iminocyclitols.

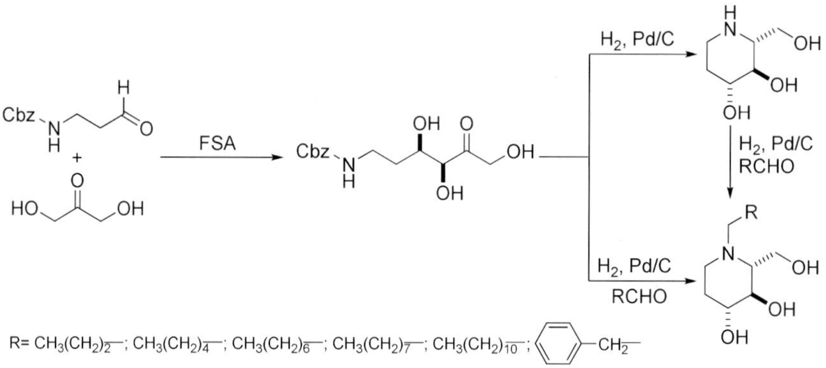

R= $CH_3(CH_2)_2$–; $CH_3(CH_2)_4$–; $CH_3(CH_2)_6$–; $CH_3(CH_2)_7$–; $CH_3(CH_2)_{10}$–; $C_6H_5$–$CH_2$–

**Scheme 4.22** FSA-mediated synthesis of D-fagomine and *N*-alkyl derivatives.

FSA is an interesting novel tool in chemoenzymatic synthesis. It use greatly simplifies the enzymatic procedures based on the use of DHAP-dependent aldolases since it is employs non-phosphorylated donors and, as a consequence, the reaction products do not require a subsequent dephosphorylation step. Unfortunatly, only one stereoconfiguration of the aldol adduct is accessible.

## 4.4 Conclusions

The examples described above, and others, demonstrate that DHAP-dependent aldolase-catalyzed reactions can be efficiently coupled in multi-step processes for the asymmetric synthesis of carbohydrates, imino sugars, cyclitols and even more complex natural products. However, the need for DHAP is still an unresolved problem. We have described several interesting strategies that facilitate the access to DHAP, but they do not prevent the need to pass through the use of this donor. Definitive solution to this drawback may come from several approaches. Directed evolution has already shown great potential to modify substrate specificity and stereospecificity of aldolases and other enzymes, but examples focused on the DHAP problem are scarce. Metagenomics as well as elucidation of new microorganism genomes are opportunities to find new aldolases with different donor specificities. In spite of the many advances that have taken place during recent years in the field of aldolases-mediated synthesis, an exciting and challenging future remains.

## Acknowledgments

Research at the authors' laboratory has been supported by grants from the Spanish Ministry of Education and Science (Grant CTQ2004-03523/BQU). L.I. acknowledges a Predoctoral Fellowship from Comunidad de Madrid.

## References


1 Corey, E.J. and Cheng, X.M. (1989) *The Logic of Chemical Synthesis*, John Wiley & Sons, Ltd, New York.

2 (a) Li, C.J. (2005) *Chemical Reviews*, **105**, 3095–166.
(b) Palomo, C., Oiarbide, M. and García, J.M. (2004) *Chemical Society Reviews*, **33**, 65–75.(c) Alcaide, B. and Almendros, P. (2003) *Angewandte Chemie (International Edition in English)*, **42**, 858–60.

3 Sukumaran, J. and Hanefeld, U. (2005) *Chemical Society Reviews*, **34**, 530–42.

4 (a) Rutter, W.J. (1964) *Federation Proceedings*, **23**, 1248–57.
(b) Horecker, B.L., Tsolas, O. and Lai, C.-Y. (1972) *The Enzymes*, 3rd edn, VII (ed. P.D. Boyer), Academic Press, New York, pp. 213–58.

5 (a) Samland, A.K. and Sprenger, G.A. (2006) *Applied Microbiology and Biotechnology*, **71**, 253–64.
(b) Hamilton, C.J. (2004) *Natural Product Reports*, **21**, 365–85.
(c) Breuer, M. and Hauer, B. (2003) *Current Opinion in Biotechnology*, **14**, 570–6.
(d) Silvestri, M.G., Desantis, G., Mitchell, M. and Wong, C.-H. (2003) *Topics in Stereochemistry* (ed. E.D. Scott), Wiley-VCH, Weinheim, Vol. 23, pp. 267–342.
(e) Fessner, W.-D. and Helaine, V. (2001) *Current Opinion in Biotechnology*, **12**, 574–86.
(f) Machajewski, T.D. and Wong, C.-H. (2000) *Angewandte Chemie (International Edition in English)*, **39**, 1352–74.
(g) Wymer, N. and Toone, E.J. (2000) *Current Opinion in Chemical Biology*, **4**, 110–19.
(h) Seoane, G. (2000) *Current Organic Chemistry*, **4**, 283–304.
(i) Liu, J.Q., Dairi, T., Itoh, N., Kataoka, M., Shimizu, S. and Yamada, H. (2000) *Journal of Molecular Catalysis B: Enzymatic*, **10**, 107–15.

6 (a) Fessner, W.-D., Sinerius, G., Schneider, A., Dreyer, M., Schulz, G.E., Badia, J. and Aguilar, J. (1991) *Angewandte Chemie (International Edition in English)*, **30**, 555–8.
(b) Lin, C.H., Sugai, T., Halcomb, R.L., Ichikawa, Y. and Wong, C.H. (1992) *Journal of the American Chemical Society*, **114**, 10138–45.
(c) Espelt, L., Parella, T., Bujons, J., Solans, C., Joglar, J., Delgado, A. and Clapés, P. (2003) *Chemistry–A European Journal*, **9**, 4887–99.

7 Fessner, W.-D. and Eyrisch, O. (1992) *Angewandte Chemie (International Edition in English)*, **31**, 56–8.

8 Williams, G.J., Domann, S., Nelson, A. and Berry, A. (2003) *Proceedings of the National Academy of Sciences of the United States of America*, **100**, 3143–8.

9 (a) Drueckhammer, D.G., Durrwachter, J.R., Pederson, R.L., Crans, D.C., Daniels, L. and Wong, C.-H. (1989) *Journal of Organic Chemistry*, **54**, 70–7.
(b) Schoevaart, R., van Rantwijk, F. and Sheldon, R.A. (2001) *Journal of Organic Chemistry*, **66**, 4559–62.

10 Sugiyama, M., Hong, Z., Whalen, L.J., Greenberg, W.A. and Wong, C.-H. (2006) *Advanced Synthesis Catalysis*, **348**, 2555–9.

11 Sugiyama, M., Hong, Z., Greenberg, W.A. and Wong, C.-H. (2007) *Bioorganic and Medicinal Chemistry*, **15**, 5905–11.

12 (a) Jung, S.H., Jeong, J.-H., Miller, P. and Wong, C.-H. (1994) *Journal of Organic Chemistry*, **59**, 7182–4.
(b) Charmantray, F., El Blidi, L., Gefflaut, T., Hecquet, L., Bolte, J. and Lemaire, M. (2004) *Journal of Organic Chemistry*, **69**, 9310–12.

13 Gefflaut, T., Lemaire, M., Valentin, M.-L. and Bolte, J. (1997) *Journal of Organic Chemistry*, **62**, 5920–2.

14 Fessner, W.-D. and Sinerius, G. (1994) *Angewandte Chemie (International Edition in English)*, **33**, 209–12.

15 (a) Schoevaart, R., van Rantwijk, F. and Sheldon, R.A. (1999) *Chemical Communications*, 2465–6.
(b) Schoevaart, R., van Rantwijk, F. and Sheldon, R.A. (2000) *Journal of Organic Chemistry*, **65**, 6940–3.

16 Charmantray, F., Dellis, P., Samreth, S. and Hecquet, L. (2006) *Tetrahedron Letters*, **47**, 3261–3.

17 van Herk, T., Hartog, A.F., Schoemaker, H.E. and Wever, R. (2006) *Journal of Organic Chemistry*, **71**, 6244–7.

18 Wong, C.-H. and Whitesides, G.M. (1983) *Journal of Organic Chemistry*, **48**, 3199–205.

19 (a) Zhao, H. and van der Donk, W.A. (2003) *Current Opinion in Biotechnology*, **14**, 583–9.
(b) Endo, T. and Koizumi, S. (2001) *Advanced Synthesis Catalysis*, **343**, 521–6.
(c) Shiba, T., Tsutsumi, K., Ishige, K. and Noguchi, T. (2000) *Biochemistry (Moscow)*, **65**, 315–23.

20 (a) Murata, K., Uchida, T., Kato, J. and Chibata, I. (1988) *Agricultural and Biological Chemistry*, **52**, 1471–7.
(b) Noguchi, T. and Shiba, T. (1998) *Bioscience, Biotechnology, and Biochemistry*, **62**, 1594–6.

21 Erni, B., Siebold, C., Christen, S., Srinivas, A., Oberholzer, A. and Baumann, U. (2006) *Cellular and Molecular Life Sciences*, **63**, 890–900.

22 Itoh, N., Tujibata, Y. and Liu, J.Q. (1999) *Applied Microbiology and Biotechnology*, **51**, 193–200.

23 Sánchez-Moreno, I., García-García, J.F., Bastida, A. and García-Junceda, E. (2004) *Chemical Communications*, 1634–5.

24 Crans, D.C. and Whitesides, G.M. (1983) *Journal of Organic Chemistry*, **48**, 3130–2.

25 Schümperli, M., Pellaux, R. and Panke, S. (2007) *Applied Microbiology and Biotechnology*, **75**, 33–45.

26 (a) Withers, S.G., Namchuk, M. and Mosi, R. (1999) *Iminosugars as Glycosidase Inhibitors: Nojirimycin and Beyond* (ed. A.E. Stütz), Wiley-VCH, Weinheim, pp. 188–206.
(b) Lillelund, V.H., Jensen, H.H., Liang, X. and Bols, M. (2002) *Chemical Reviews*, **102**, 515–53.
(c) Compain, P. and Martin, O.R. (2003) *Current Topics in Medicinal Chemistry*, **3**, 541–60.

27 Look, G.C., Fotsch, C.H. and Wong, C.-H. (1993) *Accounts of Chemical Research*, **26**, 182–90.

28 Whalen, L.J. and Wong, C.-H. (2006) *Aldrichimica Acta*, **39**, 63–71.

29 (a) Von der Osten, C.H., Sinskey, A.J., Barbas, C.F.III, , Pederson, R.L., Wang, Y.F. and Wong, C.-H. (1989) *Journal of the American Chemical Society*, **111**, 3924–7.
(b) Pederson, R.L. and Wong, C.-H. (1989) *Heterocycles*, **28**, 477–80.
(c) Hung, R.R., Straub, J.A. and Whitesides, G.M. (1991) *Journal of Organic Chemistry*, **56**, 3849–55.
(d) Romero, A. and Wong, C.-H. (2000) *Journal of Organic Chemistry*, **65**, 8264–8.

30 (a) Espelt, L., Clapés, P., Esquena, J., Manich, A. and Solans, C. (2003) *Langmuir*, **19**, 1337–46.
(b) Espelt, L., Bujons, J., Parella, T., Calveras, J., Joglar, J., Delgado, A. and Clapés, P. (2005) *Chemistry–A European Journal*, **11**, 1392–401.
(c) Calveras, J., Bujons, J., Parella, T., Crehuet, R., Espelt, L., Joglar, J. and Clapés, P. (2006) *Tetrahedron*, **62**, 2648–56.

31 (a) Card, P.J. and Hitz, W.D. (1985) *Journal of Organic Chemistry*, **50**, 891–3.
(b) Kajimoto, T., Liu, K.K.-C., Pederson, R.L., Zhong, Z., Ichikawa, Y., Porco, J.A., Jr and Wong, C.-H. (1991) *Journal of the American Chemical Society*, **113**, 6187–96.
(c) Kajimoto, T., Chen, L., Liu, K.K.-C. and Wong, C.-H. (1991) *Journal of the American Chemical Society*, **113**, 6678–80.

32 (a) Qian, X.-H., Morís-Varas, F., Fitzgerald, M.C. and Wong, C.-H. (1996) *Bioorganic and Medicinal Chemistry*, **4**, 2055–69.
(b) Qian, X.-H, Morís-Varas, F. and Wong, C.-H. (1996) *Bioorganic and Medicinal Chemistry Letters*, **6**, 1117–22.

33 Morís-Varas, F., Qian, X.-H. and Wong, C.-H. (1996) *Journal of the American Chemical Society*, **118**, 7647–52.

34 (a) El Blidi, L., Crestia, D., Gallienne, E., Demuynck, C., Bolte, J. and Lemaire, M. (2004) *Tetrahedron: Asymmetry*, **15**, 2951–4.
(b) El Blidi, L., Ahbala, M., Bolte, J. and Lemaire, M. (2006) *Tetrahedron: Asymmetry*, **17**, 2684–8.

35 (a) Fessner, W.-D., Badia, J., Eyrisch, O., Schneider, A. and Sinerius, G. (1992) *Tetrahedron Letters*, **33**, 5231–4.
(b) Fessner, W.-D., Schneider, A., Eyrisch, O., Sinerius, G. and Badia, J. (1993) *Tetrahedron: Asymmetry*, **4**, 1183–92.

36 Alajarin, R., Garcia-Junceda, E. and Wong, C.-H. (1995) *Journal of Organic Chemistry*, **60**, 4294–5.

37 Franke, D., Machajewski, T., Hsu, C.-C. and Wong, C.-H. (2003) *Journal of Organic Chemistry*, **68**, 6828–31.

**38** (a) Durrwachter, J.R., Sweers, H.M., Nozaki, K. and Wong, C.-H. (1986) *Tetrahedron Letters*, **27**, 1261–4.
(b) Durrwachter, J.R., Drueckhammer, D.G., Nozaki, K., Sweers, H.M. and Wong, C.-H. (1986) *Journal of the American Chemical Society*, **108**, 7812–18.

**39** (a) Wong, C.-H., Alajarin, R., Moris-Varas, F., Blanco, O. and García-Junceda, E. (1995) *Journal of Organic Chemistry*, **60**, 7360–3.
(b) Fessner, W.-D., Gosse, C., Jaeschke, G. and Eyrisch, O. (2000) *European Journal of Organic Chemistry*, 125–32.

**40** (a) Schultz, M., Waldmann, H., Kunz, H. and Vogt, W. (1990) *Liebigs Annalen der Chemie*, 1019–24.
(b) Schultz, M., Waldmann, H., Vogt, W. and Kunz, H. (1990) *Tetrahedron Letters*, **31**, 867–8.

**41** (a) Maliakel, B.P. and Schmid, W. (1992) *Tetrahedron Letters*, **33**, 3297–300.
(b) Maliakel, B.P. and Schmid, W. (1993) *Journal of Carbohydrate Chemistry*, **12**, 415–24.

**42** Phung, A.N., Zannetti, M.T., Whited, G. and Fessner, W.-D. (2003) *Angewandte Chemie (International Edition in English)*, **42**, 4821–4.

**43** (a) Gabrielsen, B., Monath, T.P., Huggins, J.W., Kefauver, D.F., Pettit, G.R., Groszek, G., Hollingshead, M., Kirsi, J.J., Shannon, W.M. and Schubert, E.M. (1992) *Journal of Natural Products*, **55**, 1569–81.
(b) Hoshino, O. (1998) *The Alkaloids* (ed. G.A. Cordell), Academic Press, New York, Vol. 51, pp. 324–424.
(c) Ouarzane-Amara, M., Franetich, J.-F., Mazier, D., Pettit, G.R., Meijer, L., Doerig, C. and Desportes-Livage, I. (2001) *Antimicrobial Agents and Chemotherapy*, **45**, 3409–15.

**44** (a) Fessner, W.-D. and Walter, C. (1996) *Topics in Current Chemistry*, **184**, 97–194.
(b) Fessner, W.-D. (2000) *Stereoselective Biocatalysis* (ed. R.N. Patel), Marcel Dekker, New York, pp. 239–65.

**45** Shimagaki, M., Muneshima, H., Kubota, M. and Oishi, T. (1993) *Chemical & Pharmaceutical Bulletin*, **41**, 282–6.

**46** Chênevert, R., Lavoie, M. and Dasser, M. (1997) *Canadian Journal of Chemistry – Revue Canadienne de Chimie*, **75**, 68–73.

**47** Malleron, A. and David, S. (1996) *New Journal of Chemistry*, **20**, 153–9.

**48** Midland, S.L., Keen, N.T., Sims, J.J., Midland, M.M., Stayton, M.M., Burton, V., Smith, M.J., Mazzola, E.P., Graham, K.J. and Clardy, J. (1993) *Journal of Organic Chemistry*, **58**, 2940–5.

**49** Chênevert, R. and Dasser, M. (2000) *Journal of Organic Chemistry*, **65**, 4529–31.

**50** Schürmann, M. and Sprenger, G.A. (2001) *Journal of Biological Chemistry*, **276**, 11055–61.

**51** Schürmann, M., Schürmann, M. and Sprenger, G.A. (2002) *Journal of Molecular Catalysis B: Enzymatic*, **19–20**, 247–52.

**52** Sugiyama, M., Hong, Z., Liang, P.H., Dean, S.M., Whalen, L.J., Greenberg, W.A. and Wong, C.-H. (2007) *Journal of the American Chemical Society*, **129**, 14811–17.

**53** Castillo, J.A., Calveras, J., Casas, J., Mitjans, M., Vinardell, M.P., Parella, T., Inoue, T., Sprenger, G.A., Joglar, J. and Clapes, P. (2006) *Organic Letters*, **8**, 6067–70.

# 5
# Multi-Enzyme Systems for the Synthesis of Glycoconjugates

*Birgit Sauerzapfe and Lothar Elling*

## 5.1
## Introduction

In the course of the functional characterization of genomes, glycosylation (the glycome) has become one of the most challenging biochemical modifications to elucidate the biological role of carbohydrates and to envisage future application prospects [1]. Glycoproteins, glycolipids, glycosylphosphatidylinositol-anchored proteins and glycosaminoglycans on proteoglycans are the most prominent examples of glycoconjugates. Thus, glycoconjugates comprise a vast variety of molecular structures. They are found on the surfaces of animal and plant cells, as well as on the outer surface of microorganisms. About 50% of human proteins are glycosylated with the consequence that 1–2% of the human genome encodes glycosylation pathway enzymes [2]. The challenge in understanding the biological function of glycan structures on glycoproteins lies in the diversity of glycosyl-amino acid linkages [3] and the heterogeneity of linked carbohydrate structures which is seen, for example, in different glycoforms of glycoproteins [1, 4]. From these studies it becomes more and more clear that glycosylation of proteins has an influence on protein stability [5], on biological activity [6], on protein half-life [7], as well as on a range of pharmacokinetic, pharmacodynamic, immunogenic and safety parameters [8, 9].

Human diseases are now related to changes in glycosylation patterns of glycoproteins that are caused by defects in genes and enzymes involved in glycosylation pathway enzymes [10]. In human medicine, various laboratory assay systems have been developed over the last 15 years that detect alterations in the glycan part of glycoconjugates related to cancer, infectious diseases, genetic defects of glycoconjugate biosynthesis and catabolism, autoimmunity, drug abuse, and liver disease [11]. A promising perspective of glycoconjugates is the utilization of synthetic glycans [12], neoglycopeptides [13, 14] or neoglycoproteins [15] as partial or full components of vaccines targeting tumor-associated antigens or microbe and parasite antigens.

*Multi-Step Enzyme Catalysis: Biotransformations and Chemoenzymatic Synthesis*
Edited by Eduardo Garcia-Junceda

ISBN: 978-3-527-31921-3

Analytical tools have been developed in order to identify carbohydrate structures as well as carbohydrate-binding proteins and to understand their underlying structure–function relationships of protein–carbohydrate and carbohydrate–carbohydrate interactions: lectin arrays [16], glycan microarrays [17, 18], glyco-nanoparticles [19], frontal affinity chromatography [20] and carbohydrate tools for metabolic labeling [21].

In the field of antibiotics the induction of resistance of pathogenic microorganisms is an ongoing problem [22]. Glyco-engineering of bioactive natural products is therefore also in the focus of current pharmaceutical research [23]. Biodiversity of natural products is caused by mono- and oligosaccharides that are attached to specific positions of an aglycone and involved in the molecular recognition event of its target. Many glycosylated bioactive natural compounds are produced as macrolides by actinomycetes with the unique feature to carry complex deoxy sugars. Tools to alter the glycosylation pattern and to generate glycosylated natural products have been developed, and were recently summarized [24–26].

In summary, for all of the above-mentioned applications, the synthesis of carbohydrates and derivatives thereof is necessary to foster the field of glyco-engineering. Many procedures have been established to produce synthetic (neo)glycoconjugates by chemical or enzymatic methods and these are summarized in excellent reviews [12, 27, 28].

Enzymatic methods have already become a viable alternative to organic chemistry in the synthesis of complex carbohydrate structures due to their selectivity and simplicity [25, 27, 29, 30]. Glycosyltransferases and glycosidases are utilized under environmentally benign conditions in aqueous solutions and at ambient temperature. Combinations of various enzymes in a sequence or multi-enzyme one-pot method bring additional advantages for multi-step reaction sequences and for cofactor regeneration [31–33]. The efficiency of some glycosidases for glycosylation reactions has been increased by active-site mutagenesis to generate glycosynthases [34]. First reports describe the expansion and optimization of glycosynthase [34] and glycosyltransferase [35] substrate specificity by directed evolution. However, the repertoire of available glycosynthases for enzymatic synthesis is still small, but is expected to grow. Glycosynthases and glycosyltransferases need special donor substrates, glycosyl fluorides or nucleotide sugars, which have to be synthesized in additional chemical or enzymatic reactions. The simplicity of native glycosidases lies in the use of very cheap substrates from renewable resources, monosaccharides or disaccharides, for the synthesis of glycoconjugates by condensation (reverse hydrolysis) or transglycosylation (thermodynamic approach) reactions. However, glycosidases show lower reaction yields and a mixed regioselectivity which can be improved by reaction engineering (lowering the water content) and substrate engineering (varying the aglycone part of the acceptor substrate), respectively [36]. Although glycosyltransferases are considered to have an absolute regioselectivity it could be demonstrated by substrate engineering that products with different regiospecific glycosidic linkages can be generated by variation of acceptor substrates [37].

## 5.2 *In Vitro* and *In Vivo* Multi-Enzyme Systems

In this chapter we focus on the utilization of multi-enzyme systems to demonstrate their variability and utility for glycoconjugate synthesis. Multi-enzyme systems can be regarded as the combination of enzyme modules that are put together *in vitro* or *in vivo*. Each enzyme module generates a product by one or more biocatalysts. The product can be isolated or directly transferred to the next enzyme module. The complexity of these multi-enzyme systems is determined by the number of enzyme modules and the number of enzymes included in each enzyme module. Figure 5.1 shows the general approach for the synthesis of nucleotide sugars and oligosaccharides or (neo)glycoconjugates by multi-enzyme systems. The enzyme modules are linked by the output of products that are used as substrates of the following module.

*In vitro* multi-enzyme systems are set up by the combination of enzyme modules including pathway and even pathway-unrelated enzymes. Also, the synthesis of saccharides in combination with *de novo* enzymatic sugar synthesis can be accomplished. This so-called **combinatorial biocatalysis** can be performed in sequential reactors or in a one-pot reaction vessel which challenges further reaction engineering for optimization. Even the combination of an enzyme module with a chemical

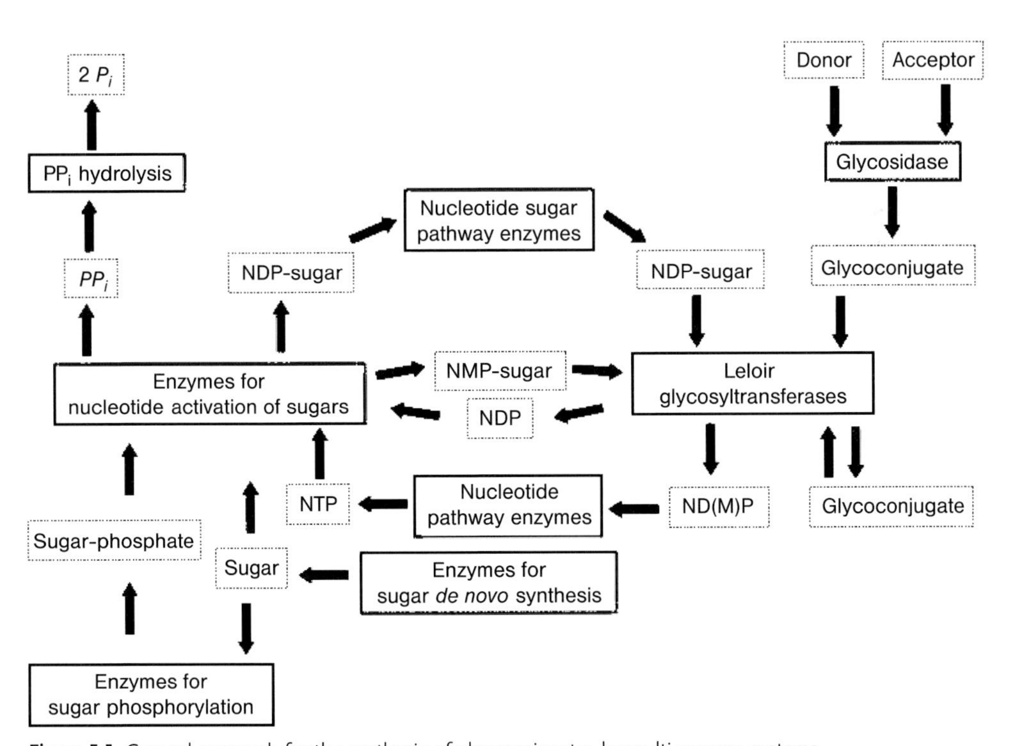


**Figure 5.1** General approach for the synthesis of glycoconjugates by multi-enzyme systems.

step in one pot is realized for glycoconjugate synthesis. Important parameters for the optimization of space–time yields and enzyme-specific productivities are the pH value, the choice of buffer, the variation and relation of enzyme activities, the enzyme kinetics, the enzymes' reaction stabilities, and cross-inhibition of intermediate products.

Taking microbial hosts as 'reaction vessels' opens the way to **combinatorial biosynthesis** by using a set of pathway-related enzymes as modules in combination with metabolic enzymes of the host. This strategy minimizes the overall number of new enzymes added to the host's metabolism, and utilizes enzymes of the host for the regeneration of donor substrates and cofactors. Optimization strategies aim mainly towards genetic manipulation of the production hosts to avoid side reactions of intermediates and the product, as well as towards the modular design and modulation of gene expression by genetic engineering methods to set up novel multi-enzyme systems in the production host. Further critical points are the substrate uptake and product export when growing or resting cells are employed in glycoconjugate synthesis.

In summary, the combination of enzymes is advantageous from an enzymology and reaction engineering point of view. Reaction yields can be increased by avoiding product inhibition of single enzymatic reactions. Product decomposition (e.g. by hydrolysis) can be overcome by further enzymatic transformations. Tedious isolation of intermediate products is not necessary. However, both strategies – combinatorial biocatalysis and combinatorial biosynthesis – have their disadvantages. The *in vitro* approach needs every enzyme to be produced by recombinant techniques and purified in high amounts, which is in some cases difficult to achieve. On the other hand, product isolation from a biotransformation with permeabilized or whole host cells can be tedious and results in low yields.

In the following sections we summarize multi-enzyme systems for the synthesis of glycoconjugates (Table 5.1) following the general scheme for the combination of enzyme modules (see Figure 5.1). Many of these multi-enzyme systems have been already reviewed – here we will highlight those which have been recently been developed.

## 5.3 Combinatorial Biocatalysis

As depicted in Figure 5.1 the routes to glycoconjugates via *in vitro* multi-enzyme systems are not restricted as long as pools of isolated and well-characterized biocatalysts are available for the enzyme module systems set up. These pools comprise Leloir glycosyltransferases, enzymes for the synthesis and *in situ* regeneration of donor nucleotide sugars, enzymes for the generation of nucleotides, glycosidases, and maybe also enzymes for the *de novo* synthesis of monosaccharides, which could themselves serve as acceptor substrates of glycosyltransferases.

**Table 5.1** Multi-enzyme systems for the synthesis of glycoconjugates by the combination of enzyme modules in combinatorial biocatalysis and combinatorial biosynthesis.

| Products | Combination of enzyme modules in | References |
|---|---|---|
| **Nucleotide sugars** | | |
| NDP-α-D-glucose (N = A, C, dT, dU, U)<br>UDP-α-D-GlcA<br>UDP-α-D-Gal<br>UDP-α-D-GlcNAc<br>UDP-α-D-GalNAc<br>GDP-α-D-Man<br>GDP-β-L-Fuc<br>CMP-Neu5Ac | combinatorial biocatalysis | [29, 38, 39] and references therein |
| dTDP-α-D-Glc, dTDP-β-L-Rha | combinatorial biocatalysis | [32] |
| dTDP-D,L-deoxy sugars | | [25] and references therein |
| UDP-α-D Glc<br>UDP-α-D-Gal<br>UDP-α-D-GlcNAc<br>GDP-α-D-Man<br>GDP-β-L-Fuc<br>CMP-Neu5Ac | combinatorial biosynthesis | [40–43] |
| ***In situ* regeneration systems for nucleotide sugars** | | |
| UDP-α-D-Glc<br>UDP-α-D-Gal<br>UDP-α-D-GlcA<br>UDP-α-D-GlcNAc<br>UDP-α-D-GalNAc<br>GDP-α-D-Man<br>GDP-β-l-Fuc<br>CMP-Neu5Ac | combinatorial biocatalysis | [44–49] |
| dTDP-α-D-Glc, dTDP-β-l-Rha | combinatorial biocatalysis | [33] |
| **Oligosaccharides** | | |
| Poly-LacNAc | combinatorial biocatalysis | |
| Blood group epitopes:<br>H(type2),H(type2)–LacNAc,<br>H(type2)–LacNAc–LacNAc | | [50] |
| Lewis epitopes: $Le^x$, $Le^x$–$Le^x$, $Le^x$–$Le^x$–$Le^x$,<br>$Le^y$, $Le^y$–$Le^x$, $Le^y$–$Le^x$–$Le^x$ | | [51] |
| Hyaluronan | | |
| Heparosan (4GlcUA(β1–4)GlcNAc(α1–)$_n$ | | [52] |

**Table 5.1** Continued

| Products | Combination of enzyme modules in | References |
|---|---|---|
| *N*-acetyllactosamine (LacNAc) | combinatorial biosynthesis | [53] |
| 3′-Sialyllactose | | [41] |
| $Le^x$ | | [42, 54] |
| Gal(α1–3)Gal(β1–4)Glc | | |
| Sialyl-$Le^x$ | | [55] |
| Lacto-*N*-neotetraose | | |
| H antigen oligosaccharides | | [56, 57] |
| **Glycolipid oligosaccharides** | | |
| GD3, GT3, GM2, GD2, GT2, GM1, | combinatorial biocatalysis | [58] |
| GD1a, GM3 | | [59] |
| Globotriose | combinatorial biosynthesis | [40] |
| Globotetraose | | [60] |
| P1 trisaccharide | | [61] |
| GM2, GM1 | | [62] |
| **Glycosylated natural products** | | |
| Glycosylated polyketides and macrolides | combinatorial biosynthesis | [24, 25] and references therein [63] |

### 5.3.1
### Synthesis and *In Situ* Regeneration of Nucleotide Sugars

The utilization of Leloir glycosyltransferases has been considered to be too expensive since they need the supply of donor substrates – the nucleotide sugars. However, enzyme modules combining different enzymes have been developed for most common nucleotide sugars and some rare nucleotide deoxy sugars (Figure 5.2 or Table 5.2). The bottleneck to produce nucleotide deoxy sugars is the efficient production of recombinant enzymes involved in the biosynthetic pathways and efficient product isolation. Considering the vast diversity of deoxy sugars attached to natural products, the enzymatic synthesis of nucleotide deoxy sugars is still a challenge [33].

In order to provide dTDP-deoxy sugars by combinatorial biocatalysis we have utilized the enzymes for the dTDP-β-L-rhamnose pathway. The successful combination of pathway enzymes with optimized enzyme productivities (amount of product per unit of enzyme) needs a concise kinetic and inhibition analysis. Scheme 5.1 depicts the biosynthetic pathway of dTDP-β-L-rhamnose with important $K_m$ and $K_i$ constants. The enzymes RmlA and RmlB are highly controlled by the intermediate, dTDP-4-keto-6-deoxy-α-D-glucose **3**, the product **5** or by

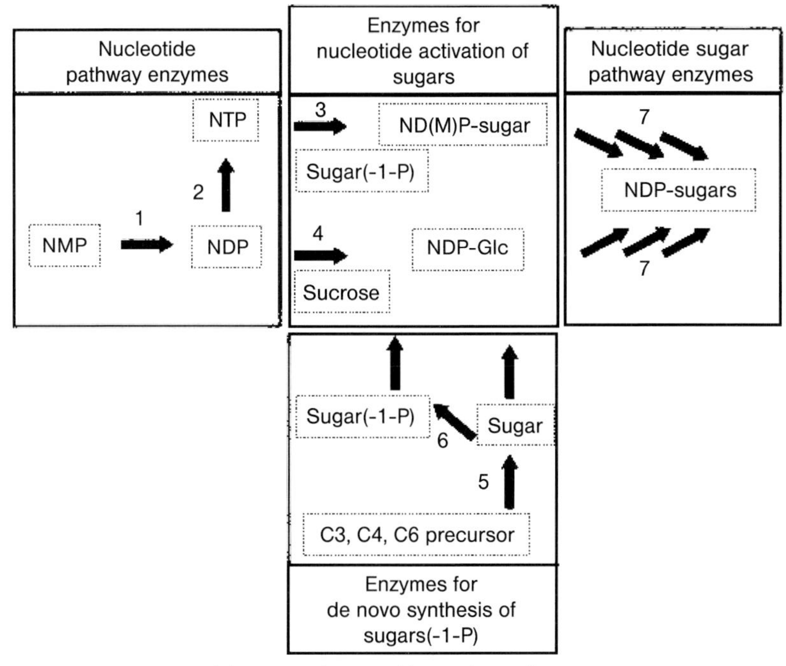


**Figure 5.2** Enzyme modules in combinatorial biocatalysis of nucleotide sugars. See Table 5.2 for enzymes.

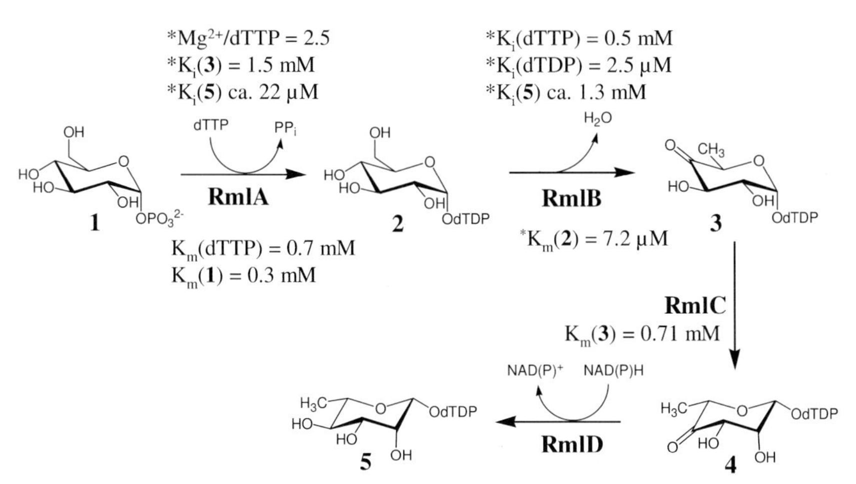


**Scheme 5.1** Biosynthetic pathway of dTDP-β-L-rhamnose with important kinetic and inhibitory constants [68, 70, 73]. RmlA, dTDP-Glc pyrophosphorylase (EC 2.7.7.24); RmlB, dTDP-Glc 4,6-dehydratase (EC 4.2.1.46); RmlC, dTDP-4-dehydrorhamnose 3,5-epimerase (EC 5.1.3.13); RmlD, dTDP-4-dehydrorhamnose reductase (EC 1.1.1.133). An asterisk indicates data from our own analysis.

**Table 5.2** Enzymes used in modules for combinatorial biocatalysis of nucleotide sugars.

| Figure 5.2 | Enzymes (EC numbers) | Products | References |
|---|---|---|---|
| 1 | Myokinase, EC 2.7.4.3<br>Nucleoside monophosphate kinase, EC 2.7.4.4<br>dTMP-kinase, EC 2.7.4.9 | ADP<br>UDP, dUDP, CDP<br>dTDP | [32, 64–66] |
| 2 | Pyruvate kinase, EC 2.7.1.40<br>Acetate kinase, EC 2.7.2.1<br>Polyphosphate kinase, EC 2.7.4.1<br>Creatine kinase, EC 2.7.3.2 | ATP, GTP, UTP, CTP<br>ATP, UTP, dTTP<br>ATP, GTP, UTP, CTP<br>UTP | [44]<br>[66]<br>[47]<br>[49] |
| 3 | UDP-Glc pyrophosphorylase, EC 2.7.7.9<br>UDP-GlcNAc pyrophosphorylase, EC 2.7.7.23<br>UDP-GalNAc pyrophosphorylase, EC 2.7.7.X<br>Gal-1-P uridylyltransferase[a], EC 2.7.7.9<br>GDP-Man pyrophoshorylase, EC 2.7.7.13<br>GDP-Fuc pyrophosphorylase, EC 2.7.7.30<br>RmlA (dTDP-glucose pyrophosphorylase), EC 2.7.7.24<br>CMP-Neu5Ac synthetase, EC 2.7.7.43 | UDP-Glc<br>UDP-GlcNAc<br>UDP-GalNAc<br>UDP-Gal<br>CMP-Neu5Ac<br>GDP-Man<br>GDP-Fuc<br>dTDP-Glc | [38] and references therein<br>[67, 68] |
| 4 | Sucrose Synthase, EC 2.4.1.13 | UDP-Glc<br>dUDP-Glc<br>dTDP-Glc<br>CDP-Glc<br>ADP-Glc | [32, 64, 65] |
| 5 | Aldolases<br>Transketolases | C6-sugars, C9-sugar | [69] |
| 6 | Hexokinase, EC 2.7.1.1<br>Phospoglucomutase, EC 5.4.2.2<br>Phosphomannomutase, EC 5.4.2.8<br>GlcNAc-6-P mutase, EC 5.4.2.3<br>Fuc-1-P kinase, EC 2.7.1.52<br>Gal-1-P kinase, EC 2.7.1.6<br>GalNAc-1-P kinase, EC 2.7.1.157 | Glc-1-P<br>Man-1-P<br>GlcNAc-1-P<br>Fuc-1-P<br>Gal-1-P<br>GalNAc-1-P | [38] and references therein<br>[67] |
| 7 | UDP-Gal(NAc) 4-epimerase, EC 5.1.3.2<br>dTDP-Glc 4,6-dehydratase, EC 4.2.1.46<br>dTDP-4-dehydrorhamnose 3,5-epimerase, EC 5.1.3.13<br>dTDP-4-dehydrorhamnose reductase, EC 1.1.1.133<br>dTDP-4-keto-6-deoxyglucose aminotransferase<br>GDP-Man 4,6,dehydratase, EC 4.2.1.47<br>GDP-Fuc synthetase, EC 1.1.1.271 | UDP-Gal(NAc)<br>dTDP-4-keto-6-deoxyglucose<br>dTDP-4-keto-2,6-dideoxyglucose<br>dUDP-4-keto-6-deoxyglucose<br>dTDP-4-amino-4,6-dideoxy-D-glucose | [38] and references therein<br>[25, 32, 70, 71]<br>[68]<br>[64]<br>[72] |

a Uses Gal-1-P and UDP-Glc as substrates.

both. Also important for the development of an enzyme module system are cross-inhibition effects of the nucleotides. The very strong inhibition of RmlB by dTDP is not obvious from the biosynthetic pathway; however, in enzymatic synthesis it is present as a decomposition product of dTDP-deoxy sugars. A more general aspect is the dependence of pyrophosphorylase activity from the optimum ratio of $Mg^{2+}$ and the nucleoside triphosphate. We have shown that this ratio is specific for individual pyrophosphorylases [68, 74]. Finally, these conditions lead to an optimum reaction time for substrate conversion allowing also the repetitive use of enzyme module systems (repetitive batch synthesis) for optimum enzyme productivity and space–time yield (amount of product per liter reaction volume and day) as applied throughout all of our applied multi-enzyme systems.

The multi-enzyme system was recently utilized for the synthesis of dTDP-4-keto-6-deoxy-α-D-glucose **3** and dTDP-β-L-rhamnose **5** generating dTTP from dTMP by dTMP kinase and acetate kinase [66, 71]. Further combination with GerB, a dTDP-4-keto-6-deoxy-D-glucose aminotransferase, gave dTDP-4-amino-4,6-dideoxy-D-glucose [72].

With 2-deoxy-D-glucose-6-phosphate **6** as starting substrate and an additional enzymatic step the synthesis of dTDP-2,6 dideoxy-4-ketoglucose **7**, an important intermediate of dTDP-deoxy sugars, and dTDP-β-L-olivose **8** was realized (Scheme 5.2) [68].

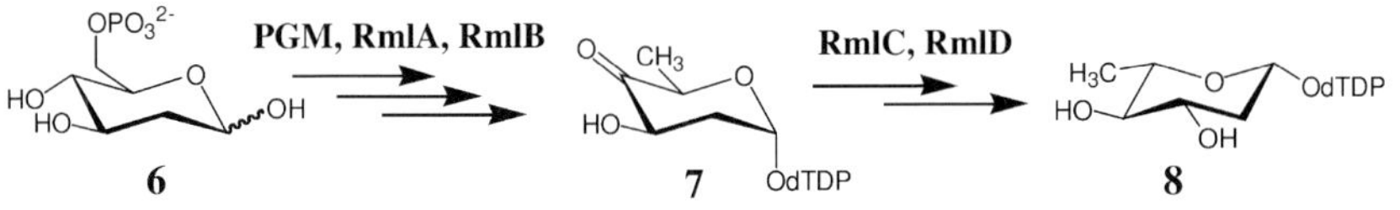

**Scheme 5.2** Enzyme module system for the synthesis of dTDP-4-keto-2,6-dideoxy-α-D-glucose **7** and dTDP-β-L-olivose **8** [68]. PGM, phosphoglucomutase (EC 5.4.2.2). See Scheme 5.1 for other enzymes [68].

A feasible approach to generate nucleotide-activated glucose from sucrose is the utilization of the cleavage reaction (back reaction) of the plant glycosyltransferase sucrose synthase (SuSy, EC 2.4.1.13) [25, 29]. In the case of dTDP-deoxy sugar synthesis the access to dTDP-α-D-Glc **2** as precursor substrate was accomplished by use of SuSy and dTMP kinase from the cheap substrates dTMP and sucrose by a novel enzyme module system (Scheme 5.3). The combination of pathway-unrelated enzymes led further to the enzymatic synthesis of dTDP-4-keto-6-deoxy-α-D-glucose **3** [32, 70] and dTDP-β-L-rhamnose **5** [32, 33]. The SusSy reaction module produced **2** and **3**, respectively, at the gram scale, with an average yield of around 94% in five to 10 repetitive batches. Efficient synthesis of **5** was accomplished by combination of two successive one-pot reactions (**A** and **B** in

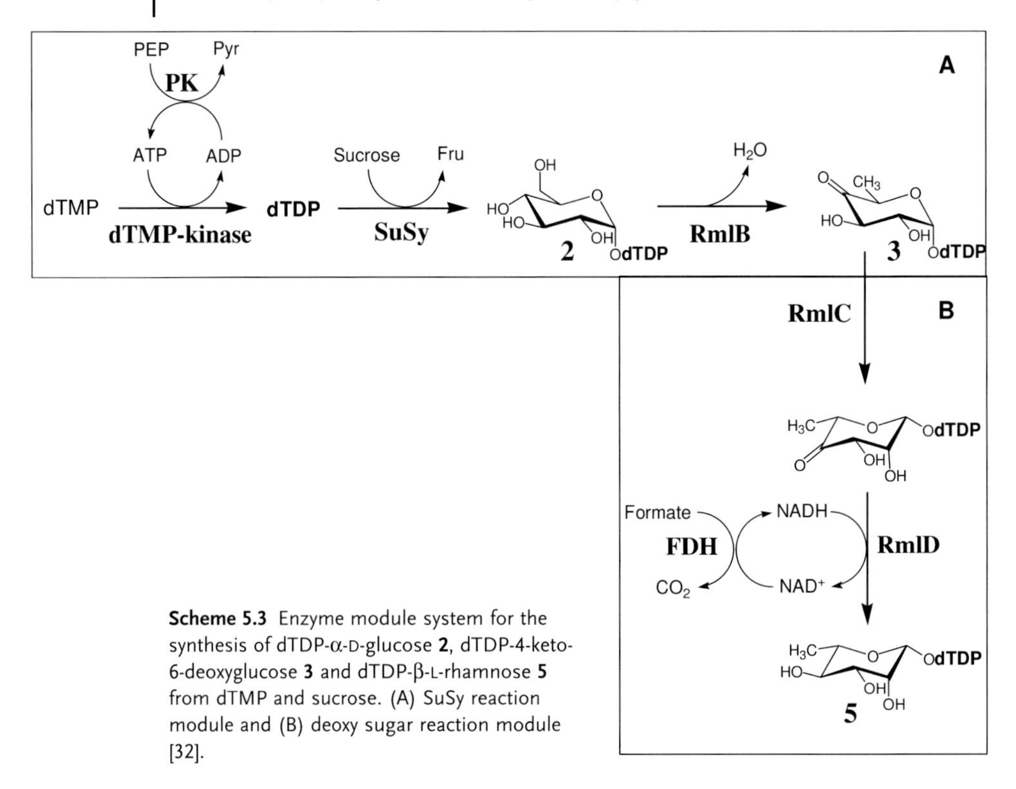


**Scheme 5.3** Enzyme module system for the synthesis of dTDP-α-D-glucose **2**, dTDP-4-keto-6-deoxyglucose **3** and dTDP-β-L-rhamnose **5** from dTMP and sucrose. (A) SuSy reaction module and (B) deoxy sugar reaction module [32].

Scheme 5.3) – the SuSy reaction module and the deoxy sugar reaction module, respectively. Product **3** from the SuSy could be utilized in the next module without prior isolation. However, a one-pot reaction was possible, but with less efficiency due to inhibition of RmlB.

In general, multi-enzyme systems for the *in situ* regeneration of nucleotide sugars recycle the nucleotide byproduct of the glycosyltransferase reaction (Figure 5.1), avoiding tedious and costly isolation of nucleotide sugars. Although these enzyme cascades have been previously developed for most of the common nucleotide sugars (Table 5.1) the synthesis of large amounts of oligosaccharides is now performed without such *in situ* regeneration systems which is mainly due to the kilogram-scale production of nucleotide sugars (UDP-Glc, UDP-Gal, UDP-GlcNAc, GDP-Man, GDP-Fuc, CMP-Neu5Ac) by biotransformation (Table 5.1). However, since large-scale synthesis of UDP-GalNAc is still missing, *in situ* regeneration systems for UDP-GalNAc are of high interest because of the recently published biocatalytic syntheses of mucin glycopeptides and glycolipid oligosaccharide structures as possible vaccines (see Section 5.3.2).

So far, the 4-epimerase reaction [44, 75] has been used to generate UDP-GalNAc from UDP-GlcNAc which is formed by GlmU, a bifunctional UDP-GlcNAc pyrophosphorylase [76] from *Escherichia coli*. Recently, Piller *et al.* developed a novel regeneration cycle for UDP-GalNAc involving the enzymes from the UDP-GalNAc

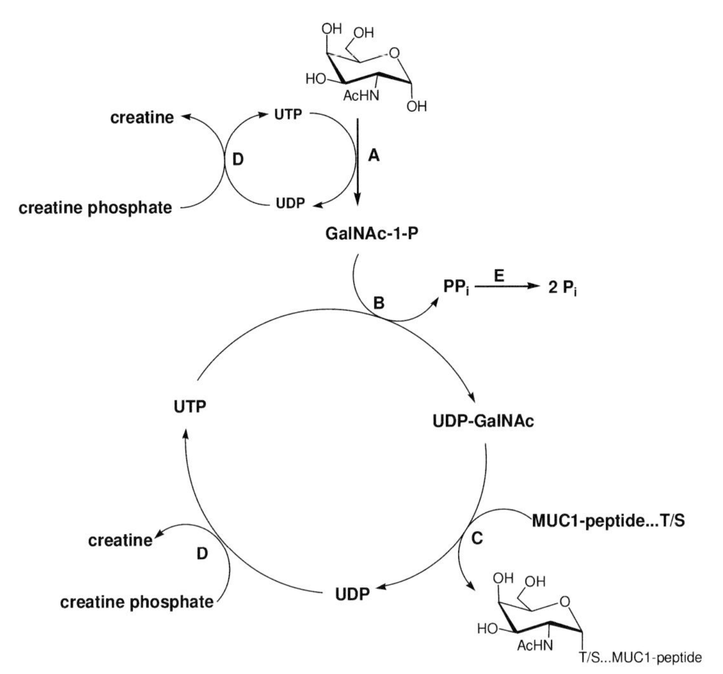


**Scheme 5.4** *In situ* regeneration of UDP-GalNAc [49]. (A) Human GalNAc-1-P kinase, (B) human UDP-GalNAc pyrophosphorylase, (C) bovine polypeptide GalNAc-transferase, (D) rabbit creatine kinase and (E) inorganic pyrophosphatase.

salvage pathway, the human GalNAc-1-P kinase 2 (GK2) and human UDP-GalNAc pyrophosphorylase (AGX1) (Scheme 5.4) [49].

The first *in situ* regeneration cycle for dTDP-deoxy sugars was recently developed [33]. The enzyme module system for dTDP-L-rhamnose was extended by a glycosyltransferase module (Scheme 5.5). Exploiting the donor substrate promiscuity of SorF [77], different sorangiosides were synthesized with *in situ* regeneration of dTDP-glucose, dTDP-4-keto-6-deoxyglucose and dTDP-L-rhamnose, respectively. The byproduct dTDP is directly recycled by SuSy and inhibition effects by dTDP-deoxy sugars onto the pathway enzymes are minimized by an optimum ratio of all six enzymes in the enzyme modules combined in one pot.

In summary, the synthesis and *in situ* regeneration of nucleotide sugars by combinatorial biocatalysis suffers from the main disadvantage that each enzyme has to be produced in sufficient amounts. This affords efficient recombinant protein production hosts being a bottleneck for some genes [25]. However, once a multi-enzyme system has been developed, the productivity can be improved by repetitive use of the biocatalysts as demonstrated for repetitive batch syntheses with soluble enzymes [25, 38] or with immobilized enzymes [48]. The advantage

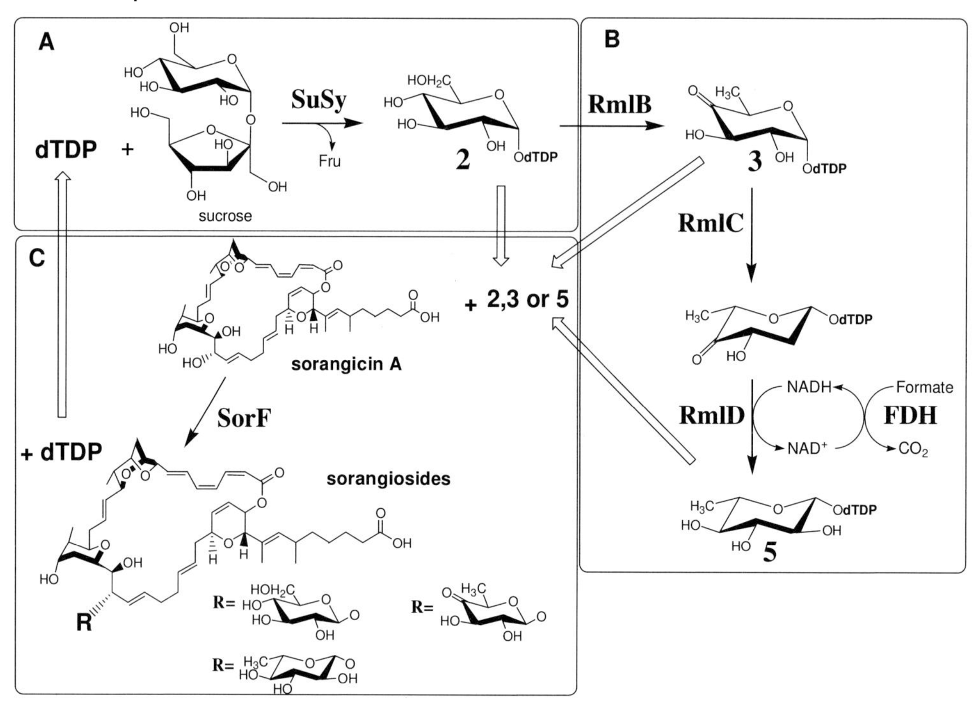


**Scheme 5.5** *In situ* regeneration of dTDP-sugars. (A) SuSy module, (B) deoxy sugar module and (C) glycosyltransferase module [33].

of both methods is the easier purification of the products in the supernatant with immobilized enzymes or by ultrafiltration with repetitive use of soluble enzymes. However, the drawbacks are still multiple production and purification steps of the desired enzymes, and the loss of activity during immobilization and repeated enzyme use.

### 5.3.2
### Synthesis of Oligosaccharides, Glycopeptides and Glycolipids Oligosaccharides

Multi-enzyme systems for the synthesis of a variety of oligosaccharides for use in glycan microarrays [18, 78] have been established by the consortium of functional glycomics (www.functionalglycomics.org). Among them are the tumor-associated sialylated Tn antigen (Neu5Ac(α2–6)GalNAc(α1–Thr/Ser), the sialylated T antigen (Neu5Ac(α2–3)Gal(β1–3)GalNAc(α1–Thr/Ser) and (Gal(β1–3)[Neu5Ac(α2–6)] GalNAc(α1–Thr/Ser) and the disialylated T antigen (Neu5Ac(α2–3)Gal(β1–3) [Neu5Ac(α2–6)]GalNAc(α1–Thr/Ser) [79]. The most impressive examples so far are the gram-scale syntheses of poly-*N*-acetyllactosamines (poly-LacNAcs) carrying blood groups and tumor-associated antigens [50], and of ganglio-oligosaccharides [58] by combinations of different Leloir glycosyltransferases. The successive action

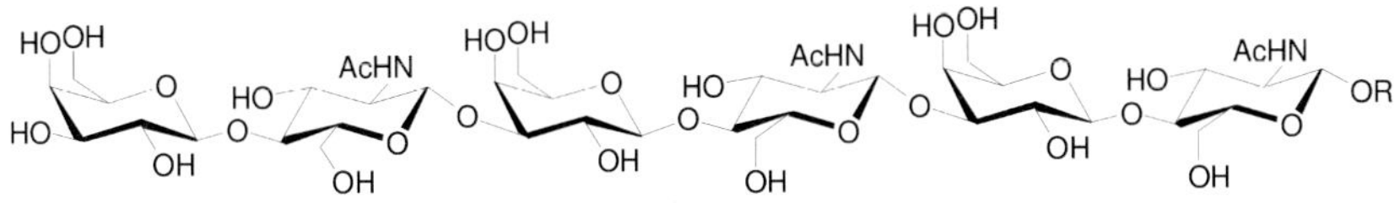

**Scheme 5.6** Poly-LacNAc-derived oligosaccharides synthesized by combinatorial biocatalysis [80].

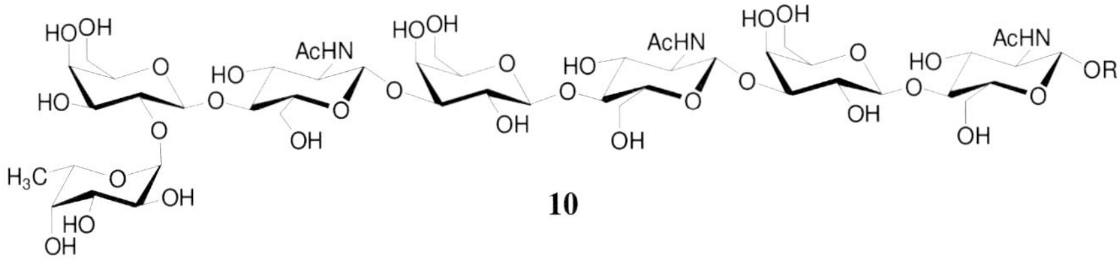

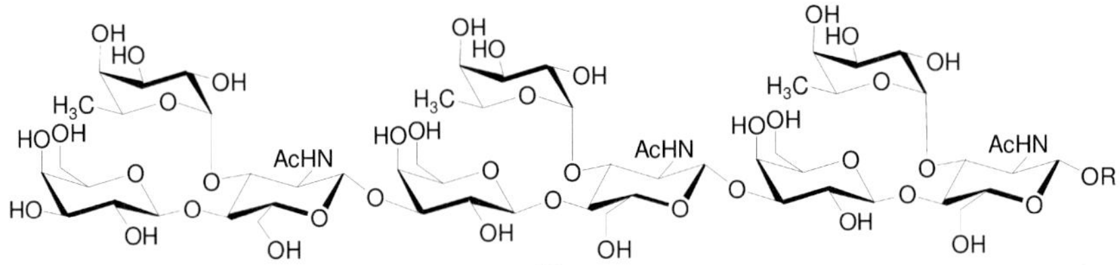

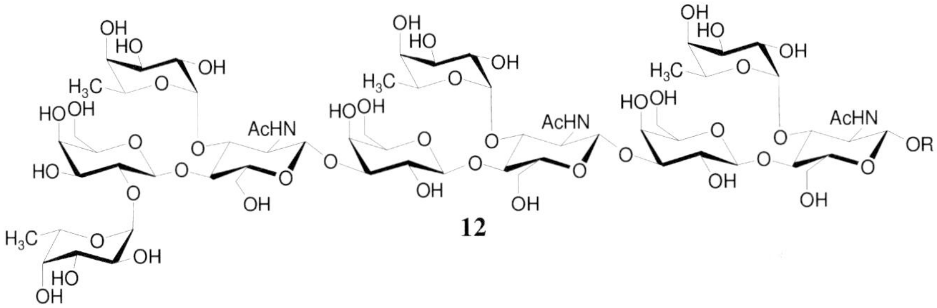

of two recombinant bacterial enzymes β1–4-galactosyltransferase (LgtB, EC 2.4.1.38) [80] and β1–3-*N*-acetylglucosaminyltransferase (LgtA, EC 2.4.1.149) [81], both from *Neisseria meningitides*, resulted in poly-LacNAc with up to four LacNAc units. Interestingly, the donor substrate UDP-Gal was generated *in situ* from inexpensive UDP-Glc by fusion of the bacterial β1–4-galactosyltransferase with UDP-Glc 4′-epimerase from *Streptococcus thermophilus* [80]. In a similar way the fusion of bovine α1–3-galactosyltransferase with GalE from *E. coli* can be exploited for formation of the Galili epitope [82]. The poly-LacNAc structures [e.g. tri-LacNAc (**9**) in Scheme 5.6] were further modified with different recombinant fucosyltransferases (human α1–3-fucosyltransferase VI, FUT-VI, EC 2.4.1.152 and human α1–2-fucosyltransferase II, FUT-II, EC 2.4.1.69) to obtain H(type2),H(type2)–LacNAc, H(type2)–LacNAc–LacNAc (**10**), $Le^x$, $Le^x$–$Le^x$, $Le^x$–$Le^x$–$Le^x$ (**11**) as well as $Le^y$, $Le^y$–$Le^x$, $Le^y$–$Le^x$–$Le^x$ (**12**) (Scheme 5.6). Sialyltransferases (ST3 from

*N. meningitides* and human ST6Gal-I, EC2.4.99.1) were exploited to add sialic acid in α2–3 and α2–6 linkages to the terminal galactose residue of mono-, di- and tri-LacNAc. The fusion of the bacterial ST3 with CMP-Neu5Ac synthetase [83] generated the donor substrate *in situ* from added Neu5Ac and CTP.

In a similar fashion glycosyltransferases from *Campylobacter jejuni* (α2–3/8-sialyltransferase, Cst-II; β1–4-*N*-acetylgalactosaminyltransferase, CgtA; β1–3-galactosyltransferase, CgtB) were combined to yield GD3 (Neu5Ac(α2–8)Neu5Ac(α2–3)Gal(β1–4)Glc(β1–R), GT3 (Neu5Ac(α2–8)Neu5Ac(α2–8)Neu5Ac(α2–3)Gal(β1–4)Glc(β1–R), GM2 (GalNAc(β1–4)[Neu5Ac(α2–3)]Gal(β1–4)Glc(β1–R), GD2 (GalNAc(β1–4)[Neu5Ac(α2–8)Neu5Ac(α2–3)]Gal(β1–4)Glc(β1–R), GT2 (GalNAc(β1–4)[Neu5Ac(α2–8)Neu5Ac(α2–8)Neu5Ac(α2–3)]Gal(β1–4)Glc(β1–R) and GM1 (Gal(β1–3)GalNAc(β1–4)[Neu5Ac(α2–3)]Gal(β1–4)Glc(β1–R), synthesized on the gram scale [58]. GD1a (Neu5Ac(α2–3)Gal(β1–3)GalNAc(β1–4)[Neu5Ac(α2–3)]Gal(β1–4)Glc(β1–R) was generated from GM1 by the addition of a mammalian α2–3-sialyltransferase (ST3Gal I). For cost saving, *in situ* conversion of less-expensive UDP-GlcNAc to UDP-GalNAc was achieved by a rat UDP-*N*-acetylglucosamine 4′-epimerase expressed in *E. coli* AD202 cells.

Multi-enzyme systems were also effectively used in the synthesis of potential glycopeptide and glycolipid vaccines that present tumor-associated carbohydrate antigens. Gangliosides like GM3, GM2, GD3 and GD2 are significantly overexpressed in melanoma and other tumors [84], and represent interesting targets for active cancer immunotherapy. However, since oligosaccharides are poorly immunogenic and subject to decomposition by serum hydrolases such structures have to be conjugated to proteins and chemically stable. In this direction glycosidase-stable *S*-glycosidic oligosaccharide derivatives of GM3 and GM2 have been synthesized involving also GalNAc and Neu5Ac glycosyltransferases as outlined above [59].

In adenocarcinoma cancer cells (breast and ovarian cancer) the mucin 1 (MUC1) is aberrantly glycosylated in tandem repeats of 20-amino-acid residues with five potential *O*-glycosylation sites. This results in the expression of the *O*-glycan carbohydrate antigens Tn (GalNAc(α1-*O*-Ser/Thr), STn (Neu5Ac(α2–6)GalNAc(α1-*O*-Ser/Thr) and T (Gal(β1–3)GalNAc(α1-*O*-Ser/Thr) [85]. Breast carcinoma cells carry the cancer-associated Tn, STn and T antigens, as well as the mono- and disialyl core 1 structure (ST, Neu5Ac(α2–3)Gal(β1–3)GalNAc(α1-*O*-Ser/Thr) and Neu5Ac(α2–3)Gal(β1–3)[Neu5Ac(α2–6)]GalNAc(α1-*O*-Ser/Thr) found widely in normal cells. The density of occupied *O*-glycosylation sites in a tandem repeat sequence is an important parameter for a strong immunogenic glycopeptide [86]. Through the availability of a whole set of polypeptide-*N*-galactosaminyltransferases [87], core 1 β1–3-galactosyltransferase [88] and *O*-glycan sialyltransferases [89] it is now possible to synthesize a variety of cancer-related glycopeptides by multi-enzyme systems [14, 86].

The site-specific introduction of glyco-poly(ethylene glycol) (PEG) chains into pharmaproteins (granulocyte colony-stimulating factor, interferon-α2b and granulocyte macrophage colony-stimulating factor) produced in *E. coli* was accomplished by successive action of polypeptide-*N*-galactosaminyltransferases, core 1 β1–3-

galactosyltransferase and sialyltransferases [90]. Selective glycosylation with GalNAc at their natural *O*-glycosylation sites with specific recombinant polypeptide-*N*-galactosaminyltransferases initiated *O*-glycan biosynthesis *in vitro*. GlycoPEGylation was achieved by specific sialyltransferases which are able to utilize a PEG-modified CMP-Neu5Ac derivative. The pharmacokinetic data of the enzymatically derived PEG-pharmaproteins were similar to those produced by chemical coupling procedures.

Hyaluronan (HA), a glycosaminoglycan composed of repeating (4GlcA(β1–3)-GlcNAc(β1–)) disaccharide units, is a polysaccharide chain with molecular masses ranging from $10^4$ to $10^7$ Da in vertebrates and bacteria. The HA polymer and HA oligosaccharides are of great interest for biomedical applications since they display important biological roles in recognition, signaling, cellular behavior and growth. A two-enzyme system has been developed to synthesize uniform size-defined HA oligosaccharides [51]. The polymerizing bifunctional HA synthase from *Pasteurella multocida* (pmHAS) was converted by mutagenesis into two single-action glycosyltransferases (glucuronic acid transferase and *N*-acetylglucosamine transferase). The immobilized biocatalysts were used to produce single-length size-defined sugar polymers containing up to 10 disaccharide units. With native pmHAS the enzymatic synthesis of monodisperse HA polymers with very narrow size distributions in the range of around 16 kDa to 2 MDa was demonstrated. This concept has been recently applied for the synthesis of monodisperse heparosan (4GlcUA(β1–4)GlcNAc(α1–)$_n$) using pmHS1 from *P. multocida* [52].

The combination of glycosidases and Leloir glycosyltransferases (see Figure 5.1) has been also widely used [91], primarily to increase product yield in the glycosidase reaction by further conversion as acceptor of a successive glycosyltransferase. Since the availability of recombinant glycosyltransferases and their nucleotide sugars has been improved in recent years these strategies have no longer been followed. A great impact to extend strategies for the synthesis of glycoconjugates by combinatorial biocatalysis lies in the *de novo* synthesis of monosaccharides by exploiting the substrate flexibility of aldolases and transketolases (see Figure 5.1) [69].

In summary, in this section selected examples were highlighted which stand for the potential and future directions of glycoconjugate synthesis by combinatorial biocatalysis with multi-enzyme systems. The use of Leloir glycosyltransferases seems to be the most advantageous method for large-scale production of oligosaccharides because of the high regioselectivity and high yields.

## 5.4 Combinatorial Biosynthesis

The utilization of whole cells as bioreactors for the synthesis of oligosaccharides and engineering cell lines for the production of glycoproteins with a human-like glycosylation pattern are highlights in the field of glycobiotechnology.

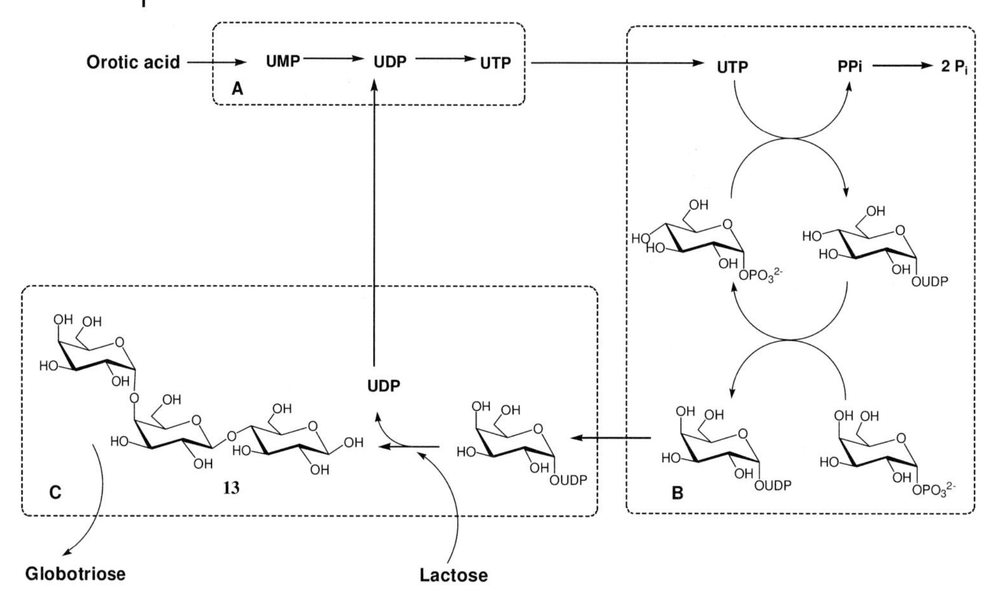


**Scheme 5.7** Combination of three bacterial strains for the production of globotriose (**13**) with regeneration of UDP-Gal [40]. Combination of *C. ammoniagenes* DN510 (A), *E. coli* NM 522 pNT25/pNT32 (B) and *E. coli* NM522 pGT5 (C).

### 5.4.1
### Synthesis of Oligosaccharides with Metabolically Engineered Cells

Metabolically engineered cells were established by combining genes encoding enzymes for glycan synthesis in appropriate expression vectors for use in heterologous expression systems. Two strategies have been developed for the production of oligosaccharides: (i) bacterial coupling using permeabilized cells, and (ii) living cells with active import of substrates and intracellular product accumulation.

The first prominent example of large-scale oligosaccharide production by bacterial coupling was described by Koizumi *et al.* (Scheme 5.7) [40]. Three engineered bacterial strains were combined for the production of globotriose (**13**) including regeneration of UDP-Gal.

The bacterial strain *Corynebacterium ammoniagenes* DN510 provides UTP for the reaction of recombinant glucose-1-phosphate-uridyltransferase generating UDP-Glc in *E. coli* NM 522 pNT25/pNT32, and formation of UDP-Gal with galactose-1-phosphate-uridyltransferase and galactokinase. *E. coli* NM522 pGT5 provides recombinant α1–4-galactosyltransferase. The combination of the three bacterial strains produced 188 g/l globotriose; however, due to the lack of efficient export/import systems for UTP, UDP-Gal and globotriose the cells have to be permeabilized. This technique was further developed for the synthesis of *N*-acetyllactosamine (Gal(β1–4)GlcNAc, LacNAc) [53], 3′-sialyllactose (Neu5Ac(α2–3)Gal(β1–4)Glc) [41] and $Le^x$ (Gal(β1–4)[Fuc(α1–3)]GlcNAc) [42] with a productivity of 107, 36 and 21 g/l, respectively.

The UDP-Gal regeneration system using SuSy and UDP-Glc 4′-epimerase [46] was also established in genetically engineered microorganisms. Permeabilized *E. coli* cells were used for the synthesis of the P1 trisaccharide (Gal(α1–4)Gal(β1–4)GlcNAc) with 5.4 g in a 200-ml reaction volume [61]. Four enzymes (*Anabaena* sp. SuSy, *E. coli* UDP-Glc 4′-epimerase, *Helicobacter pylori* βl–4-galactosyltransferase and a *N. meningitides* α1–4-galactosyltransferase) were coexpressed in a single genetically engineered *E. coli* strain. *Pichia pastoris* was genetically engineered by chromosomal integration of three genes (S11E mutant of plant SuSy from *Vigna radiata*, UDP-Glc 4′-epimerase from *S. cerevisiae* and bovine α1–3-galactosyltransferase) [54]. Permeabilized cells gave 2.8 g of Gal(α1–3)Gal(β1–4)Glc product from a 200-ml reaction solution with a space–time yield of 8.4 g/l/day.

A different approach of using metabolically engineered cells is to express the enzymes as fusion proteins bound to the cell wall of yeast cells instead of expression in the cytoplasm. The first example was established for the synthesis of sialyl-$Le^x$ by the coexpression of the catalytical domains of human α1–3-fucosyltransferase VII and rat α2–3-sialyltransferase in a *S. cerevisiae* or *P. pastoris* strain [55]. Both enzymes were fused to the Hsp150Δ carrier, which is an N-terminal fragment of a secretory glycoprotein of *S. cerevisiae*. The fusion protein was externalized and remained mostly attached to the cell wall in a non-covalent fashion. Synthesis of sialyl-$Le^x$ was accomplished by adding *N*-acteyllactosamine, CMP-Neu5Ac and GDP-Fuc to the medium. Shimma *et al.* attempted to immobilize glycosyltransferases *in vivo* at the yeast cell surface by fusion with cell wall Pir proteins [92]. The genetically engineered *S. cerevisiae* strains express glycosyltransferases that are covalently linked to cell wall glucans. Simple cultivation of the yeast cells and centrifugation of the culture broth provided large amounts of immobilized enzymes. A total of 51 human glycosyltransferases were expressed in this study as immobilized proteins and 40 of them showed enzymatic activity. Therefore, this system provides a promising approach to synthesizing a variety of oligosaccharides using immobilized human glycosyltransferases.

The production of oligosaccharides with living bacterial strains was extensively elaborated by the group of Samain. *E. coli* strains were metabolically engineered to facilitate uptake of substrates (e.g. lactose, Neu5Ac), and avoiding their metabolic degradation by knocking out the genes encoding β-galactosidase (*lacZ*) and Neu5Ac-aldolase (*nanA*) (Scheme 5.8). The combination of different expression vectors provides glycosyltransferases (bacterial sources) and enzymes for donor substrate synthesis (e.g. CMP-Neu5Ac and UDP-GalNAc). Lactose and Neu5Ac are taken up by the specific permeases LacY and NanT. Neu5Ac is converted into a nucleotide-activated form (CMP-Neu5Ac) by CMP-Neu5Ac synthase (**A**) and then transferred onto lactose by α2–3-sialyltransferase (encoded by Lst, **B**), to form 3′-sialyllactose (**14**). Use of the endogenous pool of UDP-GlcNAc gives UDP-GalNAc by the recombinant UDP-GlcNAc 4-epimerase (WbpP, **C**) and allows β1–4-GalNAc-transferase (CgtA, **D**) to form II3-α-Neu5Ac-Gg3 (GM2, **15**). This compound serves as an acceptor for β1–3-galactosyltransferase (CgtB, **E**) and reacts to yield II3-α-Neu5Ac-Gg4 (GM1, **16**).

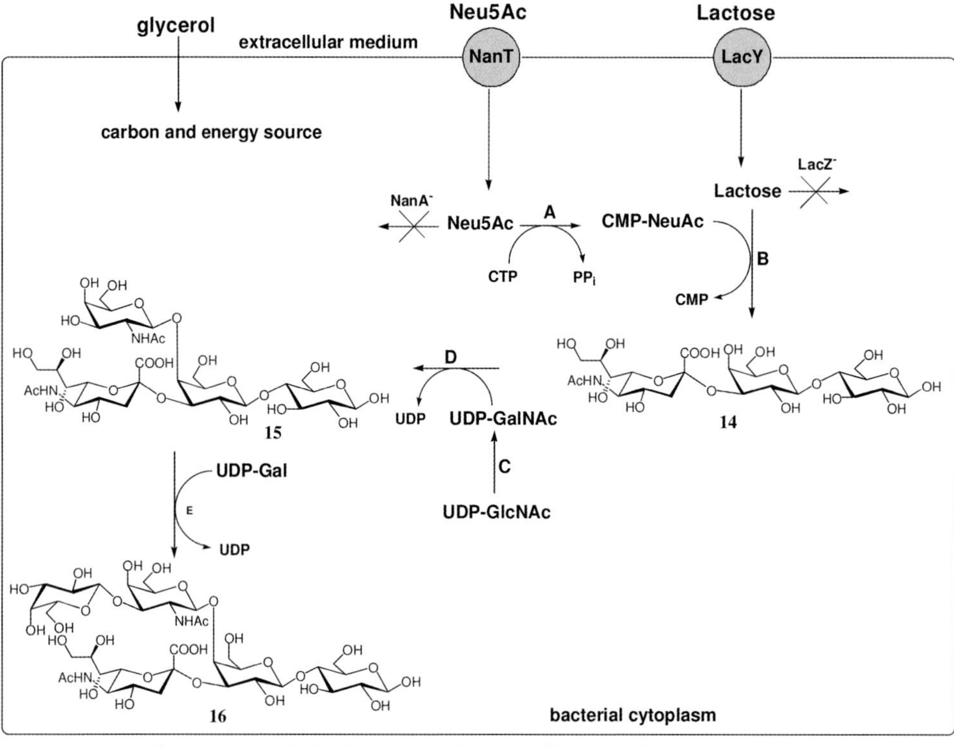


**Scheme 5.8** Metabolically engineered pathway for the production of II$^3$-α-Neu5Ac-Gg3 (GM2, **15**) and II$^3$-α-Neu5Ac-Gg4 (GM1, **16**) by combinatorial biosynthesis in *E. coli* K12 [62].

Further examples for the production of oligosaccharides on the gram scale with metabolically engineered *E. coli* cells are summarized in Table 5.3 and Scheme 5.9.

In summary, different strategies for the combinatorial biosynthesis of oligosaccharides have been established successfully. However, some systems are limited mostly to bacterial glycosyltransferases because of their better expression in *E. coli*. The yeast systems seem to be better suited for expression of mammalian glycosyltransferases; however, it remains to be demonstrated if they can be also utilized as 'living factories' for the production of oligosaccharides.

**Table 5.3** Production of oligosaccharides by metabolic engineered living *E. coli* cells (see Scheme 5.9 for product formulas).

| Introduced genes | Product | References |
|---|---|---|
| *lgtB*: β1–4-galactosyltransferase from *N. meningitides*<br>*lgtA*: β1–3-glucosaminyltransferase from *N. meningitides* | lacto-*N*-neotetraose (**17**) | [56] |
| *cmp-NeuAc synthase*: CMP-Neu5Ac synthase from *N. meningitides*<br>*lst*: α2–3-sialyltransferase from *Neisseria*<br>*wbpP*: UDP-GlcNAc 4′-epimerase from *P. aeruginosa* O6<br>*cgtA*: β1–4-GalNAc transferase from *C. jejuni*<br>*cgtB*: β1–3-galactosyltransferase from *C. jejuni* | ganglioside GM2 (**15**) (1.28 g/l)<br>ganglioside GM1 (**16**) (0.89 g/l) | [62] |
| *lgtC*: α1–4-galactosyltransferase from *N. meningitides*<br>*lgtD*: β1–3-GalNAc transferase from *H. pylori*<br>*wbpP*: UDP-GlcNAc 4-epimerase from *P. aeruginosa* O6 | globotriose (**13**) (7 g/l)<br>globotetraose (**18**) (3 g/l) | [60] |
| *manC*: GDP-Man pyrophosphorylase from *E. coli*<br>*manB*: phosphomannomutase from *E. coli*<br>*gmd*: GDP-mannose dehydratase from *E. coli*<br>*wcaG*: fucose synthase from *E. coli*<br>*lgtB*: β1–4-galactosyltransferase from *N. meningitides*<br>*lgtA*: β1–3-glucosaminyltransferase from *N. meningitides*<br>*futA*: α1–3-fucosyltransferase from *H. pylori*<br>*nodC*: *N*-acteylglucosaminyltransferase from *Azorhizobium*<br>*chiA*: chitinase from Bacillus | $Le^x$ trisaccharide:<br>bound on a GlcNAc motif (**19**) (0.62 g/l)<br>bound on a Gal motif (**20**) (1.84 g/l) | [93] |
| *lgtB*: β1–4-galactosyltransferase from *N. meningitides*<br>*lgtA*: β1–3-glucosaminyltransferase from *N. meningitides*<br>*futC*: α1–2-fucosyltransferase from *H. pylori* | H antigen oligosaccharides:<br>lacto-*N*-neofucopentaose-1 (**21**) (1.5 g/l)<br>2′-fucosyllactose (**22**) (11 g/l) | [57] |

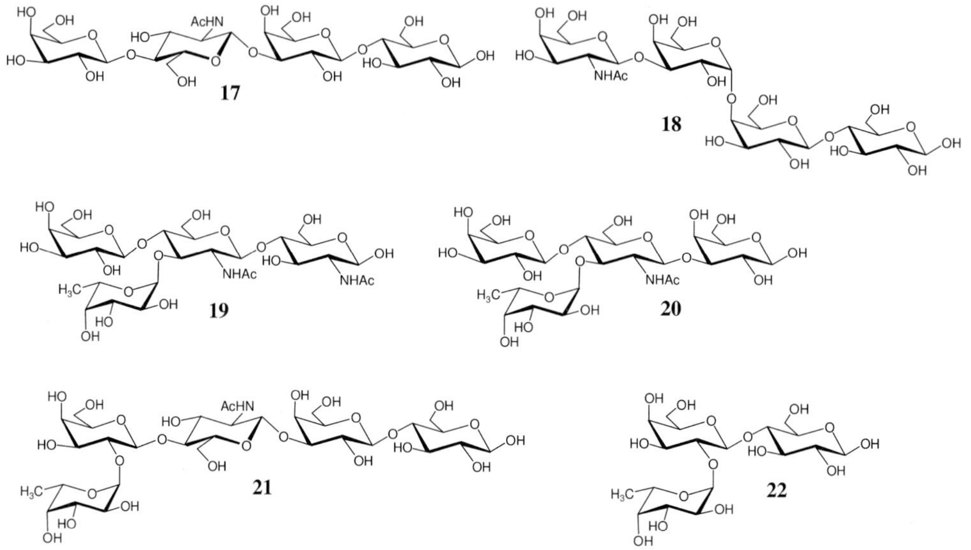

**Scheme 5.9** Examples of oligosaccharides produced by metabolically engineered living *E. coli* cells (see Table 5.3 for references). Lacto-*N*-neotetraose (LNnT, **17**), globotetraose (**18**), Le$^x$ trisaccharide bound on a GlcNAc motif (**19**), Le$^x$ trisaccharide bound on a Gal motif (**20**), lacto-*N*-neofucopentaose-1 (LNnF-1) (**21**) and 2′-fucosyllactose (**22**).

## 5.5 Conclusions

In this chapter we have restricted ourselves to the multi-enzyme synthesis of glycoconjugates including nucleotide sugars, glycosylated natural compounds, oligosaccharides, glycopeptides and glycolipid-related oligosaccharides. We have selected examples to demonstrate strategies for *in vitro* combinatorial biocatalysis and *in vivo* combinatorial biosynthesis. We think that both approaches have their advantages in specialized fields. Complex metabolic engineering of yeast strains has brought up humanized glycoproteins [9], which is an excellent example for overcoming glycosylation barriers between species. In this respect, we may also see human-like *N*-glycosylation of proteins in *E. coli* in the near future [94].

## References


1 Varki, A. (1993) *Glycobiology*, **3**, 97–130.

2 Campbell, C.T. and Yarema, K.J. (2005) *Genome Biology*, **6**, 236–44.

3 Spiro, R.G. (2002) *Glycobiology*, **12**, 43R–56R.

4 Gagneux, P. and Varki, A. (1999) *Glycobiology*, **9**, 747–55.

5 Helenius, A. and Aebi, M. (2001) *Science*, **291**, 2364–9.

**6** (a) Rudd, P.M., Elliott, T., Cresswell, P., Wilson, I.A. and Dwek, R.A. (2001) *Science*, **291**, 2370–6.
(b) Rudd, P.M., Wormald, M.R. and Dwek, R.A. (2004) *Trends in Biotechnology*, **22**, 524–30.

**7** Koury, M.J. (2003) *Trends in Biotechnology*, **21**, 462–4.

**8** (a) Saint-Jore-Dupas, C., Faye, L. and Gomord, V. (2007) *Trends in Biotechnology*, **25**, 317–23.
(b) Sethuraman, N. and Stadheim, T.A. (2006) *Current Opinion in Biotechnology*, **17**, 341–6.
(c) Li, H., Sethuraman, N., Stadheim, T.A., Zha, D., Prinz, B., Ballew, N., Bobrowicz, P., Choi, B.-K., Cook, W.J., Cukan, M., Houston-Cummings, N.R., Davidson, R., Gong, B., Hamilton, S.R., Hoopes, J.P., Jiang, Y., Kim, N., Mansfield, R., Nett, J.H., Rios, S., Strawbridge, R., Wildt, S. and Gerngross, T.U. (2006) *Nature Biotechnology*, **24**, 210–15.
(d) Walsh, G. and Jefferis, R. (2006) *Nature Biotechnology*, **24**, 1241–52.
(e) Jefferis, R. (2006) *Nature Biotechnology*, **24**, 1230–1.
(f) Jefferis, R. (2005) *Biotechnology Progress*, **21**, 11–16.
(g) Grabenhorst, E., Schlenke, P., Pohl, S., Nimtz, M. and Conradt, H.S. (1999) *Glycoconjugate Journal*, **16**, 81–97.
(h) Altmann, F., Staudacher, E., Wilson, I.B.H. and März, L. (1999) *Glycoconjugate Journal*, **16**, 109–23.

**9** Hamilton, S.R., Davidson, R.C., Sethuraman, N., Nett, J.H., Jiang, Y., Rios, S., Bobrowicz, P., Stadheim, T.A., Choi, H., Li, B.-K., Hopkins, D., Wischnewski, H., Roser, J., Mitchell, T., Strawbridge, R.R., Hoopes, J., Wildt, S. and Gerngross, T.U. (2006) *Science*, **313**, 1441–3.

**10** (a) Freeze, H.H. (2006) *Nature Reviews Genetics*, **7**, 537–51.
(b) Kobata, A. and Amano, J. (2005) *Immunology and Cell Biology*, **83**, 429–39.
(c) Marquardt, T. and Denecke, J. (2003) *European Journal of Pediatric73*, **162**, 359–379.
(d) Grewal, P.K. and Hewitt, J.E. (2003) *Human Molecular Genetics*, **12**, 259R–64.
(e) Axford, J.S. (1999) *Biochimica et Biophysica Acta – Molecular Basis of Disease*, **1455**, 219–29.
(f) Floege, J. and Feehally, J. (2000) *Journal of the American Society of Nephrology*, **11**, 2395–403.

**11** Schulz, B.L., Laroy, W. and Callewaert, N. (2007) *Current Molecular Medicine*, **7**, 397–416.

**12** Seeberger, P.H. and Werz, D.B. (2007) *Nature*, **446**, 1046–51.

**13** Liakatos, A. and Kunz, H. (2007) *Current Opinion in Molecular Therapeutics*, **9**, 35–44.

**14** Thayer, D.A. and Wong, C.-H. (2007) *Topics in Current Chemistry*, **267**, 37–63.

**15** (a) Danishefsky, S.J. and Allen, J.R. (2000) *Angewandte Chemie (International Edition in English)*, **39**, 836–63.
(b) Wan, Q., Chen, J., Chen, G. and Danishefsky, S.J. (2006) *Journal of Organic Chemistry*, **71**, 8244–9.

**16** (a) Naeem, A., Saleemuddin, M. and Khan Hasan, R. (2007) *Current Protein & Peptide Science*, **8**, 261–71.
(b) Uchiyama, N., Kuno, A., Koseki-Kuno, S., Ebe, Y., Horio, K., Yamada, M. and Hirabayashi, J. (2006) *Methods in Enzymology*, **415**, 341–51.

**17** (a) Alvarez, R.A. and Blixt, O. (2006) *Methods in Enzymology*, **415**, 292–310.
(b) de Paz, J.L., Horlacher, T. and Seeberger, P.H. (2006) *Methods in Enzymology*, **415**, 269–92.
(c) Liu, Y., Chai, W., Childs, R.A. and Feizi, T. (2006) *Methods in Enzymology*, **415**, 326–40.
(d) Biskup, M.B., Müller, J.U., Weingart, R. and Schmidt, R.R. (2005) *ChemBioChem*, **6**, 1007–15.
(e) Feizi, T. and Chai, W. (2004) *Nature Reviews Molecular Cell Biology*, **5**, 582–8.

**18** Paulson, J.C., Blixt, O. and Collins, B.E. (2006) *Nature Chemical Biology*, **2**, 238–48.

**19** (a) de la Fuente, J.M. and Penades, S. (2006) *Biochimica et Biophysica Acta – General Subjects* **1760**, 636–51.
(b) Andre, S., Maljaars, C.E.P., Halkes, K.M., Gabius, H.-J. and Kamerling, J.P.

(2007) *Bioorganic & Medicinal Chemistry Letters*, **17**, 793–8.
(c) Halkes, K.M., Maljaars, A.C., Souza, C.E.P., Gerwig, G.J. and Kamerling, J.P. (2005) *European Journal of Organic Chemistry*, **2005**, 3650–9.
(d) Carvalho de Souza, A., Halkes, K.M., Meeldijk, J.D., Verkleij, A.J., Vliegenthart, J.F.G. and Kamerling, J.P. (2005) *ChemBioChem*, **6**, 828–31.

**20** Nakamura-Tsuruta, S., Uchiyama, N. and Hirabayashi, J. (2006) *Methods in Enzymology*, **415**, 311–25.

**21** (a) Laughlin, S.T., Agard, N.J., Baskin, J.M., Carrico, I.S., Chang, P.V., Ganguli, A.S., Hangauer, M.J., Lo, A., Prescher, J.A. and Bertozzi, C.R. (2006) *Methods in Enzymology*, **415**, 230–50.
(b) Sawa, M., Hsu, T.-L., Itoh, T., Sugiyama, M., Hanson, S.R., Vogt, P.K. and Wong, C.-H., (2006) *Proceedings of the National Academy of Sciences of the United States of America*, **103**, 12371–6.

**22** Coates, A., Bax, Y., Hu, R. and Page, C. (2002) *Nature Reviews Drug Discovery*, **1**, 895–910.

**23** (a) Walsh, C.T. (2002) *ChemBioChem*, **3**, 124–34.
(b) Clardy, J., Fischbach, M.A. and Walsh, C.T. (2006) *Nature Biotechnology*, **24**, 1541–50.

**24** Mendez, C. and Salas, J.A. (2001) *Trends in Biotechnology*, **19**, 449–56.

**25** Rupprath, C., Schumacher, T. and Elling, L. (2005) *Current Medicinal Chemistry*, **12**, 1637–75.

**26** (a) Blanchard, S. and Thorson, J.S. (2006) *Current Opinion in Chemical Biology*, **10**, 263–71.
(b) Thibodeaux, C.J. and Liu, H.-W. (2007) *Pure and Applied Chemistry*, **79**, 785–99.

**27** Koeller, K.M. and Wong, C.-H. (2000) *Chemical Reviews*, **100**, 4465–93.

**28** (a) Hamilton, C.J. (2004) *Natural Product Reports*, **21**, 365–85.
(b) Hanson, S., Best, M., Bryan, M.C. and Wong, C.-H. (2004) *Trends in Biochemical Sciences*, **29**, 656–63.
(c) Rowan, A.S. and Hamilton, C.J. (2006) *Natural Product Reports*, **23**, 412–43.
(d) Murata, T. and Usui, T. (2006) *Bioscience, Biotechnology, and Biochemistry*, **70**, 1049–59.
(e) Galonic, D.P. and Gin, D.Y. (2007) *Nature*, **446**, 1000–7.

**29** Elling, L. (1997) *Advances in Biochemical Engineering./Biotechnology*, **58**, 89–144.

**30** Garcia-Junceda, E., Garcia-Garcia, J.F., Bastida, A. and Fernandez-Mayoralas, A. (2004) *Bioorganic & Medicinal Chemistry*, **12**, 1817–34.

**31** (a) Bülter, T., Schumacher, T., Namdjou, D.-J., Gutiérrez Gallego, R., Clausen, H. and Elling, L. (2001) *ChemBioChem*, **2**, 884–94.
(b) Namdjou, D.-J., Sauerzapfe, B., Schmiedel, J., Dräger, G., Bernatchez, S., Wakarchuk, W.W. and Elling, L. (2007) *Advanced Synthesis Catalysis*, **349**, 314–18.

**32** Elling, L., Rupprath, C., Günther, N., Römer, U., Verseck, S., Weingarten, P., Dräger, G., Kirschning, A. and Piepersberg, W. (2005) *ChemBioChem*, **6**, 1423–30.

**33** Rupprath, C., Kopp, M., Hirtz, D., Müller, R. and Elling, L. (2007) *Advanced Synthesis Catalysis*, **349**, 1489–96.

**34** Hancock, S.M., Vaughan, M.D. and Withers, S.G. (2006) *Current Opinion in Chemical Biology*, **10**, 509–19.

**35** Aharoni, A., Thieme, K., Chiu, C.P.C., Buchini, S., Lairson, L.L., Chen, H., Strynadka, N.C.J., Wakarchuk, W.W. and Withers, S.G. (2006) *Nature Methods*, **3**, 609–14.

**36** (a) Nilsson, K.G.I. (1990) *Annals of the New York Academy of Sciences*, **613**, 431–4.
(b) Nilsson, K.G.I. (1996) *Modern Methods in Carbohydrate Synthesis* (eds S.H. Khan and R.A. O'Neill), Harwood Academic, Amsterdam, pp. 518–47.
(c) Nilsson, K.G.I., Ljunger, G. and Melin, P.M. (1997) *Biotechnology Letters*, **19**, 889–92.
(d) Nilsson, K.G.I., Pan, H.F. and Larssonlorek, U. (1997) *Journal of Carbohydrate Chemistry*, **16**, 459–77.
(e) Crout, D.H.G. and Vic, G. (1998) *Current Opinion in Chemical Biology*, **2**, 98–111.
(f) Zervosen, A., Nieder, V., Gutiérrez Gallego, R., Kamerling, J.P., Vliegenthart,

J.F.G. and Elling, L. (2001) *Biological Chemistry*, **382**, 299–311.
(g) Nieder, V., Marx, S.P., Gutiérrez Gallego, R., Kamerling, J.P., Vliegenthart, J.F.G. and Elling, L. (2003) *Journal of Molecular Catalysis B: Enzymatic*, **21**, 157–66.

**37** (a) Lairson, L.L., Watts, A.G., Wakarchuk, W.W. and Withers, S.G. (2006) *Nature Chemical Biology*, **2**, 724–8.
(b) Nishida, Y., Tamakoshi, H., Kitagawa, Y., Kobayashi, K. and Thiem, J. (2000) *Angewandte Chemie (International Edition in English)*, **39**, 2000–3.

**38** Bülter, T. and Elling, L. (1999) *Glycoconjugate Journal*, **16**, 147–59.

**39** Bülter, T. and Elling, L. (2000) *Journal of Molecular Catalysis B: Enzymatic*, **8**, 281–4.

**40** Koizumi, S., Endo, T., Tabata, K. and Ozaki, A. (1998) *Nature Biotechnology*, **16**, 847–50.

**41** Endo, T., Koizumi, S., Tabata, K. and Ozaki, A. (2000) *Applied Microbiology and Biotechnology*, **53**, 257–61.

**42** Koizumi, S., Endo, T., Tabata, K., Nagano, H., Ohnishi, J. and Ozaki, A. (2000) *Journal of Industrial Microbiology & Biotechnology*, **25**, 213–17.

**43** (a) Tabata, K., Koizumi, S., Endo, T. and Ozaki, A. (2000) *Biotechnology Letters*, **22**, 479–83.
(b) Endo, T. and Koizumi, S. (2001) *Advanced Synthesis Catalysis*, **6–7**, 521–6.

**44** Ichikawa, Y., Wang, R. and Wong, C.H. (1994) *Methods in Enzymology*, **247**, 107–27.

**45** (a) Ichikawa, M., Schnaar, R.L. and Ichikawa, Y. (1995) *Tetrahedron Letters*, **36**, 8731–2.
(b) Hokke, C.H., Zervosen, A., Elling, L., Joziasse, D.H. and van den Eijnden, D.H.(1996) *Glycoconjugate Journal*, **13**, 687–92.

**46** Zervosen, A. and Elling, L. (1996) *Journal of the American Chemical Society*, **118**, 1836–40.

**47** Noguchi, T. and Shiba, T. (1998) *Bioscience, Biotechnology, and Biochemistry*, **62**, 1594–6.

**48** Nahalka, J., Liu, Z., Chen, X. and Wang, P.G. (2003) *Chemistry–A European Journal*, **9**, 372–7.

**49** Bourgeaux, V., Cadène, M., Piller, F. and Piller, V. (2007) *ChemBioChem*, **8**, 37–40.

**50** Vasiliu, D., Razi, N., Zhang, Y., Jacobsen, N., Allin, K., Liu, X., Hoffmann, J., Bohorov, O. and Blixt, O. (2006) *Carbohydrate Research*, **341**, 1447–57.

**51** DeAngelis, P.L., Oatman, L.C. and Gay, D.F. (2003) *Journal of Biological Chemistry*, **278**, 35199–35203.

**52** Sismey-Ragatz, A.E., Green, D.E., Otto, N.J., Rejzek, M., Field, R.A. and DeAngelis, P.L. (2007) *Journal of Biological Chemistry*, **282**, 28321–7.

**53** Endo, T., Koizumi, S., Tabata, K., Kakita, S. and Ozaki, A. (1999) *Carbohydrate Research*, **316**, 179–83.

**54** Shao, J., Hayashi, T. and Wang, P.G. (2003) *Applied and Environmental Microbiology*, **69**, 5238–42.

**55** Salo, H., Sievi, E., Suntio, T., Mecklin, M., Mattila, P., Renkonen, R. and Mararow, M. (2005) *FEMS Yeast Research*, **5**, 341–50.

**56** Priem, B., Gilbert, M., Wakarchuk, W.W., Heyraud, A. and Samain, E. (2002) *Glycobiology*, **12**, 235–40.

**57** Drouillard, S., Driguez, H. and Samain, E. (2006) *Angewandte Chemie (International Edition in English)*, **45**, 1778–80.

**58** Blixt, O., Vasiliu, D., Allin, K., Jacobsen, N., Warnock, D., Razi, N., Paulson, J.C., Bernatchez, S., Gilbert, M. and Wakarchuk, W. (2005) *Carbohydrate Research*, **340**, 1963–72.

**59** (a) Rich, J.R., Szpacenko, A., Palcic, M.M. and Bundle, D.R. (2004) *Angewandte Chemie (International Edition in English)*, **43**, 613–15.
(b) Bundle, D.R., Rich, J.R., Jacques, S., Yu, H.N., Nitz, M. and Ling, C.-C. (2005) *Angewandte Chemie (International Edition in English)*, **44**, 7725–9.
(c) Rich, J.R., Wakarchuk, W.W. and Bundle, D.R. (2005) *Chemistry–A European Journal*, **12**, 845–58.
(d) Jacques, S., Rich, J.R., Ling, C.-C. and Bundle, D.R. (2006) *Organic & Biomolecular Chemistry*, **4**, 142–54.

**60** Antoine, T., Bosso, C., Heyraud, A. and Samain, E. (2005) *Biochimie*, **87**, 197–203.

**61** Liu, Z., Zhang, Y., Lu, J., Pardee, K. and Wang, P.G. (2003) *Applied and Environmental Microbiology*, **69**, 2110–15.

**62** Antoine, T., Priem, B., Heyraud, A., Greffe, L., Gilbert, M., Wakarchuk, W. W., Lam, J.S. and Samain, E. (2003) *ChemBioChem*, **4**, 406–12.

**63** (a) Hutchinson, C.R. and McDaniel, R. (2001) *Current Opinion in Investigational Drugs*, **2**, 1681–90.
(b) Trefzer, A., Blanco, G., Remsing, L., Kuenzel, E., Rix, U., Lipata, F., Brana, A.F., Mendez, C., Rohr, J., Bechthold, A. and Salas, J.A. (2002) *Journal of the American Chemical Society*, **124**, 6056–62.

**64** Zervosen, A., Stein, A., Adrian, H. and Elling, L. (1996) *Tetrahedron*, **52**, 2395–404.

**65** Zervosen, A., Römer, U. and Elling, L. (1998) *Journal of Molecular Catalysis B: Enzymatic*, **5**, 25–8.

**66** Kang, Y.-B., Yang, Y.-H., Lee, K.-W., Lee, S.-G., Sohng, J.K., Lee, H.C., Liou, K. and Kim, B.-G. (2006) *Biotechnology and Bioengineering*, **93**, 21–7.

**67** Bourgeaux, V., Piller, F. and Piller, V. (2005) *Bioorganic & Medicinal Chemistry Letters*, **15**, 5459–62.

**68** Amann, S., Dräger, G., Rupprath, C., Kirschning, A. and Elling, L. (2001) *Carbohydrate Research*, **335**, 23–32.

**69** Takayama, S., McGarvey, G.J. and Wong, C.H. (1997) *Annual Review of Microbiology*, **51**, 285–310.

**70** Stein, A., Kula, M.R. and Elling, L. (1998) *Glycoconjugate Journal*, **15**, 139–45.

**71** Oh, J., Lee, S.-G., Kim, B.-G., Sohng, J.K., Liou, K. and Chan Lee, H.C. (2003) *Biotechnology and Bioengineering*, **84**, 452–8.

**72** Chung, Y.S., Kim, D.H., Seo, W.M., Lee, H.C., Liou, K., Oh, T.-J. and Sohng, J.K. (2007) *Carbohydrate Research*, **342**, 1412–18.

**73** (a) Melo, A. and Glaser, L. (1965) *Journal of Biological Chemistry*, **240**, 398–405.
(b) Barton, W.A., Lesniak, J., Biggins, J.B., Jeffrey, P.D., Jiang, J.Q., Rajashankar, K.R., Thorson, J.S. and Nikolov, D.B. (2001) *Nature Structural Biology*, **8**, 545–51.

**74** (a) Elling, L. and Kula, M.-R. (1994) *Journal of Biotechnology*, **34**, 157–63.
(b) Elling, L. (1996) *Phytochemistry*. **42**, 955–960.
(c) Elling, L., Ritter, J.E. and Verseck, S. (1996) *Glycobiology*, **6**, 591–7.

**75** Shao, J., Zhang, J., Kowal, P. and Wang, P.G. (2002) *Applied and Environmental Microbiology*, **68**, 5634–40.

**76** Sunthankar, P., Pastuszak, I., Rooke, A., Elbein, A.D., van de Rijn, I., Canfield, W.M. and Drake, R.R. (1998) *Analytical Biochemistry*, **258**, 195–201.

**77** Kopp, M., Rupprath, C., Irschik, H., Bechthold, A., Elling, L. and Müller, R. (2007) *ChemBioChem*, **8**, 813–19.

**78** Stevens, J., Blixt, O., Paulson, J.C. and Wilson, I.A. (2006) *Nature Reviews Microbiology*, **4**, 857–64.

**79** Blixt, O., Allin, K., Pereira, L., Datta, A. and Paulson, J.C. (2002) *Journal of the American Chemical Society*, **124**, 5739–46.

**80** Blixt, O., Brown, J., Schur, M.J., Wakarchuk, W. and Paulson, J.C. (2001) *Journal of Organic Chemistry*, **66**, 2442–8.

**81** Blixt, O., van Die, I., Norberg, T. and van den Eijnden, D.H. (1999) *Glycobiology*, **9**, 1061–71.

**82** (a) Chen, X., Liu, Z., Wang, J.Q., Fang, J.W., Fan, H.N. and Wang, P.G. (2000) *Journal of Biological Chemistry*, **275**, 31594–600.
(b) Fang, J.W., Chen, X., Zhang, W., Janczuk, A. and Wang, P.G. (2000) *Carbohydrate Research*, **329**, 873–8.

**83** Gilbert, M., Bayer, R., Cunningham, A.-M., DeFrees, S., Gao, Y., Watson, D.C., Young, N.M. and Wakarchuk, W.W. (1998) *Nature Biotechnology*, **16**, 769–72.

**84** Hakomori, S.-I. (1989) *Advances in Cancer Research*, **52**, 257–331.

**85** (a) Taylor-Papadimitriou, J., Burchell, J., Miles, D.W. and Dalziel, M. (1999) *Biochimica et Biophysica Acta – Molecular Basis of Disease*, **1455**, 301–13.
(b) Schietinger, A., Philip, M., Yoshida, B.A., Azadi, P., Liu, H., Meredith, S.C. and Schreiber, H. (2006) *Science*, **314**, 304–8.

**86** Sorensen, A.L., Reis, C.A., Tarp, M.A., Mandel, U., Ramachandran, K., Sankaranarayanan, V., Schwientek, T., Graham, R., Taylor-Papadimitriou, J., Hollingsworth, M.A., Burchell, J. and Clausen, H. (2006) *Glycobiology*, **16**, 96–107.

**87** Hagen, F.K., Ten Hagen, K.G. and Tabata, K. (2002) *Handbook of Glycosyltransferases and Related Genes* (eds N. Taniguchi,
K. Honke and M. Fukuda), Springer, Tokyo, pp. 167–73.

**88** (a) Ju, T., Brewer, K., D'Souza, A., Cummings, R.D. and Canfield, W.M. (2002) *Journal of Biological Chemistry*, **277**, 178–86.
(b) Ju, T. and Cummings, R.D. (2002) *Proceedings of the National Academy of Sciences of the United States of America*, **99**, 16613–18.
(c) Muller, R., Hulsmeier, A.J., Altmann, F., Ten Hagen, K., Tiemeyer, M. and Hennet, T. (2005) *FEBS Journal*, **272**, 4295–305.
(d) Ju, T., Zheng, Q. and Cummings, R.D. (2006) *Glycobiology*, **16**, 947–58.

**89** Taniguchi, N., Honke, K. and Fukuda, M. (2002) *Handbook of Glycosyltransferases and Related Genes*, Springer, Tokyo.

**90** DeFrees, S., Wang, Z.-G., Xing, R., Scott, A.E., Wang, J., Zopf, D., Gouty, D.L., Sjoberg, E.R., Panneerselvam, K., Brinkman-Van der Linden, E.C.M., Bayer, R.J., Tarp, M.A. and Clausen, H. (2006) *Glycobiology*, **16**, 833–43.

**91** (a) Kren, V. and Thiem, J. (1995) *Angewandte Chemie (International Edition in English)*, **34**, 893–5.
(b) Gambert, U. and Thiem, J. (1997) *Topics in Current Chemistry*, **186**, 21–43.

**92** Shimma, Y.-i., Saito, F., Oosawa, F. and Jigami, Y. (2006) *Applied and Environmental Microbiology*, **72**, 7003–70012.

**93** Dumon, C., Bosso, C., Utille, J.P., Heyraud, A. and Samain, E. (2006) *ChemBioChem*, **7**, 359–65.

**94** (a) Feldman, M.F., Wacker, M., Hernandez, M., Hitchen, P.G., Marolda, C.L., Kowarik, M., Morris, H.R., Dell, A., Valvano, M.A. and Aebi, M. (2005) *Proceedings of the National Academy of Sciences of the United States of America*, **102**, 3016–21.
(b) Kowarik, M., Numao, S., Feldman, M.F., Schulz, B.L., Callewaert, N., Kiermaier, E., Catrein, I. and Aebi, M. (2006) *Science*, **314**, 1148–50.

# 6
# Enzyme-Catalyzed Cascade Reactions

*Roger A. Sheldon*

## 6.1
## Introduction

In the last decade the concepts of green chemistry and sustainable development have become a strategic focus in both the chemical industry and the academic community at large [1]. A prominent feature of this drive towards sustainability is the widespread application of chemo- and biocatalytic methodologies in chemicals manufacture [2]. The key to successful implementation of catalytic methodologies in fine chemicals manufacture is the integration of catalytic steps in multi-step organic syntheses and downstream processing. The ultimate in integration is to combine several catalytic steps into a one-pot, multi-step catalytic cascade process (Figure 6.1). This is truly emulating Nature where metabolic pathways conducted in living cells involve an elegant orchestration of a series of enzymatic steps into an exquisite multi-catalyst cascade, without the need for isolation of intermediates [3].

The 'telescoping' of multi-step syntheses into a one-pot catalytic cascade has several advantages – fewer unit operations, less solvent and reactor volume, shorter cycle times, higher volumetric and space–time yields, and less waste (lower $E$ factor) [4], which translates to substantial economic and environmental benefits. Furthermore, coupling of reactions together can be used to drive equilibria towards product, thus avoiding the need for excess reagents. On the other hand, there are several problems associated with the construction of catalytic cascades: catalysts are often incompatible with each other, rates are very different, and it is difficult to find optimum conditions of pH, temperature, solvent and so on. Catalyst recovery and recycle is complicated, and downstream processing is difficult. Nature solves this problem by compartmentalization of the various enzymes. Hence, compartmentalization via immobilization is conceivably a way of solving these problems in cascade processes. We also note that enzymatic processes generally proceed under roughly the same conditions – in water at around ambient temperature and pressure – that facilitate the cascading process.

*Multi-Step Enzyme Catalysis: Biotransformations and Chemoenzymatic Synthesis*
Edited by Eduardo Garcia-Junceda

ISBN: 978-3-527-31921-3

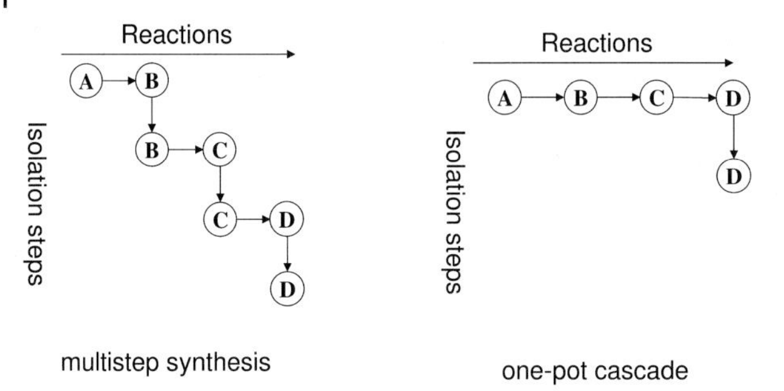


**Figure 6.1** Multi-step organic synthesis and one-pot cascade processes.

Strictly speaking a catalytic cascade process is one in which all of the catalysts (enzymes or chemocatalysts) are present in the reaction mixture from the outset. A one-pot process, on the other hand, is one in which several reactions are conducted sequentially in the same reaction vessel, without the isolation of intermediates. However, not all of the reactants or catalysts are necessarily present from the outset. Hence, a cascade process is by definition a one-pot process, but the converse is not necessarily true. Clearly a cascade process is a more elegant solution, but a one-pot process that is not, according to the strict definition, a cascade reaction may have equal practical utility. In this chapter we shall be primarily concerned with enzymatic cascade processes, but the occasional chemocatalytic step may be included where relevant and sometimes a sequential one-pot procedure may slip through the net.

Cascade processes involving multiple enzymes, many of which involve a cofactor regeneration system, can be performed as whole-cell biotransformations, whereby all the enzymes needed are present in the wild-type organism. This is less expensive as it circumvents the need for isolation and purification of the individual enzymes, but often suffers from lower selectivities owing to competition from other, similar enzymes present in the cell. Baker's yeast reductions, for example, often give suboptimum enantioselectivities owing to the presence of several alcohol dehydrogenases. This problem can be avoided by using the isolated enzymes in concert. There is a third possibility: the wild-type enzymes are cloned and heterologously coexpressed in a suitable host microorganism that does not contain competing homologous enzymes. A major advantage of this 'designer bug' approach [5] is that only one fermentation is needed to produce all the required enzymes, whereas with the isolated enzyme approach each individual enzyme has to be produced in a separate fermentation step (or purchased, if commercially available).

## 6.2 Enzyme Immobilization

As noted in the preceding section, successful compartmentalization by immobilization may be the key to the compatibility of different enzymes in a cascade process. Hence, before going on to discuss different types of enzymatic cascades we shall briefly discuss the immobilization of enzymes. Industrial application of enzymes is often hampered by a lack of long-term operational stability, and difficult recovery and reuse of the enzyme. These drawbacks can often be overcome by immobilization of the enzyme [6] (For a recent comprehensive review of enzyme immobilization, see [7].) Improved enzyme performance via enhanced stability and repeated reuse is reflected in higher catalyst productivities (kilogram of product/kilogram of enzyme) which directly influence the enzyme costs per kilogram of product.

Methods for the immobilization of enzymes can be divided into three main categories: (i) binding to a support (carrier), (ii) entrapment (encapsulation) and (iii) cross-linking [6, 7]. Binding to a support can be physical (such as hydrophobic interactions), ionic or covalent in nature. Physical bonding is generally too weak to keep the enzyme fixed to the carrier under industrial conditions of high reactant and product concentrations and high ionic strength. Ionic binding (e.g. to an ion-exchange resin) is generally stronger and covalent binding of the enzyme to the support even more so, which has the advantage that the enzyme cannot be leached from the surface. However, if the enzyme is irreversibly deactivated both the enzyme and the (often costly) support are rendered unusable. The support can be a synthetic resin, a biopolymer or an inorganic polymer such as silica. Entrapment involves inclusion of an enzyme in a polymer network, such as an organic polymer or a silica sol-gel, or a membrane device, such as a hollow fiber or a microcapsule. The physical restraints generally are too weak, however, to prevent enzyme leakage entirely and additional covalent attachment is often required. Cross-linking involves the reaction of reactive groups, generally free amino groups in lysine residues, on the surface of the enzyme with a bifunctional reagent such as glutaraldehyde, affording carrierless macroparticles.

The use of a carrier inevitably leads to dilution of catalytic activity, owing to the introduction of a large portion of non-catalytic ballast, generally ranging from 90 to more than 99%, resulting in lower space–time yields and productivities. Consequently, there is an increasing interest in carrier-free immobilized enzymes, such as cross-linked enzyme crystals [8] and cross-linked enzyme aggregates (CLEAs) [9]. The cross-linking approach offers clear advantages: highly concentrated enzyme activity in the catalyst, high stability and low production costs owing to the exclusion of an additional (expensive) carrier.

## 6.3 Reaction Types: General Considerations

A major advantage of enzymes as catalysts is that they are capable of inducing very high degrees of enantioselectivity and, consequently, they are particularly useful in the synthesis of enantiomerically pure compounds. In cases where the enantioselectivity is less than optimum it can generally be improved using protein engineering techniques such as *in vitro* evolution [10]. Hence, in this chapter we shall be mainly concerned with the application of enzymatic cascade processes to the

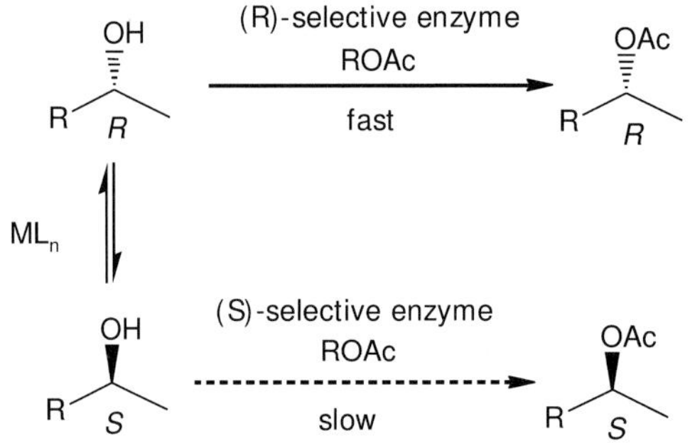

**Scheme 6.1** DKR of an alcohol.

synthesis of enantiomerically pure compounds. From the viewpoint of industrial applications, enantiomerically pure alcohols, amines and carboxylic acid derivatives probably constitute more than 90% of the target molecules of interest. Consequently, we have divided the material on this basis. With regard to the enzymatic conversion involved, these generally fall into three categories: hydrolysis (hydrolases), oxidation or reduction (oxidoreductases) and addition/elimination reactions (lyases), including C—C bond formation. As we shall see, the three different classes of products (see above) can be synthesized using the different classes of enzymes.

## 6.4 Chiral Alcohols

There are basically two approaches to the synthesis of enantiomerically pure alcohols: (i) kinetic resolution of the racemic alcohol using a hydrolase (lipase, esterase or protease) or (ii) reduction mediated by a ketoreductase (KRED). Both of these processes can be performed as a cascade process. The first approach can be performed as a dynamic kinetic resolution (DKR) by conducting an enzymatic transesterification in the presence of a redox metal [e.g. a Ru(II) complex] to catalyze *in situ* racemization of the unreacted alcohol isomer [11] (Scheme 6.1). We shall not discuss this type of process in any detail here since it forms the subject of Chapter 1.

In the second approach the reducing equivalents are supplied by a nicotinamide cofactor (NADH or NADPH) and for commercial viability it is necessary to regenerate the cofactor using a sacrificial reductant [12]. This can be achieved in two ways: substrate coupled or enzyme coupled (Scheme 6.2). Substrate-coupled regeneration involves the use of a second alcohol (e.g. isopropanol) that can be accommodated by the KRED in the oxidative mode. A problem with this approach is that it affords an equilibrium mixture of the two alcohols and two ketones. In order to obtain a high yield of the desired alcohol product a large excess of the sacrificial alcohol needs to be added and/or the ketone product (acetone) removed

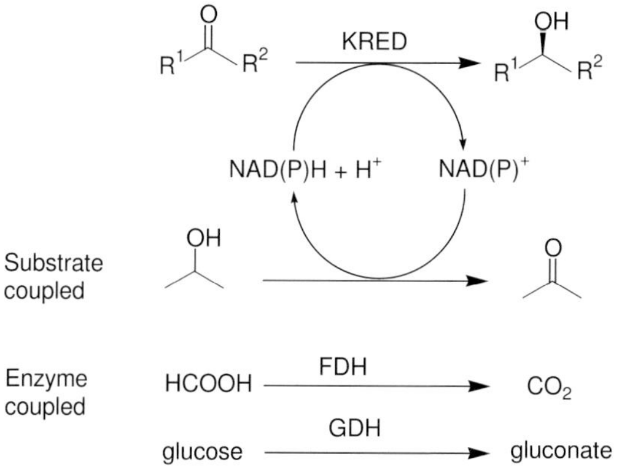

KRED = Ketoreductase
FDH = Formate dehydogenase
GDH =Glucose dehydrogenase

**Scheme 6.2** Biocatalytic reduction of ketones with cofactor regeneration.

by fractional distillation. In enzyme-coupled regeneration a second oxidoreductase is used, in an oxidation mode, together with a sacrificial reductant.

The most commonly used combinations are glucose/glucose dehydrogenase (GDH) and formate/formate dehydrogenase (FDH) or, more recently, phosphite/phosphite dehydrogenase [13]. Reactions can be conducted with the isolated enzymes [14] or as whole-cell processes [15]. An example of the former is provided by the Codexis process for the production of the atorvastatin (Lipitor) intermediate by KRED-catalyzed reduction of a β-keto ester coupled with a GDH-catalyzed conversion of glucose to gluconate for cofactor regeneration (Scheme 6.3). A third

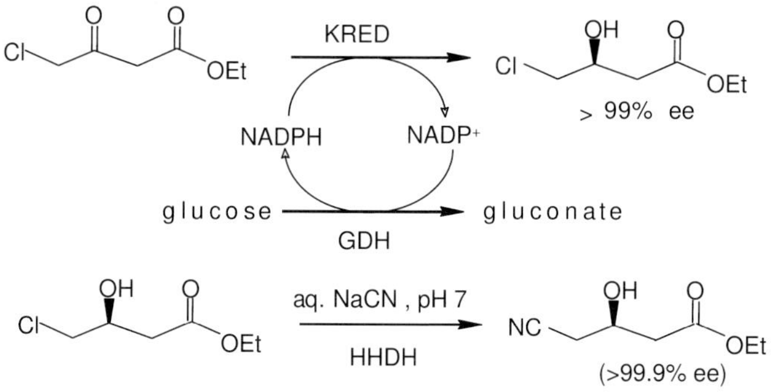

KRED = keto reductase
GDH = glucose dehydrogenase
HHDH = halohydrin dehalogenase

**Scheme 6.3** Codexis process for atorvastatin intermediate.

enzyme, halohydrin dehalogenase, is employed for the subsequent conversion of the chlorohydrin product to the corresponding cyano compound, although it is not clear if this is conducted as a one-pot process. In order to obtain a process with commercially viable volumetric and space–time yields and catalyst productivities all three enzymes were optimized by *in vitro* evolution using gene shuffling [16]. Codexis received a 2006 presidential Green Chemistry Challenge Award for this three-enzyme process.

An example of a whole-cell process is the two-step synthesis of an enantiopure epoxide by asymmetric reduction of an α-chloro ketone (Scheme 6.4), catalyzed by recombinant whole cells of an *Escherichia coli* sp. overexpressing an (*R*)-KRED from *Lactobacillus kefir* and GDH from *Thermoplasma acidophilum*, to the corresponding chlorohydrin, followed by non-enzymatic base-catalyzed ring closure to the epoxide [17].

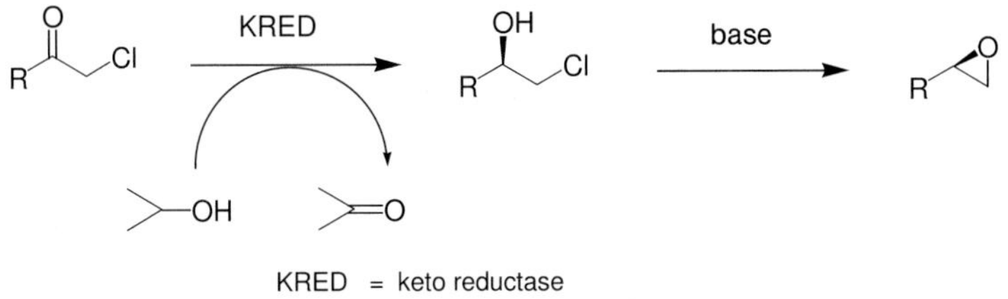

KRED = keto reductase

**Scheme 6.4** Two-step synthesis of an (*S*)-epoxide with designer cells.

Kroutil *et al.* have recently reported [18] an elegant one-pot oxidation/reduction sequence for the deracemization of a chiral secondary alcohol using a single biocatalyst. Lyophilized cells of a *Rhodococcus* sp. CBS 717.73 converted racemic 2-decanol into the (*S*)-enantiomer in 82% yield and 92% enantiomeric excess (e.e.). via a non-specific oxidation followed sequentially by an (*S*)-selective reduction (Scheme 6.5). Acetone was used as the hydrogen acceptor in the first step and isopropanol as the hydrogen donor in the second step.

## 6.5
## Chiral Amines

Analogous to the reactions of chiral alcohols, enantiomerically pure amines can be prepared by (D)KR of the racemate via enzymatic acylation. In the case of alcohols the subsequent hydrolysis of the ester product to the enantiomerically pure alcohol is trivial and is generally not even mentioned. In contrast, the product of enzymatic acylation of an amine is an amide and hydrolysis of an amide is by no means trivial, often requiring forcing conditions.

BASF successfully developed a process, which is operated on a multi-thousand-ton scale, for the resolution of chiral primary amines by lipase-catalyzed acylation

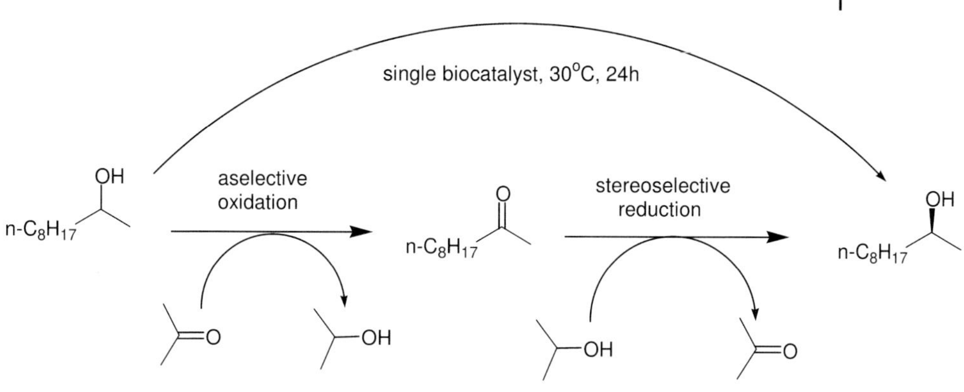


**Scheme 6.5** Biocatalytic one-pot deracemization of 2-decanol.

[19]. Commercially acceptable rates were obtained by using a methoxyacetic acid ester as the acyl donor, which is attributed to hydrogen bond formation with the ether oxygen in the active site of the enzyme [20]. As shown in Scheme 6.6, the process affords an amide of the reacting enantiomer together with the other enantiomer as the free amine.

In order to recover both amines in optically active form the amide is hydrolyzed chemically by reaction with NaOH in aqueous ethylene glycol at 150 °C. This 'brute force' method would certainly lead to problems with amines containing other functional groups and is in stark contrast to the elegant enzymatic procedure used for the first step. Hence, an overall greener process can be obtained by employing an enzymatic deacylation step in what we have called an 'easy-on/easy-off' process

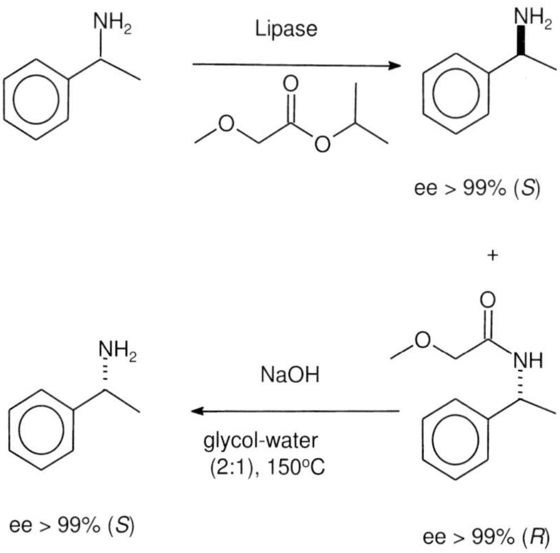

$NH_2$ — Lipase → $NH_2$, ee > 99% (*S*) + amide (NH), ee > 99% (*R*); NaOH, glycol-water (2:1), 150°C → $NH_2$, ee > 99% (*S*)

**Scheme 6.6** BASF process for enzymatic resolution of amines.

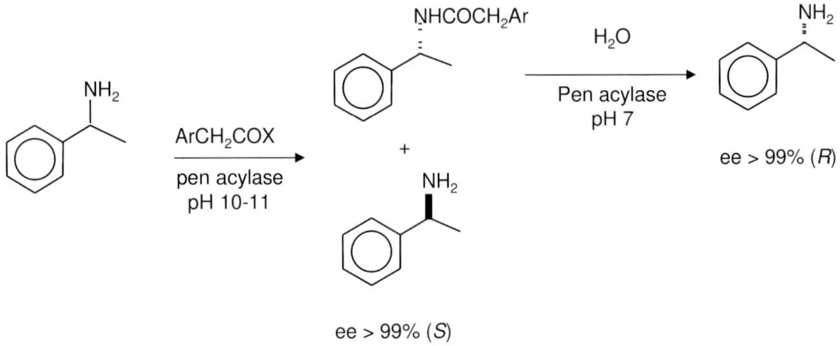

ee > 99% (*R*)

ee > 99% (*S*)

**Scheme 6.7** Easy-on/easy-off resolution of amines with penicillin G amidase.

(Scheme 6.7). Penicillin G amidase from *Alcaligenes faecalis,* which is used in the manufacture of semisynthetic penicillins and cephalosporins, was used in both steps to afford a one-pot cascade process [21]. The acylation was performed in an aqueous medium at pH 10–11 and, after separation of the remaining amine enantiomer, the acylated amine was hydrolyzed with the same enzyme by lowering the pH to 7.

This process has many benefits in the context of green chemistry: it involves two enzymatic steps, in a one-pot procedure, in water as solvent at ambient temperature. It has one shortcoming, however – penicillin acylase generally works well only with amines containing an aromatic moiety and poor enantioselectivities are often observed with simple aliphatic amines. Hence, for the easy-on/easy-off resolution of aliphatic amines a hybrid form was developed in which a lipase [*Candida antarctica* lipase B (CALB)] was used for the acylation step and penicillin acylase for the deacylation step [22]. The structure of the acyl donor was also optimized to combine a high enantioselectivity in the first step with facile deacylation in the second step. It was found that pyridyl-3-acetic acid esters gave optimum results (see Scheme 6.8).

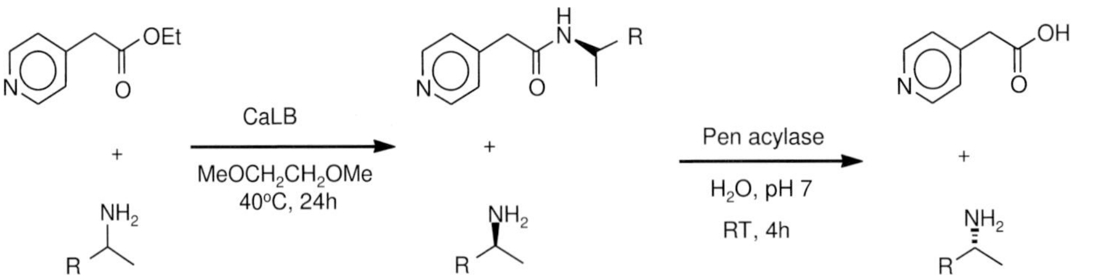

| R | conv.(%) | ee(%) | E |
|---|---|---|---|
| $C_3H_7$ | 51 | 92 | 46 |
| $C_5H_{11}$ | 50 | 97 | >100 |

| R | conv.(%) | ee(%) |
|---|---|---|
| $C_3H_7$ | 99 | 90 |
| $C_5H_{11}$ | 92 | 97 |

**Scheme 6.8** Hybrid process with lipase/pen acylase.

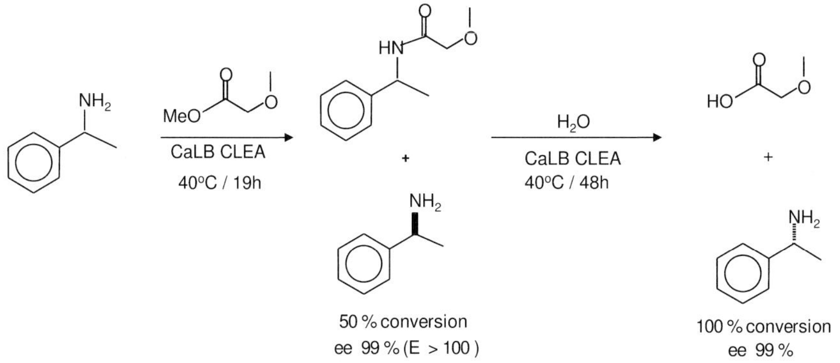

**Scheme 6.9** Acylation and deacylation with a lipase.

Both acylation and deacylation can be performed with a lipase as shown in Scheme 6.9 in which a CALB CLEA was used in the deacylation step that, not surprisingly, was rather slow compared to deacylation with pen acylase.

In order to obtain a commercially viable process it is necessary to racemize the unwanted amine enantiomer, preferably *in situ* in a so-called DKR. The palladium-on-charcoal-catalyzed racemization of amines was first reported by Murahashi *et al.* [23] and was later combined with lipase-catalyzed acylation, to afford a DKR, by Reetz [24] and others [25]. We were able to achieve a DKR of α-methyl benzylamine by performing the lipase-catalyzed acylation in the presence of a palladium nanoparticle catalyst (Scheme 6.10).

By analogy with the enantioselective reduction of prochiral ketones to chiral alcohols an attractive method for producing enantiomerically pure amines would be enantioselective reductive amination of a ketone via enzymatic reduction of an imine intermediate (Scheme 6.11). Unfortunately the required enzymes–amine

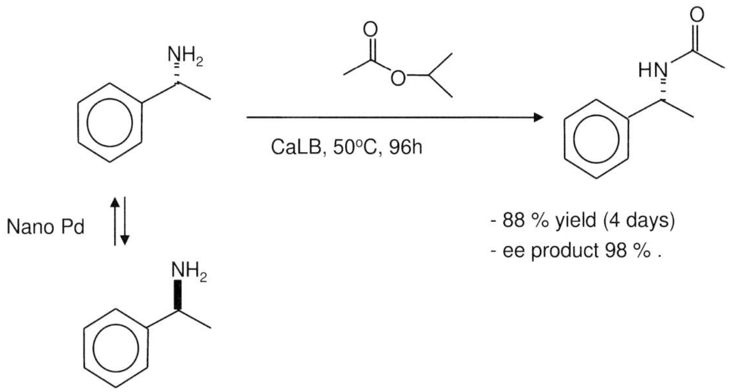

**Scheme 6.10** DKR of α-methyl benzylamine.

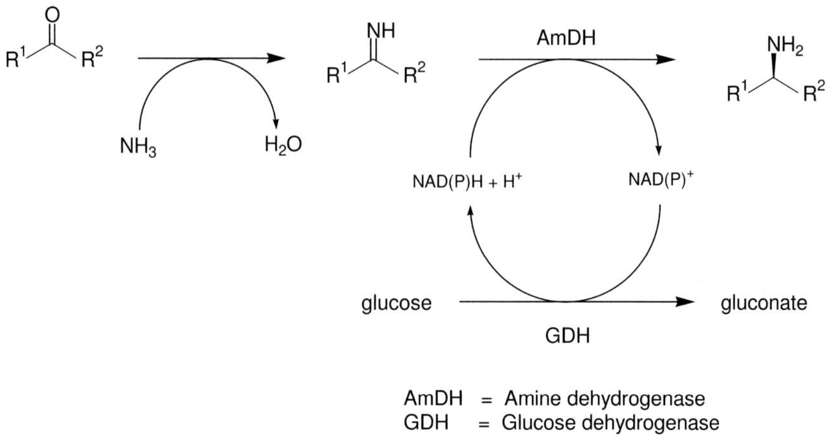

AmDH = Amine dehydrogenase
GDH = Glucose dehydrogenase

**Scheme 6.11** The holy grail: an NAD(P)-dependent amine dehydrogenase.

dehydrogenases–are not known (but see [26]). Nature has chosen a different pathway for dehydrogenation of an amine to an imine, involving molecular oxygen as a hydrogen acceptor and an amine oxidase as the catalyst (see Chapter 2). Hence, enantioselective, NAD(P)-dependent amine dehydrogenases remain a holy grail in biocatalysis.

In contrast, amino acid dehydrogenases comprise a well-known class of enzymes with industrial applications. An illustrative example is the Evonik (formerly Degussa) process for the synthesis of (*S*)-*tert*-leucine by reductive amination of trimethyl pyruvic acid (Scheme 6.12) [27]. The NADH cofactor is regenerated by coupling the reductive amination with FDH-catalyzed reduction of formate, which is added as the ammonium salt.

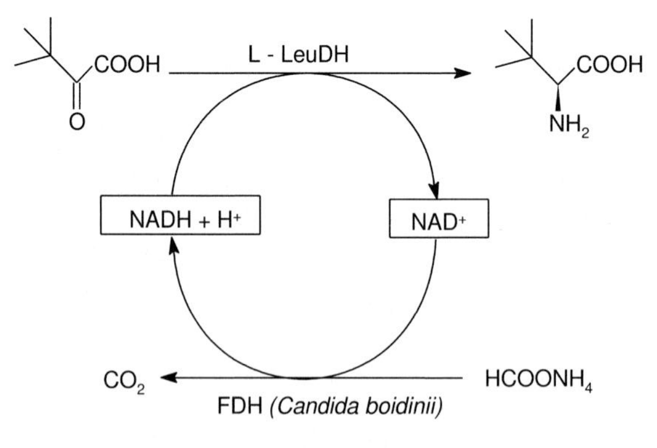


**Scheme 6.12** Synthesis of (*S*)-*tert*-leucine by enzymatic reductive amination.

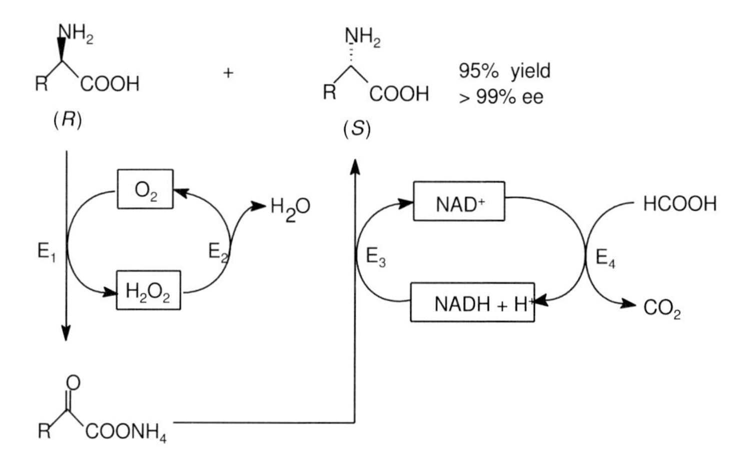


$E_1$ = D-Amino acid oxidase (E.C.1.4.3.3) $E_3$ = LeuDH (E.C.1.4.1.9)

$E_2$ = Catalase (E.C.1.11.1.6) $E_4$ = FDH (E.C.1.2.1.2)

**Scheme 6.13** Enzymatic deracemization of an amino acid.

An elegant four-enzyme cascade process was described by Nakajima *et al.* [28] for the deracemization of an α-amino acid (Scheme 6.13). It involved amine oxidase-catalyzed, (*R*)-selective oxidation of the amino acid to afford the ammonium salt of the α-keto acid and the unreacted (*S*)-enantiomer of the substrate. The keto acid then undergoes reductive amination, catalyzed by leucine dehydrogenase, to afford the (*S*)-amino acid. NADH cofactor regeneration is achieved with formate/FDH. The overall process affords the (*S*)-enantiomer in 95% yield and 99%e.e. from racemic starting material, formate and molecular oxygen, and the help of three enzymes in concert. A fourth enzyme, catalase, is added to decompose the hydrogen peroxide formed in the first step which otherwise would have a detrimental effect on the enzymes.

In an alternative approach an α-hydroxy acid first undergoes dehydrogenation to the corresponding α-keto acid, catalyzed by an α-hydroxy acid dehydrogenase, followed by reductive amination in the presence of an amino acid dehydrogenase. The reducing equivalents generated in the first step are consumed in the second step to afford an elegant redox neutral process (Scheme 6.14).

Another class of enzymes that can be used for the enantioselective synthesis of amines and amino acids is the aminotransferases or transaminases (TAs) [29]. As shown in Scheme 6.15, they can be employed in a kinetic resolution or an asymmetric synthesis mode.

From a practical viewpoint they suffer from serious product inhibition and/or equilibrium constraints. One way of avoiding the latter issue is to couple the transamination step to a second enzymatic step which removes the coproduct. In the example shown in Scheme 6.16 acetophenone undergoes TA-catalyzed transamination using L-alanine as the amine donor. The unfavorable equilibrium and

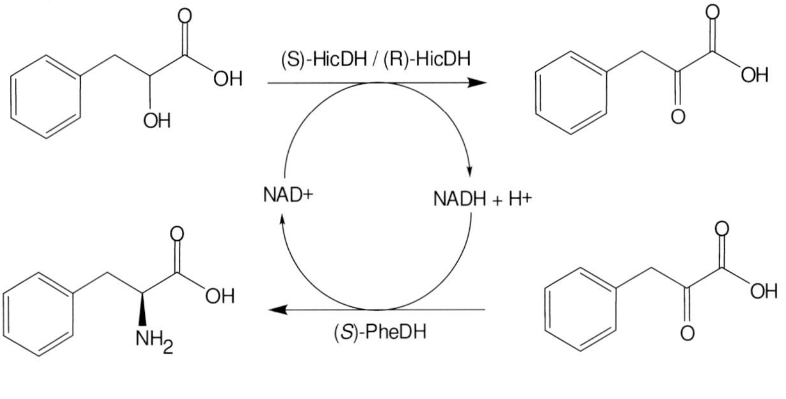

HicDH = Hydroxycaproate dehydrogenase

PheDH = Phenylalanine dehydrogenase

**Scheme 6.14** Enzymatic conversion of an α-hydroxy acid to an α-amino acid.

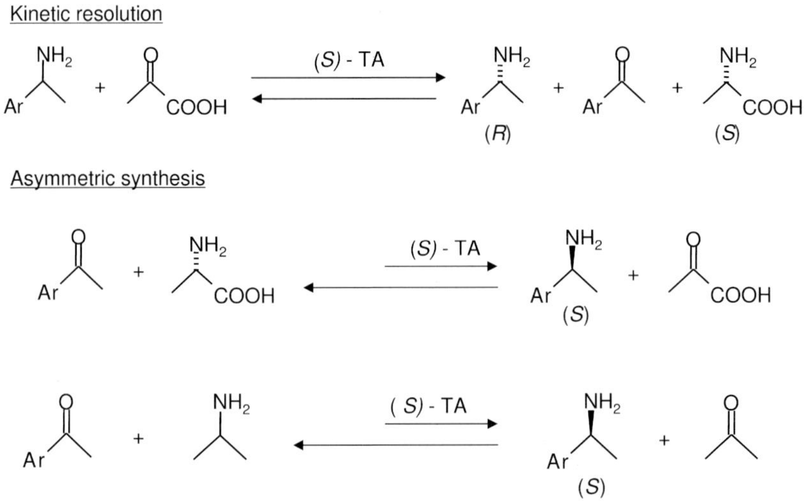

Kinetic resolution

Asymmetric synthesis

**Scheme 6.15** TA-catalyzed syntheses of chiral amines.

severe product inhibition by pyruvate were overcome by *in situ* removal of the latter by lactate dehydrogenase-catalyzed reduction to lactate [30].

Sung *et al.* [31] employed a four-enzyme cascade process for the synthesis of aromatic D-amino acids (Scheme 6.17). Glutamate racemase catalyzed the equilibration of L-glutamate to racemic glutamate. A thermostable D-amino acid transferase catalyzed the transamination of, for example, phenylpyruvate to afford D-phenylalanine and α-keto glutarate. The latter then undergoes glutamate dehydrogenase-catalyzed reduction back to L-glutamate in concert with formate/FDH as the nicotinamide cofactor-regenerating system. The overall process

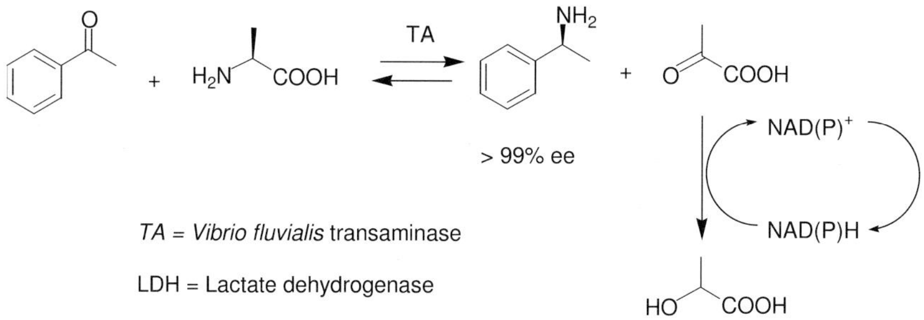
TA
NH2
H2N
COOH
+
COOH
> 99% ee
NAD(P)+
NAD(P)H
TA = Vibrio fluvialis transaminase
LDH = Lactate dehydrogenase
HO
COOH

**Scheme 6.16** TA-catalyzed transamination coupled with pyruvate reduction.

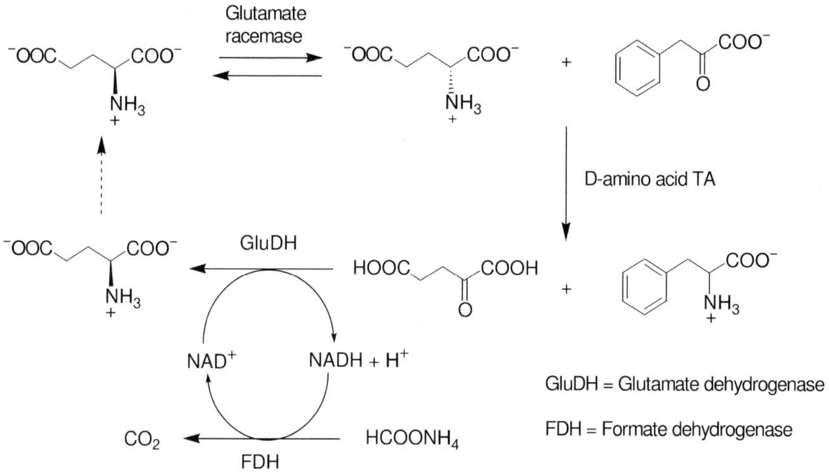
Glutamate
racemase
-OOC
COO-
NH3
+
D-amino acid TA
GluDH
HOOC
COOH
NAD+
NADH + H+
CO2
HCOONH4
FDH
GluDH = Glutamate dehydrogenase
FDH = Formate dehydrogenase

**Scheme 6.17** Production of a D-amino acid from an α-keto acid and ammonia using a four-enzyme cascade.

constitutes an elegant reductive amination of phenyl pyruvate with ammonium formate to afford D-phenylalanine together with carbon dioxide as the sole coproduct.

## 6.6 Chiral Carboxylic Acid Derivatives

The third group of target molecules comprises chiral carboxylic acid and their derivatives: esters, amides and nitriles. Enantiomerically pure esters are prepared in an analogous manner to the enantiomerically pure alcohols discussed earlier [i.e. by esterase- or lipase-catalyzed hydrolysis or (trans)esterification]. However, these reactions are not very interesting in the present context of cascade reactions. Amides can be produced by enantioselective ammoniolysis of esters or even the

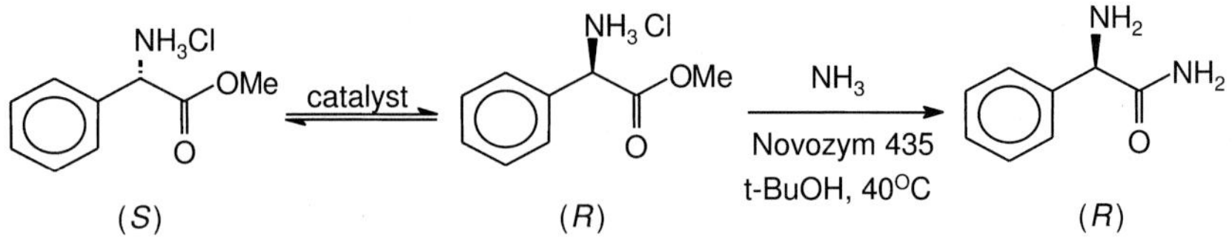

| Rac. cat | Ester/cat | Time (h) | Conv (%) | ee amide (%) |
|---|---|---|---|---|
| none | — | 4 | 46 | 78 |
| pyridoxal | 100 | 17 | 90 | 56 |
| salicylaldehyde | 100 | 17 | 86 | 56 |
| pyridoxal * | 50 | 66 | 85 | 88 |

* At -20°C

**Scheme 6.18** DKR of an amino acid ester by enzymatic ammoniolysis.

free carboxylic acid [32]. An interesting example of a DKR of an amino acid ester by ammoniolysis (Scheme 6.18) in the presence of an aldehyde, to catalyze the racemization of the remaining ester enantiomer, has been described [33].

Nitriles are interesting precursors of both amides and carboxylic acids. *In vivo* there are two pathways for the bioconversion of nitriles to carboxylic acids (Scheme 6.19). In the first method a nitrilase catalyzes the enantioselective hydrolysis of a racemic or prochiral nitrile. The second pathway involves a two-enzyme cascade in which an aselective nitrile hydratase (NHase) catalyzes the hydration of the racemic nitrile to the racemic amide followed by an amidase-catalyzed enantioselective hydrolysis to the carboxylic acid. The amidase is generally, but not always, (*S*)-selective, resulting in the formation of a 1 : 1 mixture of the (*S*)-acid

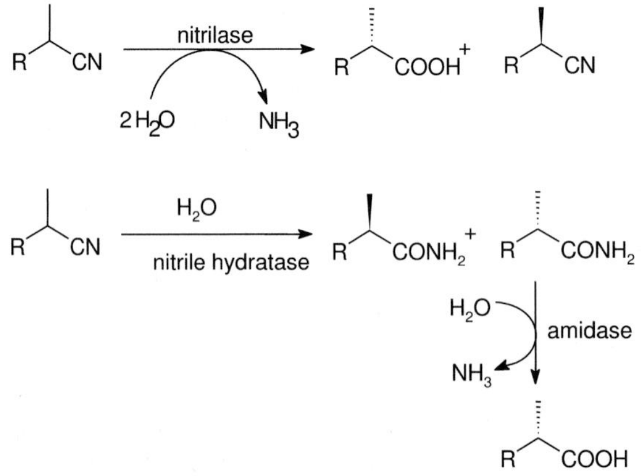

**Scheme 6.19** Nitrile-converting enzymes.

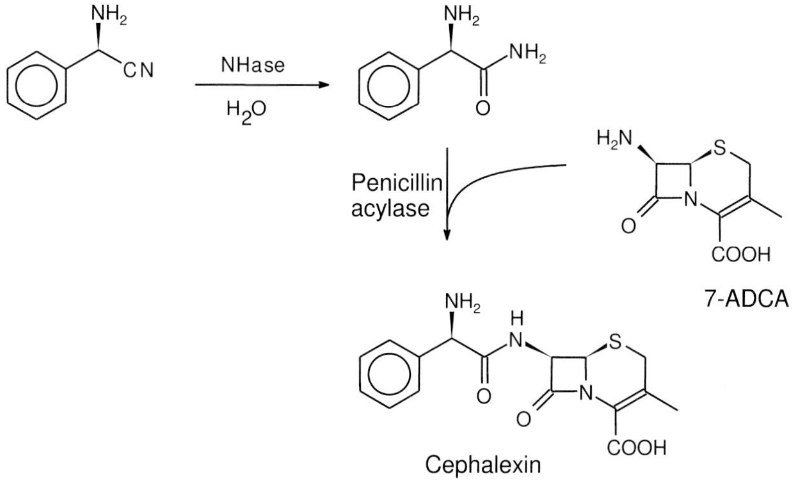

NHase
$H_2O$
Penicillin acylase
7-ADCA
Cephalexin

**Scheme 6.20** One-pot conversion of phenylglycine nitrile and 7-aminodeacetoxycephalosporanic acid to cephalexin.

and the (*R*)-amide. Nitrile-converting enzymes, in particular NHases, are used on a very large scale for the synthesis of, for example, acrylamide and nicotinamide, usually as whole-cell biotransformations mainly owing to the low stability of these enzymes outside the cell. The appearance of several reviews in recent years attests to the synthetic potential of nitrile-converting enzymes [34–38].

Since amides are often used as acyl donors in the enantioselective acylation of amines, a combination of a NHase and an amidase could, in principle, be used for the direct acylation of an amine by the nitrile precursor of the amide. An example of such a one-pot reaction is shown in Scheme 6.20. The (*R*)-enantiomer of phenylglycine nitrile undergoes NHase-catalyzed conversion to the corresponding (*R*)-amide and the latter reacts *in situ* with 7-aminodeacetoxycephalosporanic acid, in the presence of penicillin amidase, to afford the cephalosporin antibiotic, cephalexin [39].

Reaction of hydrogen cyanide with aldehydes leads to the reversible formation of cyanohydrins, which can be hydrolyzed to the corresponding α-hydroxy acids. Both steps can be performed enzymatically, using a hydroxynitrile lyase (HnL) and a nitrilase, respectively. This leads to two strategies for the synthesis of enantiomerically pure α-hydroxy acids as is illustrated for (*R*)-mandelic acid in Scheme 6.21. In the first strategy benzaldehyde is allowed to react with hydrogen cyanide, in the presence of an (*R*)-selective HnL, to afford the (*R*)-mandelonitrile. Competing non-enzymatic hydrocyanation is suppressed by performing the reaction in an organic solvent with a minimum amount of water present (microaqueous environment). Acid-catalyzed hydrolysis of the latter, in a one-pot sequential procedure, affords (*R*)-mandelic acid. In the second strategy racemic mandelonitrile is produced by non-enzymatic hydrocyanation of benzaldehyde. This reaction is performed in the presence of an (*R*)-selective nitrilase which catalyzes the

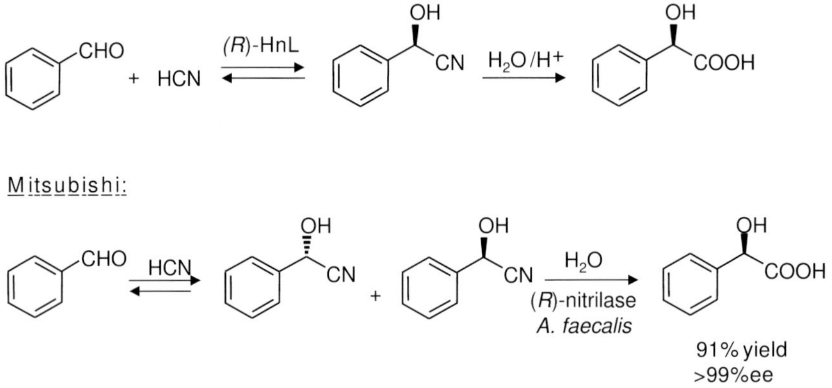

CHO + HCN (R)-HnL OH CN $H_2O/H^+$ OH COOH

Mitsubishi:

CHO HCN OH CN + OH CN $H_2O$ (R)-nitrilase *A. faecalis* OH COOH

91% yield
>99%ee

**Scheme 6.21** Two routes to (*R*)-mandelic acid.

*in situ* hydrolysis of the (*R*)-mandelonitrile. As the hydrocyanation step is reversible, the remaining (*S*)-mandelonitrile spontaneously racemizes via dehydrocyanation to benzaldehyde and hydrolysis of the (*R*)-enantiomer shifts the equilibrium towards product. Overall this results in an elegant cascade process for the synthesis of the single enantiomer in 100% theoretical yield. The process has been commercialized by Mitsubishi for the manufacture of (*R*)-mandelic acid, in 91% yield and greater than 99% e.e., using an (*R*)-selective nitrilase from *Alcaligenes faecalis* [40].

Similarly, the use of an (*S*)-selective nitrilase would afford an equally elegant process for (*S*)-mandelic acid. Unfortunately, nitrilases that are (*S*)-selective for this type of substrate are not available. Consequently, we used a different approach for the one-pot synthesis of (*S*)-mandelic acid [41]. The production of (*S*)-mandelonitrile catalyzed by (*S*)-HnL from *Manihot esculenta* (cassava) was coupled with its *in situ* hydrolysis catalyzed by an aselective nitrilase from *Pseudomonas flurescens*. Both enzymes were immobilized in a combiCLEA (see Section 6.2) and (*S*)-mandelic acid was obtained in 98% e.e., the enantioselectivity being directed by the HnL (Scheme 6.22). It was also shown that the combiCLEA gave superior results to a mixture of the individual CLEAs, which was attributed to the forced proximity of the two enzymes in the combiCLEA providing for a rapid removal of the mandelonitrile intermediate before it has a chance to racemize. However, (*S*)-mandelamide was also formed in 40% yield. It is well documented that the nitrilase-catalyzed hydrolysis of nitriles often leads to substantial amounts of the corresponding amides, depending on the structure of the nitrile. In principle, competing amide formation, which proceeds through a common intermediate, could probably be suppressed by directed evolution of the nitrilase [42]. We chose a different approach. We added a third enzyme, penicillin amidase, to catalyze the hydrolysis of the (*S*)-mandelamide. Thus, when benzaldehyde was allowed to react with hydrogen cyanide, in the presence of a triple-decker combiCLEA, containing the (*S*)-HnL, the nitrilase and penicillin amidase, (*S*)-mandelic acid was formed in 96% yield and greater than 99% e.e.

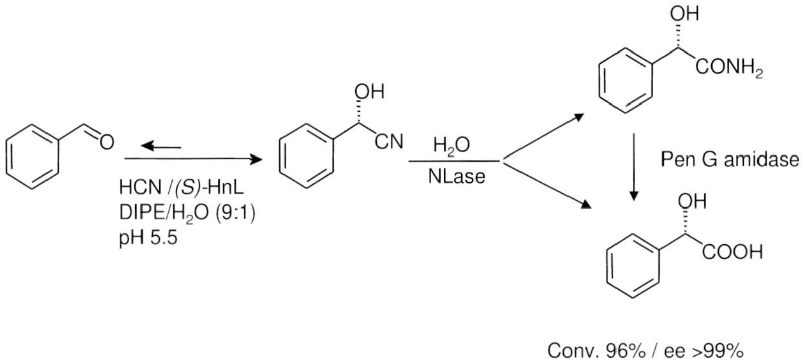

OH
CONH$_2$
OH
O
CN
H$_2$O
NLase
Pen G amidase
HCN /(*S*)-HnL
DIPE/H$_2$O (9:1)
pH 5.5
OH
COOH

Conv. 96% / ee >99%

**Scheme 6.22** A trienzymatic cascade process using a triple-decker combiCLEA.

By analogy with the synthesis of α-hydroxy acids one can envisage a one-pot synthesis of α-hydroxy amides from aldehydes via hydrocyanation and *in situ* NHase-catalyzed hydrolysis to the amide. Since enantioselective NHases are very rare, the enantioselectivity should be derived from HnL-catalyzed hydrocyanation. The second step has been described for the *Rhodococcus erythropolis* NHase-catalyzed hydration of (*R*)-mandelonitrile to give the (*R*)-amide with retention of enantiopurity [43].

Another approach to preparing enantiomerically pure carboxylic acids and related compounds is via enantioselective reduction of conjugated double bonds using NAD(P)H-dependent enoate reductases (EREDs; EC 1.3.1.X), members of the so-called 'Old Yellow Enzyme' family [44]. EREDs are ubiquitous in nature and their catalytic mechanism is well documented [45]. They contain a catalytic flavin cofactor and a stoichiometric nicotinamide cofactor which must be regenerated (Scheme 6.23).

EREDs have recently become the focus of attention because the reaction they catalyze – the asymmetric reduction of a C=C bond – is one of the most widely employed strategies for the synthesis of enantiomerically pure products from readily available prochiral substrates. Asymmetric hydrogenation catalyzed by chiral metal complexes is a well-developed methodology for performing such transformations on an industrial scale [46]. ERED-catalyzed reductions are their biological counterparts, just as KRED-catalyzed reductions of C=O bonds are the biological equivalent of the asymmetric hydrogenation of ketones. We also note that the two methodologies are complementary: asymmetric hydrogenation involves *cis*-addition of two hydrogens while ERED reduction results in *anti*-addition. They can be used for the enantioselective reduction of a C=C double bond conjugated to a variety of electron-withdrawing groups (e.g. aldehyde, ketone, carboxylic acid, ester, amide, anhydride, nitro) [47–52].

Analogous to the KRED reductions they can be performed as whole-cell biotransformations [48, 49] (baker's yeast, for example, contains a number of EREDs) or with isolated enzymes [50–52]. In the latter case the nicotinamide cofactor can

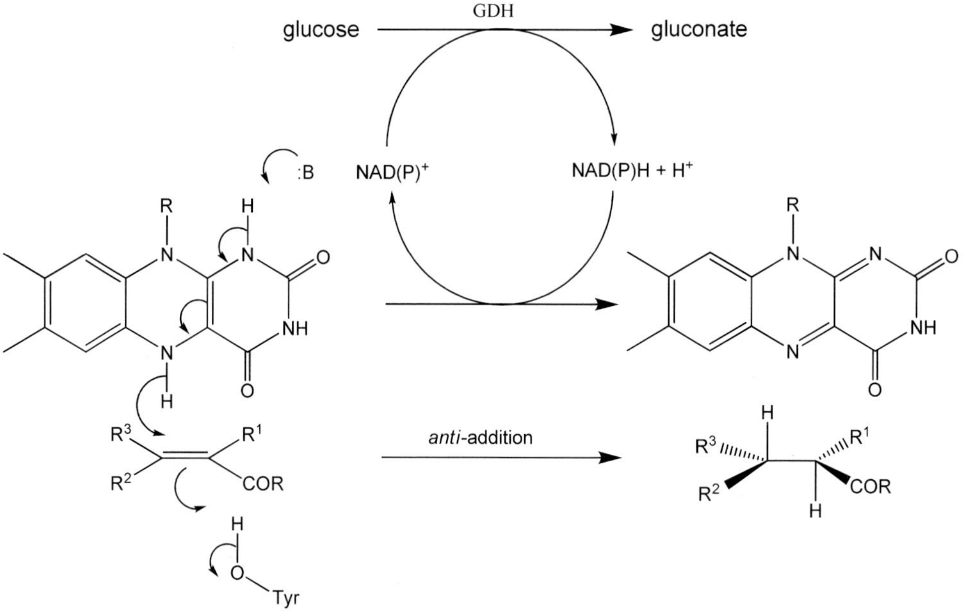
GDH
glucose
gluconate
:B
NAD(P)+
NAD(P)H + H+
R
H
N
N
O
NH
N
H
O
R3
R1
R2
COR
H
O
Tyr
anti-addition
R
N
N
O
NH
N
O
H
R3
R1
R2
COR
H

**Scheme 6.23** Mechanism of an ERED.

be regenerated by adding a cofactor regenerating system, such as glucose/GDH or formate/FDH [50–52]. Alternatively, 'designer bugs' containing both the ERED and a matching redox enzyme, usually GDH, have been used in preparative-scale biotransformations [53–55]. Two transformations of practical interest are shown in Scheme 6.24.

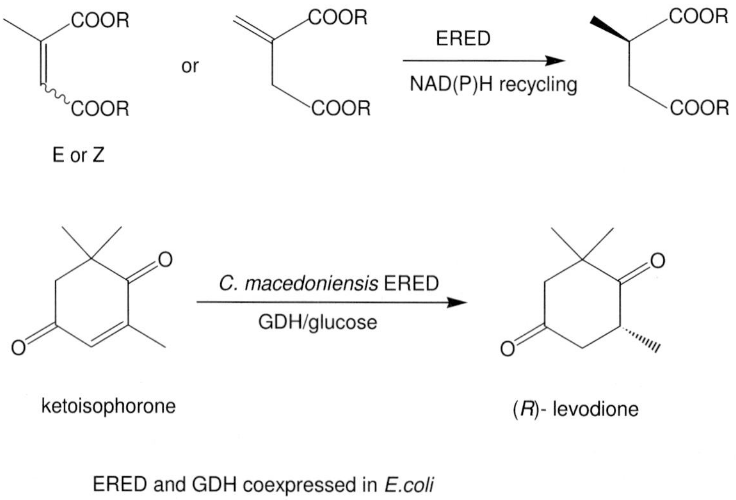
COOR
COOR
E or Z
or
COOR
COOR
ERED
NAD(P)H recycling
COOR
COOR
O
O
C. macedoniensis ERED
GDH/glucose
O
O
ketoisophorone
(R)- levodione
ERED and GDH coexpressed in E.coli

**Scheme 6.24** ERED-catalyzed reductions.

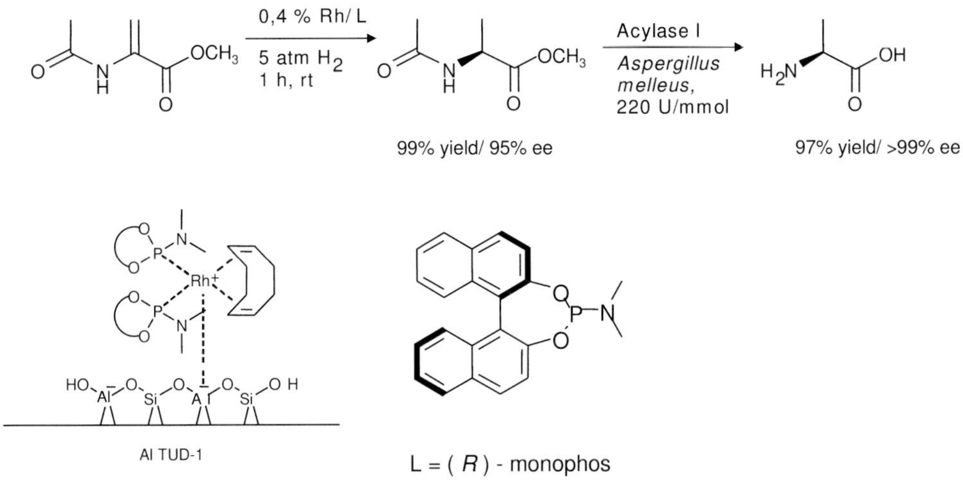

0,4 % Rh/ L
5 atm $H_2$
1 h, rt

99% yield/ 95% ee

Acylase I
*Aspergillus melleus*,
220 U/mmol

97% yield/ >99% ee

Al TUD-1

L = ( *R* ) - monophos

**Scheme 6.25** Chemoenzymatic synthesis of an amino acid.

Another interesting possibility is to combine catalytic asymmetric hydrogenation, using a chiral metal complex as the catalyst, with subsequent enzymatic hydrolysis steps to remove protecting and/or activating functional groups. An illustrative example of this is shown in Scheme 6.25. An (*S*)-amino acid is produced in a one-pot chemoenzymatic procedure, in water as solvent, by combining catalytic asymmetric hydrogenation, over a Rh(I) monophos catalyst supported on a zeolite (Al-TUD-1), with aminoacylase-catalyzed hydrolysis of the amide and ester moieties in the product [56]. The product of the hydrogenation step is produced in 95% e.e., but the subsequent enzymatic hydrolysis results in an upgrading of the optical purity, with very little loss in yield. This can be attributed to the fact that the aminoacylase is highly (*S*)-selective.

## 6.7 C–C Bond Formation: Aldolases

C–C bond formation is a pivotal transformation in organic synthesis and, at the same time, often presents a formidable stereochemical challenge in that it can lead to the formation of two new, contiguous stereogenic centers. This is the case with aldol reactions. Chemical methods for achieving such stereoselective formation of C–C bonds are generally circuitous, involving various activation, protection and deprotection steps. Enzymes, in stark contrast, are able to achieve this transformation in a single step. Enzyme-catalyzed aldol reactions are ubiquitous in nature and are intricately involved in many metabolic pathways. However, notwithstanding the enormous potential of these enzymes, there are few examples of their use on an industrial scale. The main reason for this is that many aldolase-catalyzed processes require expensive cosubstrates. A case in point is provided by

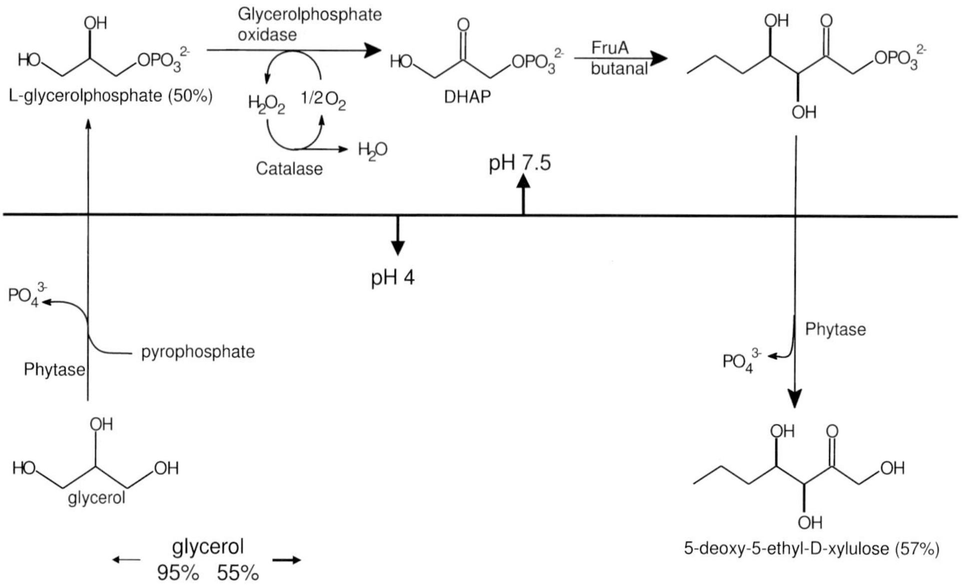

**Scheme 6.26** Catalytic cascade process for the synthesis of a non-natural carbohydrate.

the dihydroxy acetone (DHAP)-dependent aldolases which catalyze the formation of two new contiguous stereogenic centers in a single step. This reaction has tremendous synthetic potential, but DHAP is expensive and is required in stoichiometric quantities. One approach to solving this problem is to use enzymatic reactions to produce the DHAP *in situ* by employing other enzymes. This was achieved in a one-pot four enzyme catalytic cascade as shown in Scheme 6.26 [57].

Racemic glycerol-1-phosphate is formed by phytase-catalyzed phosphorylation of glycerol catalyzed by the inexpensive acid phosphatase, phytase, which is added to pig feed. This is followed by L-glycerol phosphate oxidase-catalyzed oxidation of the L-enantiomer to DHAP and DHAP-dependent aldolase-catalyzed aldol reaction with *n*-butyraldehyde to afford the phosphorylated aldol adduct. Catalase is added to decompose the hydrogen peroxide formed in this step, which would otherwise be detrimental for the enzyme(s). The final step is phytase-catalyzed dephosphorylation to the non-natural carbohydrate. The whole process is made possible by utilizing a pH switch to 'turn enzymes on and off'. Thus, the phosphorylation of glycerol is performed at pH 4 where phytase exhibits optimum activity. The pH is then raised to 7.5 and the glycerol decreased from 95 to 55 wt%, conditions which are conducive for the oxidation and the C—C bond-forming steps to afford the phosphorylated aldol adduct. The pH is subsequently reduced to 4 and the phytase effects hydrolysis of the phosphate ester of the aldol adduct, and the D-glycerol phosphate, to afford the non-natural carbohydrate product, in high stereoselectivity, together with glycerol. Overall, this provides an elegant one-pot,

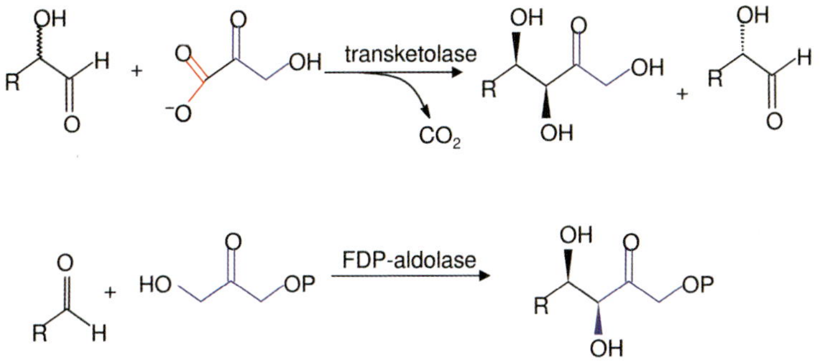

Scheme 6.27 Transketolase versus DHAP-dependent aldolase.

four-enzyme catalytic cascade process for the synthesis of a non-natural carbohydrate from inexpensive glycerol, phosphate and an aldehyde.

Another enzyme which catalyzes C–C bond-forming reactions of synthetic interest is transketolase. As shown in Scheme 6.27, transketolase can be used to make the same products that can be made using a DHAP-dependent aldolase, but from different starting materials. Transketolase is a robust, potentially readily available enzyme, with synthetic potential, but suffering from the same shortcoming as the DHAP-dependent aldolases – the requirement for a rather expensive cosubstrate, in this case hydroxy pyruvate (Scheme 6.26). In order to have industrial utility an inexpensive source of this cosubstrate is needed. One possibility could be *in situ* production from serine using a serine dehydrogenase. To our knowledge this has not been described.

In contrast to transketolase and the DHAP-dependent aldolases, deoxyribose aldolase (DERA) catalyzes the aldol reaction with the simple aldehyde, acetaldehyde. *In vivo* it catalyzes the formation of 2-deoxyribose-5-phosphate, the building block of DNA, from acetaldehyde and D-glyceraldehyde-3-phosphate, but *in vitro* it can catalyze the aldol reaction of acetaldehyde with other non-phosphorylated aldehydes. The example shown in Scheme 6.28 involves a tandem aldol reaction

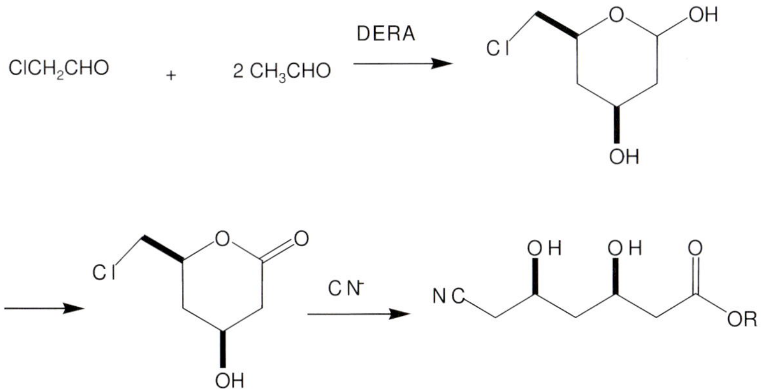

Scheme 6.28 DERA-catalyzed double aldol reaction.

of two molecules of acetaldehyde with chloroacetaldehyde to give a cyclic hemiacetal which is subsequently converted, in two steps, to an advanced intermediate for atorvastatin (Lipitor) [58]. The activity and selectivity in the DERA-catalyzed aldol reaction with the non-natural substrate, chloroacetaldehyde, were optimized by directed evolution of the wild-type DERA [58].

## 6.8
## Oxidations with $O_2$ and $H_2O_2$

Many biocatalytic oxidations involve the consumption or formation of the green oxidants oxygen and hydrogen peroxide. Peroxidases utilize hydrogen peroxide as an oxidant, while most oxidases generate one equivalent of it as a coproduct. The presence of hydrogen peroxide has a detrimental effect on the stability of most enzymes, which can be attributed to it oxidizing reactive functional groups in the enzyme. Consequently, in reactions involving oxidases a small amount of catalase is added in order to rapidly degrade the hydrogen peroxide before it can do any damage. We have already seen an example of this in Scheme 6.26, where catalase was added in order to destroy the hydrogen peroxide formed in the oxidase-catalyzed aerobic oxidation of glycerol-3-phosphate. This can also be achieved by using a combiCLEA containing the oxidase and catalase, as we have shown [59] for glucose oxidase- and galactose oxidase-catalyzed oxidation of glucose and galactose, respectively (Scheme 6.29). In both cases the combiCLEA could be recycled with no loss of activity.

In reactions catalyzed by peroxidases one cannot easily avoid the presence of hydrogen peroxide as this is the natural oxidant employed by the enzyme. However, it is necessary to maintain the steady-state concentration of hydrogen peroxide as

OH HO HO O OH OH $O_2$ / water / pH 7 combiCLEA 1 HO HO O OH O OH

HO OH HO O OR OH $O_2$ / water / pH 7 combiCLEA 2 HO O HO O OR OH

combiCLEA ! = Glucose oxidase / catalase

combiCLEA 2 = Galactose oxidase / catalase

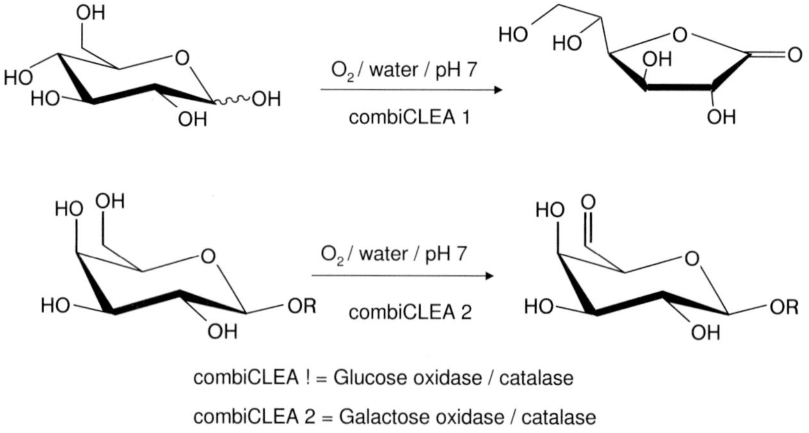

**Scheme 6.29** Glucose oxidase/catalase and galactose oxidase/catalase combiCLEAs.

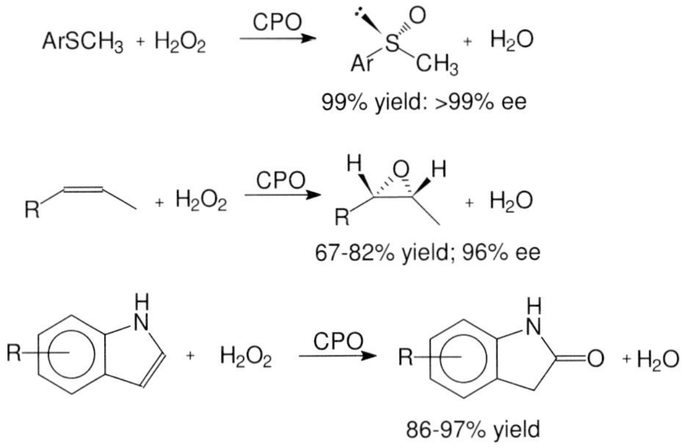

$ArSCH_3 + H_2O_2 \xrightarrow{CPO} Ar\text{-}S(O)\text{-}CH_3 + H_2O$

99% yield: >99% ee

$R\text{-}CH{=}CH\text{-}CH_3 + H_2O_2 \xrightarrow{CPO}$ epoxide $+ H_2O$

67-82% yield; 96% ee

R-indole $+ H_2O_2 \xrightarrow{CPO}$ R-oxindole $+ H_2O$

86-97% yield

**Scheme 6.30** CPO-catalyzed oxidations with hydrogen peroxide.

low as possible. A very effective way to achieve this is to generate hydrogen peroxide *in situ* at such a rate that it reacts away as fast as it is formed. For example, the heme-dependent chloroperoxidase (CPO) from *Caldariomyces fumago* catalyzes a variety of synthetically interesting oxidations with hydrogen peroxide, in the absence of chloride ion (Scheme 6.30) [60]. However, CPO has very limited stability towards hydrogen peroxide, owing to the sensitivity of its heme moiety towards oxidative degradation, which places severe limitations on its synthetic utility [61]. Much improved results could be obtained by employing *in situ* generation of hydrogen peroxide from glucose and glucose oxidase which were coimmobilized in polyurethane foam [62]. The total turnover number in the sulfoxidation of thioanisole, for example, was increased to a new optimum of 250 000 [62]. We are currently investigating the use of a CPO/glucose oxidase combiCLEA for catalytic oxidations with *in situ* generation of hydrogen peroxide [63].

## 6.9 Conclusions and Prospects

Only a few years ago it was widely accepted that the 'cofactor regeneration problem' represented a serious obstacle with respect to the commercial viability of enzymatic redox processes. Hopefully it is clear from the preceding discussion that there is no longer a cofactor regeneration 'problem' anymore than there is an 'enzyme problem'. The number of readily available enzymes has increased dramatically in the last decade and advances in *in vitro* evolution have made it possible to routinely optimize the performance of enzymes. The coupling of enzymes in multi-enzyme cascade processes is an attractive way to regenerate cofactors, shift equilibria towards products and remove intermediate products that cause inhibition. Hence, we expect that multi-enzyme cascade processes will become much more common in the future.

## References


1 Sheldon, R.A. (2007) *Green Chemistry*, **9**, 1273.

2 Sheldon, R.A., Arends, I.W.C.E. and Hanefeld, U. (2007) *Green Chemistry and Catalysis*, Wiley-VCH, Weinheim.

3 (a) Bruggink, A., Schoevaart, R. and Kieboom, T. (2003) *Organic Process Research and Development*, **7**, 622–40.
(b) see also Veum, L. and Hanefeld, U. (2006) *Chemical Communications*, 825–31.
(c) Glueck, S.M., Mayer, S.F., Kroutil, W. and Faber, K. (2002) *Pure and Applied Chemistry*, **74**, 2253–7.

4 Sheldon, R.A. (2008) *Chem. Commun.*, DOI 10.1039/G803584a.

5 (a) Menzel, A., Werner, H., Altenbuchner, J. and Groeger, H. (2004) *Engineering in Life Sciences*, **4**, 573–6.
(b) Wandrey, C. (2004) *The Chemical Record*, **4**, 254–65.

6 Sheldon, R.A. (2007) *Advanced Synthesis Catalysis*, **349**, 1289–307.

7 Cao, L. (2005) *Carrier-Bound Immobilized Enzymes, Principles, Applications and Design*, Wiley-VCH, Weinheim.

8 (a) Margolin, A.L. and Navia, M.A. (2001) *Angewandte Chemie (International Edition in English)*, **40**, 2204.
(b) Lalonde, J. (1997) *CHEMTECH*, **27**, 38–45.
(c) Roy, J.J. and Abraham, T.E. (2004) *Chemical Reviews*, **104**, 3705–21.
(d) Vaghjiani, J.D., Lee, T.S., Lye, G.J. and Turner, M.K. (2000) *Biocatalysis and Biotransformation*, **18**, 151–75.

9 (a) Cao, L., van Rantwijk, F. and Sheldon, R.A. (2000) *Organic Letters*, **2**, 1361–4.
(b) Sheldon, R.A. (2007) *Biochemical Society Transactions*, **35**, 1583.
(c) Sheldon, R.A., Sorgedrager, M. and Janssen, M.H.A. (2007) *Chimica Oggi–Chemistry Today*, **25**, 48.
(d) Sheldon, R.A. (2006) *Biocatalysis in the Pharmaceutical and Biotechnological Industries* (ed. R. Patel), CRC Press, Boca Raton, pp. 351–62.
(e) Sheldon, R.A., Schoevaart, R. and van Langen, L.M. (2005) *Biocatalysis and Biotransformation*, **23**, 141.

10 (a) Reetz, M.T. (2006) *Advanced Catalysis*, **49**, 1–69.
(b) Montiel, C. and Bustos-Jaimes, I. (2008) *Current Chemical Biology*, **2**, 50–9.
(c) Sen, S., Dasu, V.V. and Mandal, B. (2007) *Applied Biochemistry and Biotechnology*, **143**, 212–23.
(d) Joyce, G.F. (2007) *Angewandte Chemie (International Edition in English)*, **46**, 6420–36.
(e) Zhao, H. (2007) *Biotechnology and Bioengineering*, **98**, 313–17.
(f) Sylvestre, J., Chautard, H., Cedrone, F. and Delcourt, M. (2006) *Organic Process Research and Development*, **10**, 562–71.
(g) Powell, K.A., Ramer, S.W., Del Cardayre, S.B., Stemmer, W.P.C., Tobin, M.B., Longchamp, P.F. and Huisman, G.W. (2001) *Angewandte Chemie (International Edition in English)*, **40**, 3948–59.
(h) Tobin, M.B., Gustafsson, C. and Huisman, G.W. (2000) *Current Opinion in Structural Biology*, **10**, 421–7.

11 (a) Pamies, O. and Bäckvall, J.E. (2004) *Trends in Biotechnology*, **22**, 130–5.
(b) Larsson, A.L.E., Persson, B.A. and Bäckvall, J.E. (1997) *Angewandte Chemie (International Edition in English)*, **36**, 1211–12.
(c) Martin Matute, B., Edin, M., Bogar, K. and Bäckvall, J.E. (2004) *Angewandte Chemie (International Edition in English)*, **43**, 6535–9.
(d) Martin Matute, B., Edin, M., Bogar, K., Kaynak, F.B. and Bäckvall, J.E. (2005) *Journal of the American Chemical Society*, **127**, 8817–25.
(e) Choi, J.H., Kim, Y.H., Nam, S.H., Shin, S.T., Kim, M.J. and Park, J. (2002) *Angewandte Chemie (International Edition in English)*, **41**, 2373–6.
(f) Kim, N., Ko, S.-B., Keon, M.S. and Kim, M.-J. (2005) *Organic Letters*, **7**, 4523–6.
(g) van Nispen, S.F.G.M., van Buijtenen, J., Vekemans, J.A.J.M., Meuldijk, J. and Hulshof, L.A. (2006) *Tetrahedron: Asymmetry*, **17**, 2299–305.

(h) Riermeier, T.H., Gross, P., Monsees, A., Hoff, M. and Trauthwein, H. (2005) *Tetrahedron Letters*, **46**, 3403–6.
(i) Dijksman, A., Elzinga, J.M., Li, Y.X., Arends, I.W.C.E. and Sheldon, R.A. (2002) *Tetrahedron Asymmetry*, **13**, 879–84.

**12** (a) Moore, J.C., Pollard, D.J., Kosjek, B. and Devine, P.N. (2007) *Accounts of Chemical Research*, **40**, 1412–19.
(b) De Wildeman, S.M.A., Sonke, T., Schoemaker, H.E. and May, O. (2007) *Accounts of Chemical Research*, **40**, 1260–6.
(c) Nakamura, K. and Matsuda, T. (2002) *Enzyme Catalysis in Organic Synthesis*, 2nd edn (eds K. Drauz and H. Waldmann), Wiley-VCH, Weinheim, pp. 991–1047.
(d) Nakamura, K., Yamanaka, R., Matsuda, T. and Harada, T. (2003) *Tetrahedron Asymmetry*, **14**, 2659–81.
(e) Kroutil, W., Mang, H., Eddegger, K. and Faber, K. (2004) *Advanced Synthesis Catalysis*, **346**, 125–42.

**13** Johannes, T.W., Woodyer, R.D. and Zhao, H. (2006) *Biotechnology and Bioengineering*, **96**, 18–26.

**14** Goldberg, K., Schroer, K., Lütz, S. and Liese, A. (2007) *Applied Microbiology and Biotechnology*, **76**, 237–48.

**15** (a) Kataoka, M., Kita, K., Wada, M., Yasohara, Y., Hasegawa, J. and Shimizu, S. (2003) *Applied Microbiology and Biotechnology*, **62**, 437–45.
(b) Goldberg, K., Schroer, K., Lütz, S. and Liese, A. (2007) *Applied Microbiology and Biotechnology*, **76**, 249–55.
(c) Büchholz, S. and Gröger, H. (2006) *Biocatalysis in the Pharmaceutical and Biotechnology Industries* (ed. R.N. Patel), CRC Press, Boca Raton, FL, pp. 757–90.

**16** (a) Fox, R.J., Davis, C.S., Mundorff, E.C., Newman, L.M., Gavrilovic, V., Ma, S.K., Chung, L.M., Ching, C., Tam, S., Muley, S., Grate, J., Gruber, J., Whitman, J.C., Sheldon, R.A. and Huisman, G.W. (2007) *Nature Biotechnology*, **25**, 338.
(b) See also Stemmer, W.P.C. (1994) *Nature*, **370**, 389.

**17** (a) Berkessel, A., Rollmann, C., Chamouleau, F., Labs, S., May, O. and Gröger, H. (2007) *Advanced Synthesis Catalysis*, **349**, 2697–704.
(b) See also Gröger, H., Chamouleau, F., Orologas, N., Rollmann, C., Drauz, K., Hummel, W., Weckbecker, A. and May, O. (2006) *Angewandte Chemie (International Edition in English)*, **45**, 5677–81.
(c) Poessel, T.M., Kosjek, B., Ellmer, U., Gruber, C.C., Edegger, K., Hildebrandt, P., Faber, K., Bornscheuer, U.T. and Kroutil, W. (2005) *Advanced Synthesis Catalysis*, **347**, 1827–34.

**18** Voss, C.V., Gruber, C.C. and Kroutil, W. (2007) *Tetrahedron Asymmetry*, **18**, 276–81.

**19** Hieber, G. and Ditrich, K. (2001) *Chimica Oggi – Chemistry Today*, **19** (6), 16.

**20** Cammenberg, M., Hult, K. and Park, S. (2006) *ChemBioChem*, **7**, 1745.

**21** Guranda, D.T., Khimiuk, A.I., van Langen, L.M., van Rantwijk, F., Sheldon, R.A. and Svedas, V.K. (2004) *Tetrahedron Asymmetry*, **15**, 2901.

**22** Ismail, H., Madeira Lau, R., van Langen, L.M., van Rantwijk, F., Svedas, V.K. and Sheldon, R.A. (2008) *Green Chemistry*, **10**, 415–18.

**23** Murahashi, S.-I., Yoshimura, N., Tsumiyama, T. and Kojima, T. (1983) *Journal of the American Chemical Society*, **105**, 5002.

**24** Reetz, M.T. and Schimossek, K. (1996) *Chimia*, **50**, 668.

**25** (a) Ahn, Y., Ko, S.-B., Kim, M.-J. and Park, J. (2008) *Coordination Chemistry Reviews*, **252**, 647–58.
(b) Parvelescu, A., De, D. and Jacobs, P. (2005) *Chemical Communications*, 5307–8.
(c) Paetzold, J. and Backvall, J.E. (2005) *Journal of the American Chemical Society*, **127**, 17620–1.

**26** Itoh, N., Yachi, C. and Kudome, T. (2000) *Journal of Molecular Catalysis B Enzymatic*, **10**, 281–90.

**27** (a) Bommarius, A.S., Schwarm, M. and Drauz, K. (1998) *Journal of Molecular Catalysis B Enzymatic*, **5**, 1–11.
(b) Bommarius, A.S., Drauz, K., Hummel, W., Kula, M.R. and Wandrey, C. (1994) *Biocatalysis*, **10**, 37–47.
(c) Wichmann, R., Wandrey, C., Bückmann, A.F. and Kula, M.R. (1981) *Biotechnology and Bioengineering*, **23**, 2789–96.

**28** Nakajima, N., Esaki, N. and Soda, K. (1990) *Chemical Communications*, 947–8.

**29** (a) Stewart, J.D. (2001) *Current Opinion in Chemical Biology*, **5**, 120–9.
(b) Shin, J.S., Kim, B.-G., Liese, A. and Wandrey, C. (2001) *Biotechnology and Bioengineering*, **73**, 179–87.

**30** Shin, J.S. and Kim, B.G. (1999) *Biotechnology and Bioengineering*, **65**, 206–11.

**31** Bae, H.-S., Lee, S.-G., Hong, S.-P., Kwak, M.-S., Esaki, N., Soda, K. and Sung, M.-H. (1999) *Journal of Molecular Catalysis B Enzymatic*, **6**, 241–147.

**32** (a) de Zoete, M.C., Kock-van Dalen, A.C., van Rantwijk, F. and Sheldon, R.A. (1993) *Journal of the Chemical Society Chemical Communications*, 1831.
(b) de Zoete, M.C., Kock-van Dalen, A.C., van Rantwijk, F. and Sheldon, R.A. (1994) *Biocatalysis*, **10**, 307–16.
(c) de Zoete, M.C., Ouwehand, A.A., van Rantwijk, F. and Sheldon, R.A. (1995) *Recueil des Travaux Chimiques des Pays-Bas*, **114**, 171–4.
(d) de Zoete, M.C., Kock-van Dalen, A.C., van Rantwijk, F. and Sheldon, R.A. (1996) *Journal of Molecular Catalysis B: Enzymatic*, **2**, 141–5.
(e) de Zoete, M.C., Kock-van Dalen, A.C., van Rantwijk, F. and Sheldon, R.A. (1996) *Journal of Molecular Catalysis B: Enzymatic*, **2**, 19–25.

**33** Wegman, M.A., Hacking, M.A.P.J., Rops, J., van Rantwijk, F. and Sheldon, R.A. (1999) *Tetrahedron Asymmetry*, **10**, 1739.

**34** DiCosimo, R. (2007) *Biocatalysis in the Pharmaceutical and Biotechnology Industries* (ed. R.N. Patel), CRC Press, Boca Raton, pp. 1–26.

**35** Wieser, M. and Nagasawa, T. (2000) *Stereoselective Biocatalysis* (ed. R.N. Patel), Marcel Dekker, New York, pp. 463–5.

**36** Martinkova, L. and Mylerova, V. (2003) *Current Organic Chemistry*, **7**, 1279.

**37** Martinkova, L. and Kren, V. (2002) *Biocatalysis and Biotransformation*, **20**, 73.

**38** Banerjee, A., Sharma, R. and Banerjee, U.C. (2002) *Applied Microbiology and Biotechnology*, **60**, 33.

**39** Wegman, M.A., van Langen, L.M., van Rantwijk, F. and Sheldon, R.A. (2002) *Biotechnology and Bioengineering*, **79**, 356.

**40** Kaul, P., Banerjee, A., Mayilraj, S. and Banerjee, U.C. (2004) *Tetrahedron Asymmetry*, **15**, 207.

**41** Chmura, A. manuscript in preparation.

**42** DeSantis, G., Wong, K., Farwell, B., Chatman, K., Zhu, Z., Tomlinson, G., Huang, H., Tan, X., Bibbs, L., Chen, P., Kretz, K. and Burk, M.J. (2003) *Journal of the American Chemical Society*, **125**, 11476.

**43** Reisinger, C., Osprian, I., Glieder, A., Schoemaker, H.E., Griengl, H. and Schwab, H. (2004) *Biotechnology Letters*, **26**, 1675–80.

**44** Warburg, O. and Christian, W. (1933) *Biochemische Zeitschrift*, **266**, 377–411.

**45** Williams, R.E. and Bruce, N.C. (2002) *Microbiology*, **148**, 1607–14.

**46** (a) Blaser, H.U. and Schmidt, E. (eds) (2004) *Asymmetric Catalysis on Industrial Scale: Challenges, Approaches and Solutions*, Wiley-VCH, Weinheim.
(b) Blaser, H.U. (2007) *Asymmetric Synthesis* (eds M. Christmann and S. Braese), Wiley-VCH, Weinheim, pp. 296–300.
(c) Blaser, H.U., Pugin, B. and Spindler, F. (2005) *Journal of Molecular Catalysis A–Chemical*, **231**, 1–20.
(d) Noyori, R. (2002) *Angewandte Chemie (International Edition in English)*, **41**, 2008–22.
(e) Knowles, W.S. (2002) *Angewandte Chemie (International Edition in English)*, **41**, 1998–07.

**47** For a recent review see: Stuermer, R., Hauer, B., Hall, M. and Faber, K. (2007) *Current Opinion in Chemical Biology*, **11**, 203–13.

**48** Hall, M., Hauer, B., Stuermer, R., Kroutil, W. and Faber, K. (2006) *Tetrahedron Asymmetry*, **17**, 3058–62.

**49** Müller, A., Hauer, B. and Rosche, B. (2007) *Biotechnology and Bioengineering*, **98**, 22–9.

**50** (a) Chaparro-Riggers, J.F., Rogers, T.A., Vasquez-Figueroa, E., Polizzi, K.M. and Bommarius, A.S. (2007) *Advanced Synthesis Catalysis*, **349**, 1521–31.
(b) Hall, M., Stueckler, C., Ehammer, H., Pointner, E., Oberdorfer, G., Gruber, K., Hauer, B., Stuemer, R., Kroutil, W., Macheroux, P. and Faber, K. (2008) *Advanced Synthesis Catalysis*, **350**, 411–18.

**51** Stueckler, C., Hall, M., Ehammer, H., Pointner, E., Kroutil, W., Macheroux, P.

and Faber, K. (2007) *Organic Letters*, **9**, 5409–541.

**52** Hall, M., Stueckler, C., Ehammer, H., Pointner, E., Oberdorfer, G., Gruber, K., Hauer, B., Stuemer, R., Kroutil, W., Macheroux, P. and Faber, K. (2008) *Advanced Synthesis Catalysis*, **350**, 411–18.

**53** Endo, T. and Kozumi, S. (2001) *Advanced Synthesis Catalysis*, **343**, 521–6.

**54** Wada, M., Yoshizumi, A., Noda, Y., Kataoka, M., Shimizu, S., Takagi, H. and Nakamori, S. (2003) *Applied and Environmental Microbiology*, **69**, 933–7.

**55** Kataoka, M., Kotaka, A., Thiwthong, R., Wada, M., Nakamori, S. and Shimizu, S. (2004) *Journal of Biotechnology*, **114**, 1–9.

**56** (a) Simons, C., Hanefeld, U., Arends, I.W.C.E., Maschmeyer, Th. and Sheldon, R.A. (2006) *Topics in Catalysis*, **40**, 35–44.
(b) Simons, C., Hanefeld, U., Arends, I.W.C.E., Maschmeyer, Th. and Sheldon, R.A. (2006) *Advanced Synthesis Catalysis*, **348**, 471–5.

**57** (a) Schoevaart, R., van Rantwijk, F. and Sheldon, R.A. (1999) *Chemical Communications*, 2465–6.
(b) Schoevaart, R., van Rantwijk, F. and Sheldon, R.A. (2000) *Journal of Organic Chemistry*, **65**, 6940–3.

**58** Jennewein, S., Schuermann, M., Wolberg, M., Hilker, I., Luiten, R., Wubbolts, M. and Mink, D. (2006) *Biotechnology Journal*, **1**, 537–48.

**59** Schoevaart, R., Wolbers, M.W., Golubovic, M., Ottens, M., Kieboom, A.P.G., van Rantwijk, F., van der Wiele, L.A.M. and Sheldon, R.A. (2004) *Biotechnology and Bioengineering*, **87**, 754–62.

**60** van Deurzen, M.P.J., van Rantwijk, F. and Sheldon, R.A. (1997) *Tetrahedron*, **53**, 13183–220.

**61** van de Velde, F., van Rantwijk, F. and Sheldon, R.A. (2001) *Trends in Biotechnology*, **19**, 73–80.

**62** van de Velde, F., Lourenço, N.D., Bakker, M., van Rantwijk, F. and Sheldon, R.A. (2000) *Biotechnology and Bioengineering*, **69**, 286–91.

**63** Perez, D. manuscript in preparation.

# 7
# Multi-modular Synthases as Tools of the Synthetic Chemist

*Michael D. Burkart and Junhua Tao*

## 7.1
## Introduction

The chemical diversity demonstrated by natural products has inspired scientists for centuries and propelled the advent of modern-day synthetic organic chemistry. The tools available to the synthetic chemist today have made possible the construction of practically any molecules, natural or designed. Yet looking at the pharmacopeia of contemporary drugs, patterns emerge concerning the identity and source of these compounds. In general, pharmaceuticals that are fully synthetic are typically small, rigid compounds. Conversely, natural products are often comprised of large, complex structures that must be produced through fermentation of microorganisms. These simple observations have held true for decades, but process chemistry has been slowly changing to incorporate new techniques established from natural product biosynthesis. In general, enzymes catalyze specific reactivity with very high selectively and often with rapid turnover. For this reason, enzymes make excellent catalysts for synthetic chemical reactions, usually when selectivity is needed. The other general benefit to adopting enzymes in chemistry is the use of non-toxic, environmentally friendly processes to replace heavy metals and toxic solvents. The result of these benefits suggests the use of enzymes in chemistry primarily as an application for process-scale synthesis. The appeal of bioprocess chemistry – a field that incorporates the disparate tools of chemoenzymatic synthesis, fermentative semisynthesis, directed fermentation and metabolic engineering – has begun to acquire a foothold in industrial settings, where cost benefits and green technologies contain significant value. This field incorporates the accumulated knowledge of synthetic organic chemistry with that of enzymology, metabolic biochemistry and microbiology for the benefit of improved production-scale synthesis. In this chapter we will explore the use of modular synthases and associated natural product biosynthetic enzymes within the sphere of chemical synthesis. Modular synthases include the biosynthetic pathways of fatty acids (FAs), polyketides (PKs) and non-ribosomal peptides (NRPs). These three related classes of natural products are produced by a wide

*Multi-Step Enzyme Catalysis: Biotransformations and Chemoenzymatic Synthesis*
Edited by Eduardo Garcia-Junceda

ISBN: 978-3-527-31921-3

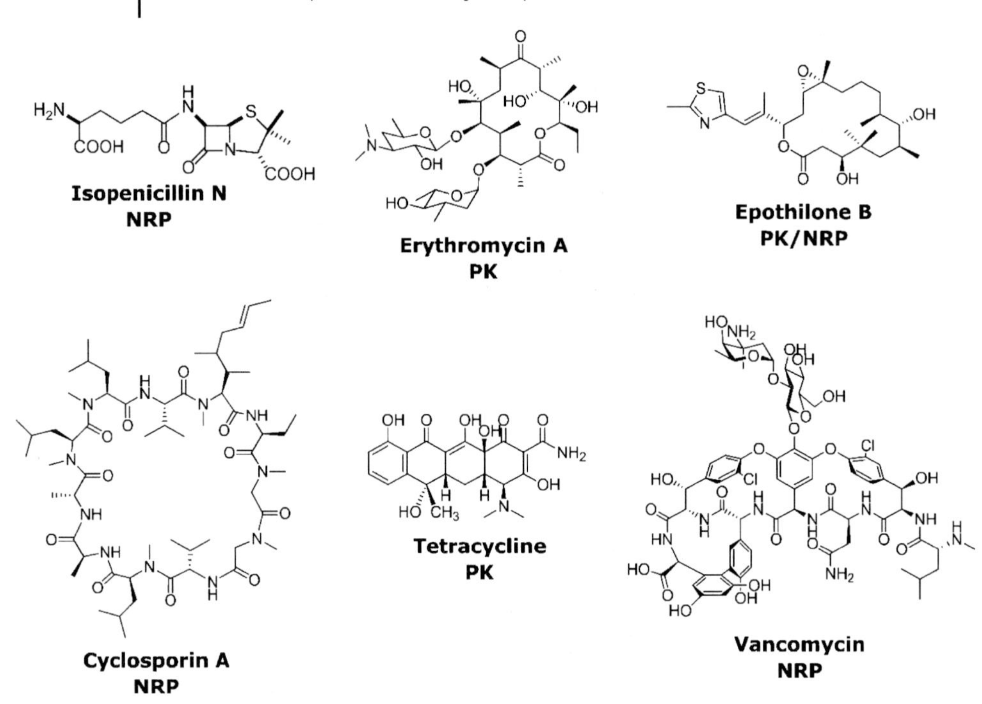


**Figure 7.1** Exemplary PK, NRP and hybrid (PK/NRP) natural products.

range of prokaryotic and eukaryotic microorganisms [1–4]. Within the vast collection of known PK and NRP molecules are many important pharmaceuticals and agrochemicals, including (to name only a few) the antibiotics penicillin, vancomycin and erythromycin A; immunosuppresants cyclosporine A and FK506; the anticholesterolemic lovastatin; the antitumor compounds doxorubicin and bleomycin; the antiparaisitic avermectin; and the antifungal amphotericin B [5]. A wide range of molecular diversity is demonstrated between these classes (Figure 7.1) despite the modular nature of their construction and seemingly facile interchangeability [6].

To date, over 7000 PKs and 10 000 NRPs have been identified. [7–11] Characterization of the mechanisms creating PKs and NRPs is a complicated process requiring years of research [2, 3, 7–17]. Study of PK synthases (PKSs) or NRP synthetases (NRPSs) is currently cued by identification of potent biological activity in culture of microbe, fungi, plants or marine organisms. Once activity is observed, the natural product(s) responsible are isolated and characterized using a combination of analytical methods (high-performance liquid chromatography, mass spectroscopy and nuclear magnetic resonance) and crystallographic analysis. Once the structure of a natural product is determined, its biosynthesis is proposed based on evidence from known pathways. The proposed mechanism is then verified by isolation and sequencing of the participating genes, recombinant expression of their proteins, and pendant biochemical characterization by *in vitro* reconstitution of domain and/or module activity. From start to finish, this process often takes in excess of 5 years.

The natural diversity in molecular structure found within these groups points to the evolutionary logic behind modular systems for the synthesis of these types of secondary metabolites. The theoretical simplicity of this modular organization coupled with the near infinite permutations of possible molecular structures has fueled enormous academic and industrial efforts to understand and modify modular biosynthetic systems, and new pathways continue to be revealed in the pursuit of this objective [7–10, 17].

## 7.2
## Excised Domains for Chemical Transformations

### 7.2.1
### Function of Individual Domains and Domain Autonomy

The past decade has seen an explosion in our outstanding of the enzyme pathways and mechanisms through which natural products and secondary metabolites are derived [18, 19]. For example, at the first level of molecular diversity, either NRP or PK scaffolds are assembled, respectively, from monomeric amino acids or acyl-CoAs by multi-modular synthases. Each module is responsible for the activation, loading and modification of an individual monomer. Interactions between upstream and downstream modules trigger the elongation of a linear NRP or PK. At the second level of molecular diversity, these linear molecules, which are either covalently or non-covalently bound to synthases, are modified by heterocyclization, macrocyclization, oxidation, reduction, glycosylation, halogenation, alkylation, acylation, sulfonation and many other transformations. In addition, the third level of molecular diversity can be created through communications among different families of synthases to produce natural product hybrids such as the PK–isoprenoid hybrid novobiocin and PK–NRP hybrid epothilone. Despite the complexity and diversity of individual molecules, each transformation is often catalyzed by a discrete enzyme with high selectivity. In this chapter, domains or enzymes responsible for macrocyclization, halogenation, glycosylation, heterocyclization and aromatization, methylation, and oxygenation will be analyzed to discuss their utility as catalysts for chemical syntheses.

### 7.2.2
### Heterocyclization and Aromatization

Many natural products contain heterocyclic motifs, which afford additional recognition elements for interaction with nucleotide and protein targets. Within a given molecule, heterocycles can occur as single, tandem and multiple moieties within a given molecule (Figure 7.2).

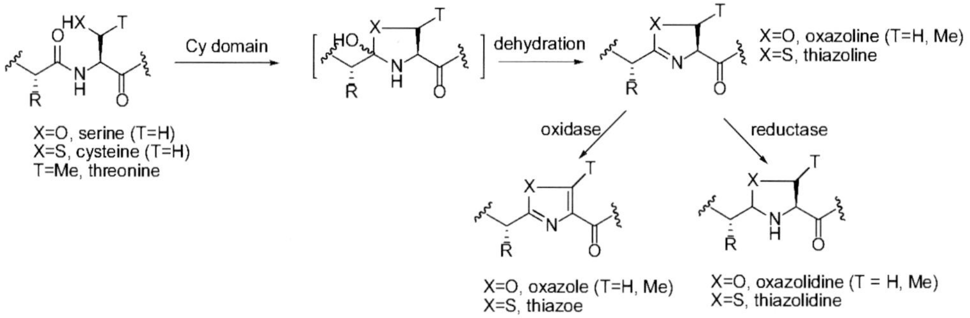

Cy domain
dehydration
X=O, oxazoline (T=H, Me)
X=S, thiazoline
X=O, serine (T=H)
X=S, cysteine (T=H)
T=Me, threonine
oxidase
reductase
X=O, oxazole (T=H, Me)
X=S, thiazoe
X=O, oxazolidine (T = H, Me)
X=S, thiazolidine

**Scheme 7.1** Biosynthesis of heterocycles by cyclization–dehydration. Cy = cyclization domain.

In NRPS, the cyclization domain catalyzes cyclization of the side-chain nucleophile from a dipeptide moiety such as AA-Ser or AA-Cys (AA = amino acids) to form a tetrahedral intermediate, followed by dehydration to form oxazolines and thiazolines (Scheme 7.1) [20]. The synthesis of a 2-methyl oxazoline from threonine follows a similar mechanism. Once a heterocycle is formed, it can be further modified by reductase to form tetrahydro thiazolidine in the case of pyochelin biosynthesis. Conversely, oxidation of the dehydroheterocycles lead to heteroaromatic thiazoles or oxazoles as in the case of epothilone D (Figure 7.2) [21].

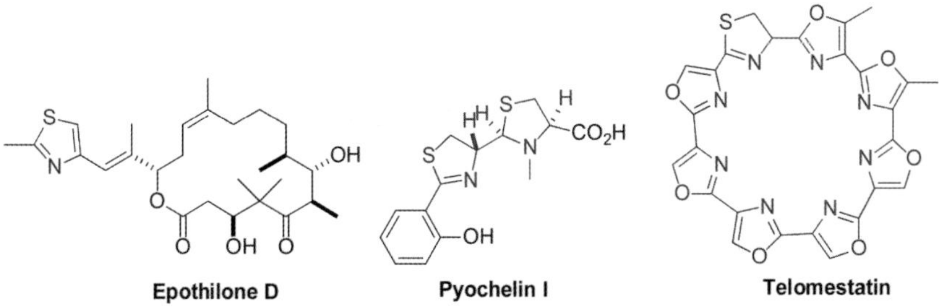

Epothilone D
Pyochelin I
Telomestatin

**Figure 7.2** Examples of heterocyclic natural products.

An alternative mechanism to form thiazoles and oxazoles is through oxidation of a dipeptide followed by cyclization from an enolate or thienolate precursor and subsequent dehydration (Scheme 7.2). This represents a higher-energy pathway and there is no accumulation of thiazoline or oxazoline intermediates [22–24].

Tandem heterocycles are formed by posttranslational conversion of oligopeptide sequences consisted of serine, cysteine and threonine. A striking example of tandem heterocycles is telomestatin (Figure 7.2), a nanomolar telomerase inhibitor, which contains eight heterocycles. In its biosynthesis, five serines were converted to five oxazoles, two threonines to two methyloxazoles and one cysteine to one thiazoline [25].

An additional paradigm of enzymatic heterocyclization can be found in cyclic polyethers such as momensin. In these antibiotics, characteristic tetrahydrofuran

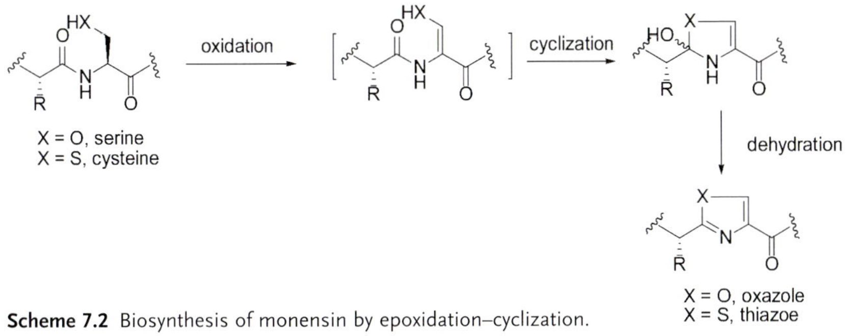

Scheme 7.2 Biosynthesis of monensin by epoxidation–cyclization.

and tetrahydropyran rings and terminal carboxylates are essential for chelating alkali metal cations. These PK-derived polycycles are formed in a cascade fashion with other enzymatic transformations. For example, in the biosynthesis of monensin (Scheme 7.2), the cascade polyether formation is triggered by an epoxidation of a polyene template derived from a PKS assembly line (Scheme 7.3) [26]. Similar mechanisms are probably used by nature to produce other polyether antibiotics such as non-actin and nanchangmycin [27].

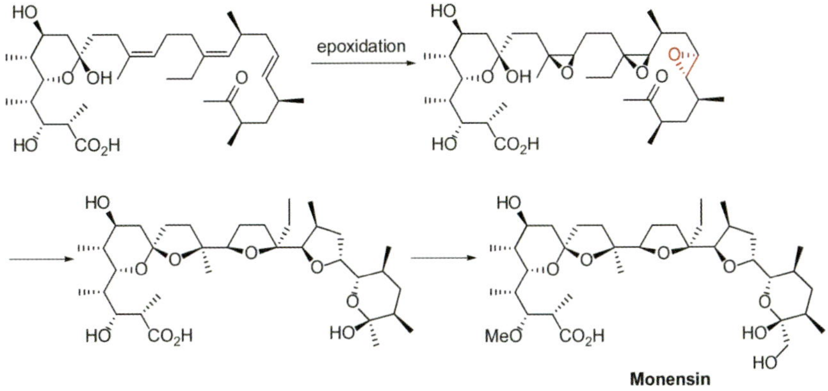

Scheme 7.3 Domain fusion to prepare heterocyclic carboxylic acids.

Several current efforts are focusing on the portability of enzymatic heterocyclization. For example, novel chiral heterocyclic carboxylic acids were produced from corresponding amino acids by domain fusion, where release of the end product was achieved by the thioesterase (TE) (Scheme 7.4) [28]. By this methodology, not only can methyl oxazoline and thiazoline be prepared, thiazole can also be created by fusing similar enzyme construction to an oxidation domain.

Stimulated by the biosynthetic pathways, biomimetic heterocyclization methods have also been developed with high efficiency. For example, using an oxo-bridged diphosphonium salt prepared *in situ* from $Ph_3PO$ and $Tf_2O$, a variety of hetero-

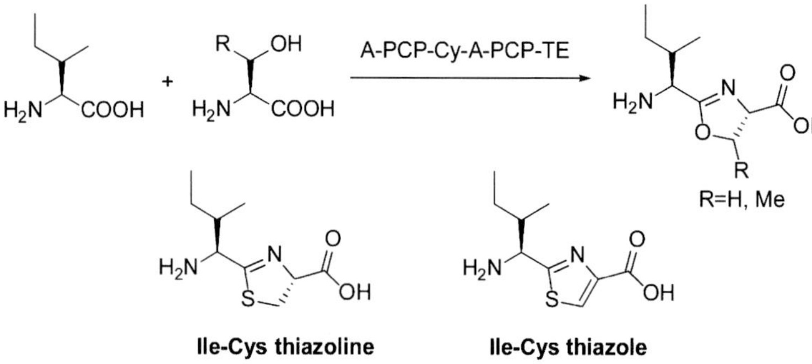

A-PCP-Cy-A-PCP-TE

R=H, Me

**Ile-Cys thiazoline** **Ile-Cys thiazole**

**Scheme 7.4** Biomimetic synthesis of heterocyclic carboxylic esters.

$Ph_3PO$ (3.0 equiv)
$Tf_2O$ (1.5 equiv)

R=Bn, Me, *i*-Pr, *s*-Bu
yield 84-95%, ee >99.5%

eq. 1

$Ph_3PO$ (6.0 equiv)
$Tf_2O$ (3.0 equiv)

yield 63%, ee >99.5%

eq. 2

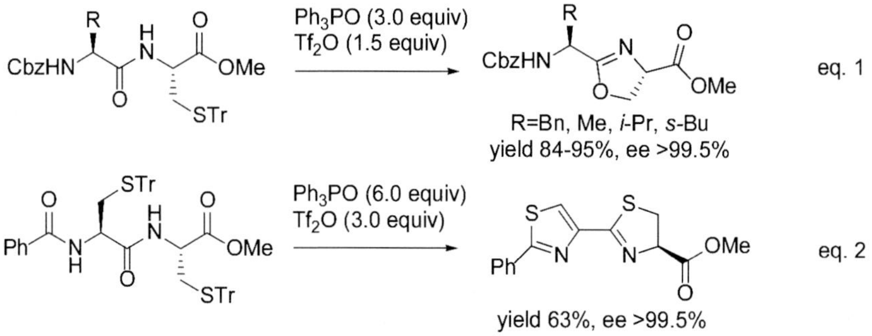

**Scheme 7.5** Biosynthesis of heterocycles by oxidation–cyclization.

cycles were produced in good yields and high optical purity from the amino acids (eq. 1 in Scheme 7.5) [29]. This method was also extended to the synthesis of tandem biheterocyclic carboxylic esters (eq. 2 in Scheme 7.5).

In addition to oxazoles and thiazoles, non-heteroaromatics are also common motifs in natural products. For example, aromatic PKs are some of the mostly commonly used antibiotic and anticancer drugs, such as oxytetracycline, tetracenomycin and doxorubicin. In contrast to reduced PKs such as macrolides, polyethers and polyenes often catalyzed by type I PKSs using a set of non-iteratively used domains arranged in a linear order, the general logic of aromatic PK biosynthesis is to form a linear precursor through chain extension by type II PKSs consisting of discrete proteins carrying out a set of iterative enzyme activities, reminiscent of bacterial FA synthase. The linear PK is then subject to single or cascade aromatization. For example, in the proposed biosynthesis of griseorhodin, a telomerase inhibitor, a linear template is created after 12 elongation catalyzed by a PKS, which is then aromatized to an angular intermediate (Scheme 7.6) [30]. The griseorhodin is finally produced after 11 redox transformations.

The aromatization often takes place after linear precursors have been modified as in the biosynthesis of actinorhodin, where condensation followed by ketone reduction takes place before aromatization (Scheme 7.7) [31].

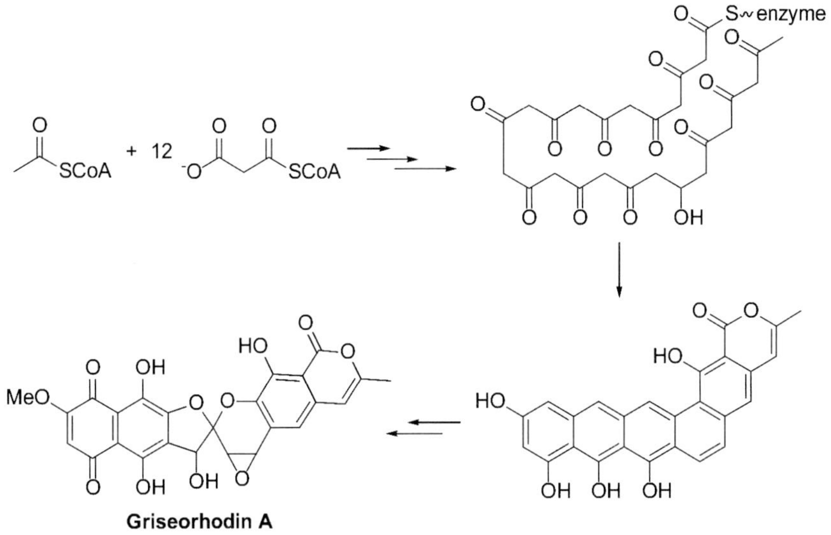

Scheme 7.6 Aromatization in biosynthesis of griseorhodin.

Although genetic studies have identified a number of putative aromatic PK aromatases and/or cyclases, *in vitro* characterization of these enzymes is usually difficult since the postulated linear or monocyclic poly β-ketone templates are often unstable to be isolated or synthesized. For these reasons, the generality and synthetic applications of these enzymes have not been studied in most cases. However, there is plenty of evidence to suggest that enzymatic aromatization and cyclization take place for a variety of linear precursors with various chain lengths [32].

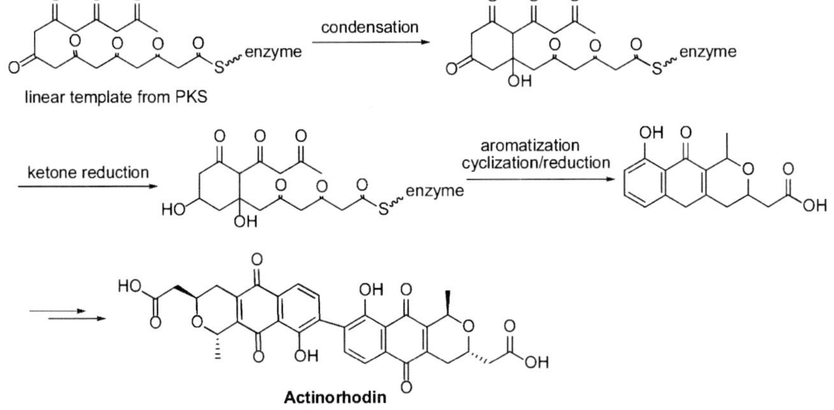

Scheme 7.7 Aromatization in biosynthesis of actinorhodin.

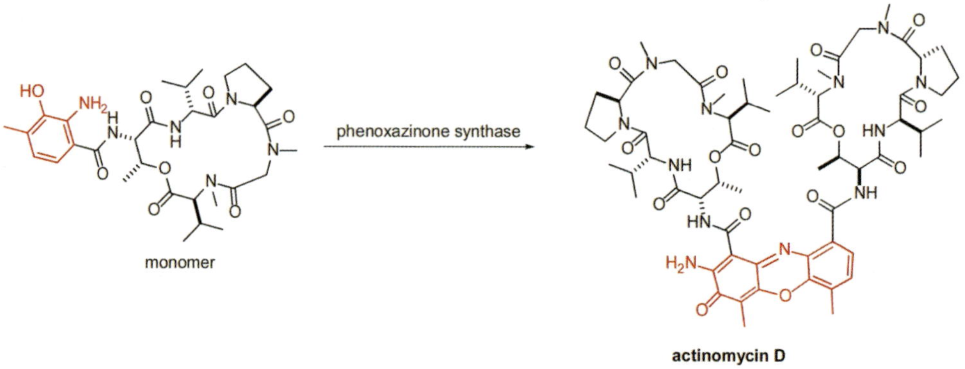

**Scheme 7.8** Oxidative cascades in biosynthesis of actinomycin.

Aromatization can also be accomplished by oxidative cascade reactions – a strategy often utilized by natural product biosynthesis [33]. For example, the phenoxazinone chromophore of actinomycin was proposed to be generated by a six-electron oxidation of aminophenols catalyzed by a copper-containing oxidase (Scheme 7.8) [34].

## 7.2.3
## Macrocyclization

Many natural products are constrained by macrocyclic motifs, which are often essential for natural products to possess the desired biological properties. In the biosynthesis of macrocyclic NRPs and PKs, linear peptides or PKs are often macrocyclized by a TE domain located at the C-terminal of multi-modular synthases. For example, in the biosynthesis of the antibiotic tyrocidine A (Tyc A), a linear enzyme-bound decapeptide, which is transferred from the last carrier protein (or thiolation) domain of the Tyc A synthase, is cyclized by an intramolecular $S_N2$ reaction between the N-terminal amine nucleophile and the C-terminal ester, which is covalently linked to serine residual in the TE domain prior to macrocyclization (Scheme 7.9) ([35] and references therein).

Remarkably, the excised TE domain (28 kDa) from the Tyc A synthatase was found to have great versatility. Not only can it catalyze the formation of macrolactams of ring sizes from 18 to 42 atoms from linear *N*-acetyl cystemine (SNAC) with excellent catalytic efficiency ($K_{cat}$ 6–30 min$^{-1}$ and $K_M$ 3–5 μM), all residues except two may be replaced by other functionalities ([36] and references therein). In one study, four amino acids (Asn, Gln, Tyr, Val in blue; left) could be replaced by a PK moiety (Scheme 7.10), allowing the production of a hybrid peptide/PK cyclic molecules ((*S,S*)-PP/PK, PP = peptide, cyclization region in red; right) [37]. Using the same methodology, other stereoisomers (*S,R*)-PP/PK, (*R,S*)-PP/PK and (*R,R*)-PP/PK could also be produced from corresponding linear SNAC substrates

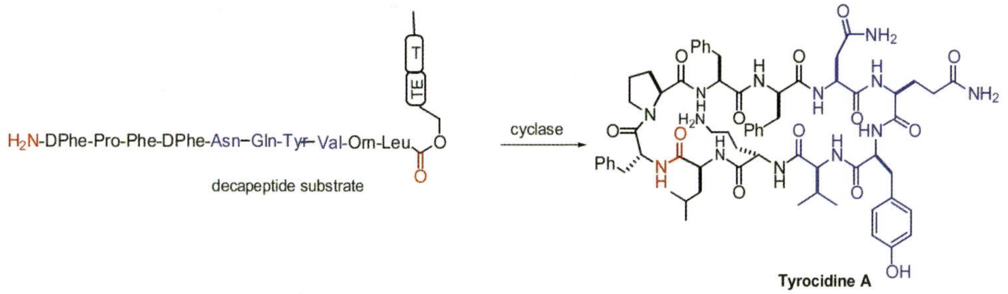

Scheme 7.9 Cyclase in biosynthesis of Tyc A.

(Scheme 7.10). In contrast to chemical macrocyclization, no protection was required for the linear precursors in enzymatic catalysis.

As the enzyme is highly flexible with regard to the template, this chemoenzymatic macrocyclization strategy can be expanded to synthesize many other natural product-like molecules. For example, by incorporation of propargyl amino acids into the linear template, a library of over 247 glycosylated macrocyclic Tyc A analogs was produced by Tyc A TE, where the sugar moieties were introduced through copper-catalyzed 1,3-dipolar cycloaddition to the alkynes [38]. Interestingly, several compounds including Tyc4PG-14 from this library were found to have a 6-fold increase in therapeutic index while retaining antibacterial efficacy

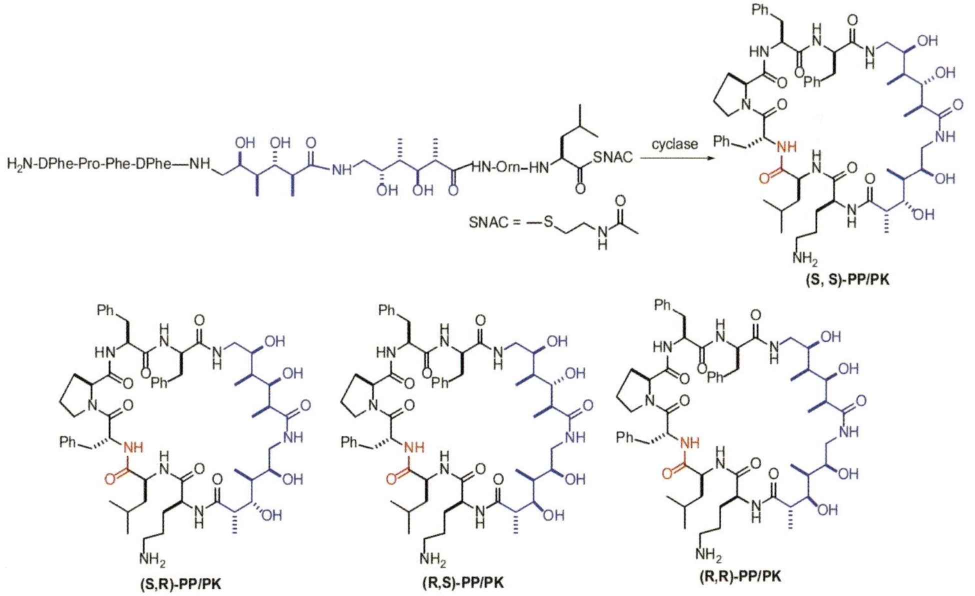

Scheme 7.10 Enzymatic macrocyclization to create novel hybrid peptide/PKs.

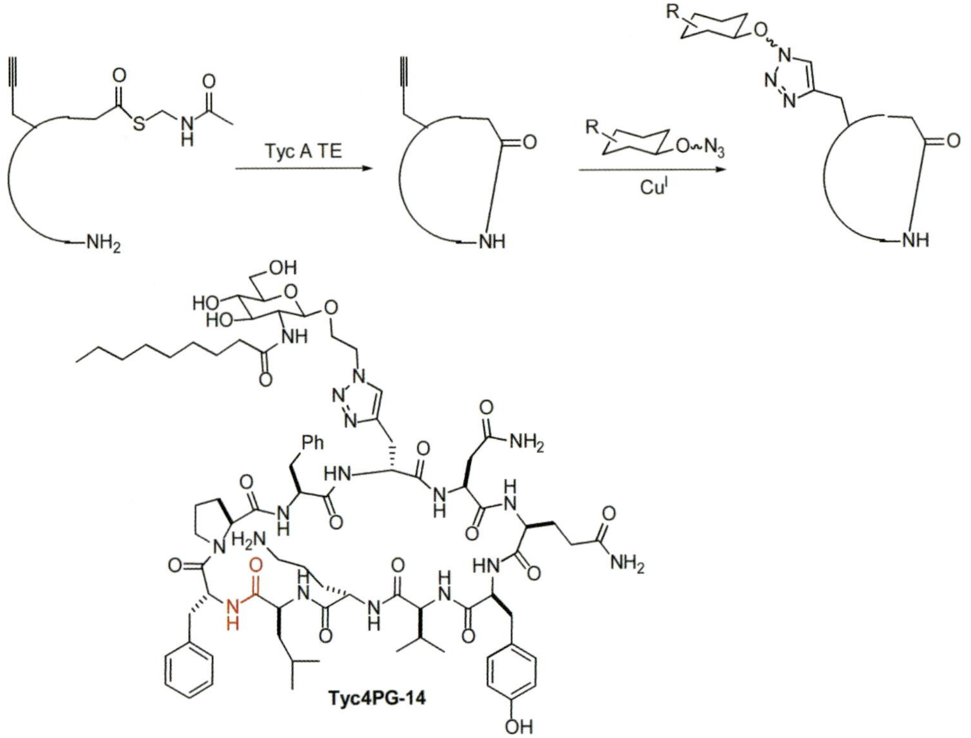

Scheme 7.11 TE-catalyzed synthesis of glycosylated Tyc A analogs.

(Scheme 7.11). Here the therapeutic index was calculated as the minimal hemolytic concentration/minimal inhibitory concentration of bacteria.

It should be noted that TE-catalyzed cyclization is not limited to the synthesis of macrocyclic peptides by catalyzing the formation of a C—N bond. These enzymes are also responsible for the cyclization of NRP depsipeptide and PK lactone. Indeed, a didomain excised from fengycin synthase was able to catalyze the formation of a macrolactone through the formation of a C—O bond [39]. Several cyclases from PKSs have also been characterized to be functional. For example, when a TE from picromycin synthase was fused to an erythromycin module (DEBS module 3), the resulting hybrid was able to convert a diketide and 2-methylmalonyl-CoA to a triketide ketolactone (Scheme 7.12) [40]. However, their *in vitro* activity is in

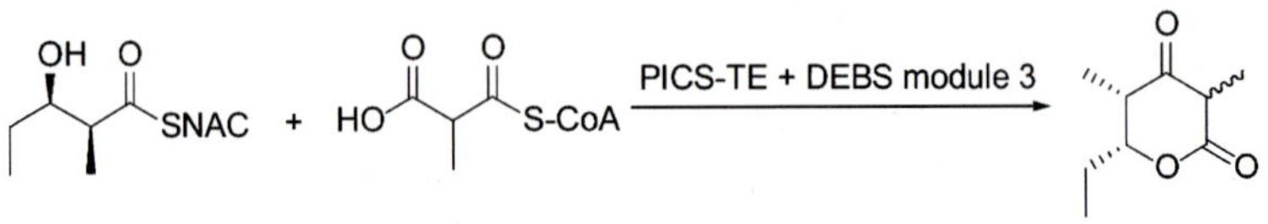

Scheme 7.12 Cyclization activities of TEs from PKS. PICS = picromycin synthase.

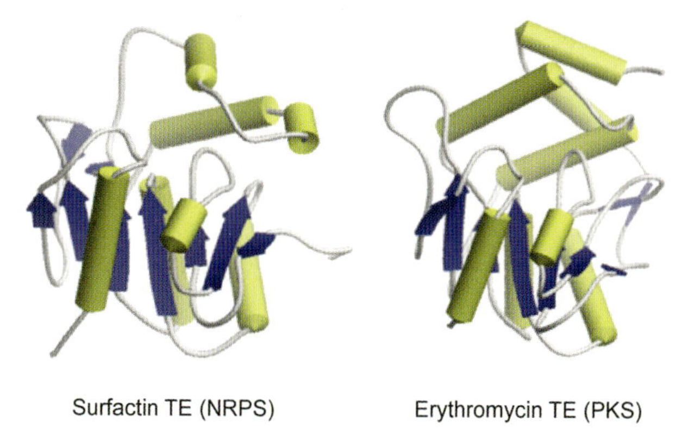


**Figure 7.3** Structures of TEs from NRPS and PKS.

general much lower than the NRPS TEs and hydrolysis tends to dominate over cyclization activity as seen in the example of epothilone D [41].

The PKS TE from erythromycin biosynthesis and the NRPS TE from surfactin biosynthesis are remarkably similar in three-dimensional topology (Figure 7.3) [42, 43]. It is expected that the application of enzyme-directed evolution will be able to further broaden the synthetic applications of TEs by increasing their activity and stability [44].

It should be noted that while TE domains represent the most common solution in releasing macrocyclic NRPs and PKs, other pathways are known. For instance, in the biosynthesis of cyclosporine, the cyclization is proposed to be catalyzed by the most downstream C-domain [45]. Macrocyclization can also occur under reduction of a carbonyl group mediated by a reduction domain as proposed in the synthesis of the macrocyclic imine nostocyclopeptide [46]. The synthetic utility of these cyclization strategies has not yet been reported.

### 7.2.4
### Halogenation

In drug discovery, halogenation is one of the most popular tools to fine-tune a drug candidate's pharmacological properties. It is estimated that up to 20% of drugs on the market and close to a quarter of those in the development pipeline are halogenated [47]. Similarly, halogenation is often essential to achieve the desired biological activities of natural products, where halogens have to be strategically placed onto organic molecules (Figure 7.4). For example, one of the chlorines in vancomycin enhances its binding affinity, while the other responsible for binding selectivity [48].

For electron-rich substrates, nature often uses flavin-dependent halogenases (e.g. chlorotetracycline), vanadium haloperoxidases (snyderol, Figure 7.4) or heme–iron haloperoxidases (tetraiodothyronine, Figure 7.4) for this role (Figure 7.5). For electron-deficient molecules such as alkanes, mononuclear iron

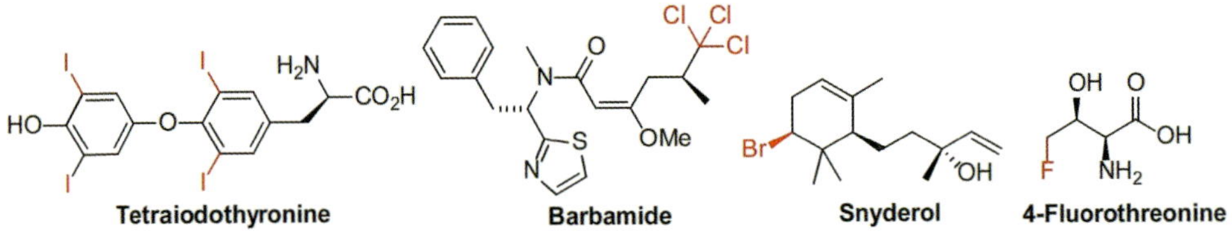

Tetraiodothyronine Barbamide Snyderol 4-Fluorothreonine

**Figure 7.4** Representative halogenated natural products.

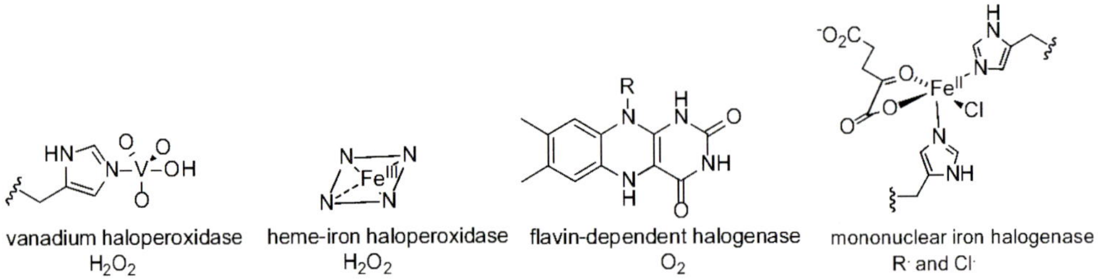

vanadium haloperoxidase $H_2O_2$ heme-iron haloperoxidase $H_2O_2$ flavin-dependent halogenase $O_2$ mononuclear iron halogenase R· and Cl·

**Figure 7.5** Halogenases: active sites and cofactors.

halogenases are utilized to catalyze halogenation (barbamide, Figure 7.4) [49]. For haloperoxidases, the halogenation often takes place with poor regio- and stereoselectivity, since the activated halogen source (hypohalous acids) are freely diffusible within and away from enzymes. As a result, their synthetic utility is largely limited. However, both flavin-dependent and mononuclear iron halogenase elicit high regio- and stereoselectivity [50, 51].

$FADH_2$ halogenases such as tryptophan 7-halogenase have been shown to catalyze regioselective halogenation of a wide range of indole derivatives and aromatic heterocycles, where a cofactor regeneration system has recently been developed to use only a catalytic amount of cofactors (Scheme 7.13) [52, 53].

Using the same enzyme, a library of antitumor indolocarbazoles was generated in a combinatorial fashion *in vivo* upon halogenation and glycosylation, producing halogenated analogs of rebeccamycin (Scheme 7.14) [54].

Much was unknown about the halogenation for unreactive substrates until very recently, when the biosynthesis of the cyclopropyl amino acid side-chain of coro-

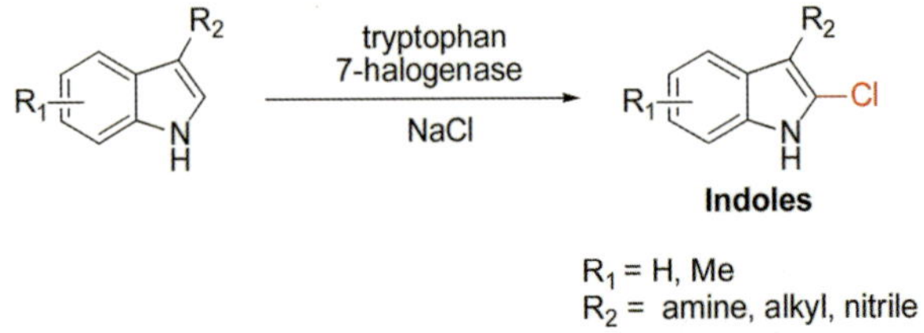

tryptophan 7-halogenase, NaCl

Indoles

$R_1$ = H, Me
$R_2$ = amine, alkyl, nitrile

**Scheme 7.13** Halogenation of indoles by tryptophan 7-halogenase.

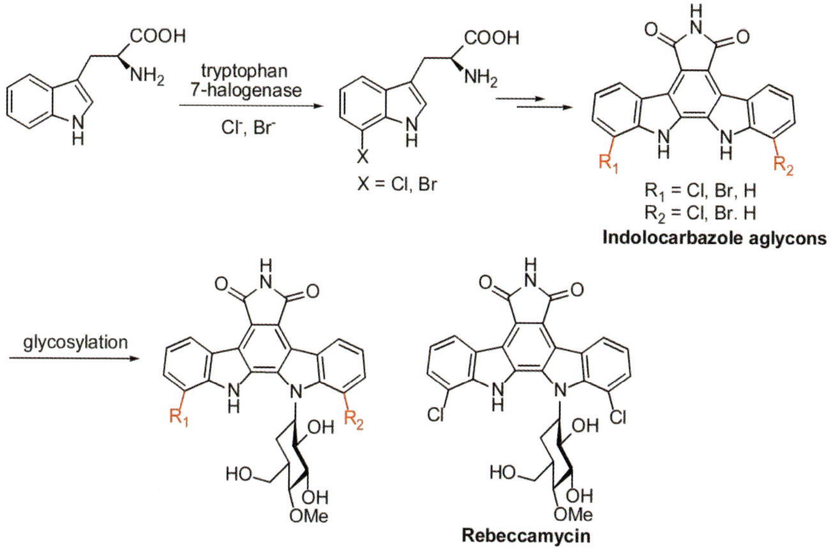

Scheme 7.14 Tryptophan 7-halogenases to create novel indolocarbazoles.

natine was elucidated. This intriguing pathway, which involves γ-chlorination of an enzyme-bound L-isoleucine by a chlorinase (Scheme 7.15), homologous to α-ketoglutarate-dependent non-heme $Fe^{2+}$ enzymes that use Asp/Glu and two His side-chains as ligands to oxygen-labile $Fe^{2+}$ [55]. Subsequently, a cyclopropyl intermediate is produced via chloride displacement by the α-carbon and then incorporated into coronatine. It is believed that a variety of halogenation takes place by this mechanism for electron-deficient molecules such as barbamide and syringomycin [56, 57].

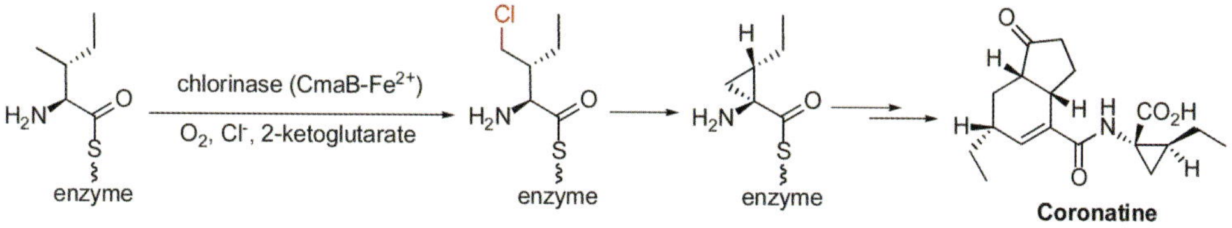

Scheme 7.15 Chlorinase in biosynthesis of coronatine.

It is yet to be determined if the halogenation can take place on substrates that are not covalently linked to natural product synthases, which would significantly broaden the synthetic applications of these enzyme for *in vitro* halogenation of non-reactive molecules – a challenge still faced by existing chemical methods.

In addition to chlorinated and brominated natural products, a few dozen fluorinated and iodinated metabolites have also been characterized. Since the energy to oxidize a fluoride is prohibitively high, it is not surprisingly that nature uses a non-oxidative halogenation strategy. So far only one fluorinase from *Streoptomyces cattleya* has been cloned and characterized, which produces several organo fluorine metabolites such as 4-flurothreonine (see Figure 7.4) [58]. The enzyme uses *S*-adenosylmethionine (SAM) as a substrate for fluorine displacement, where L-methionine serves as the leaving group to form 5-fluoroadenosine. This fluorinase was recently used to prepare $^{18}F$-containing positron emission reagents [59], where one of the current challenges is to discover novel synthetic methods to introduce fluorine into organic molecules using HF or $F^-$.

### 7.2.5
### Glycosylation

While over 50% human proteins are glycosylated, a wide range of bioactive natural products are also glycosylated through *O*-, *N*- or *C*-linkages to their respective aglycons. These sugars often endow essential biological activities to the parent molecules [60]. For example, both erythromycin and pikromycin lose their antibiotic activity when the D-desosamine is removed [61].

Recent discovery of a variety of glycosyltransferases has greatly facilitated the synthesis of glycosylated molecules. Long a target of synthetic applications, enzymatic glycosylation eliminates the need for extensive protection and deprotection often required by chemical approaches. Enzymatic glycosylation exploits the inherent flexibility of glycosyltransferases, often through processes complementary or superior to chemical approaches. These enzymes are found to display variable degrees of flexibility with respect to the aglycons, but are more often with great promiscuity toward NDP-sugars [62]. For example, one glycosyltransferase from the glycopeptide A-40926 biosynthesis gene cluster accepts a variety of sugar donors allowing the creation of novel glycopeptides (Scheme 7.16). While sugar moieties are essential for the antibiotic activity of this class of glycopeptides [63], a critical difference between lipoglycopeptides and glycopeptides is the existence of an acyl chain, which may contribute to their different antimicrobial profiles. By integrating the synthesis with a promiscuous acyltransferases, a library of novel lipoglycopeptides was generated (Scheme 7.16) [64]. By incorporating azido sugars, the glycol moiety can be further randomized through copper-catalyzed 1,3-dipolar cycloaddiiton [65].

Glycosyltransferases have also been found to be able to accomplish sugar exchange. For example, four glycosyltransferases from calicheamicin and vancomycin can catalyze reversible exchange of sugars and aglycons (Scheme 7.17) [66].

The availability of a variety of glycosyltransferases and sugar biosynthesis genes provides the basis for a general strategy to produce glycosylated molecules *in vivo* through pathway engineering and combinatorial biosynthesis [67]. For example, a macrolide glycosyltransferase from the picromycin biosynthesis pathway of *Streptomyces venezualae* was found to have broad substrate specificity. With its ability to take a wide range of both PK aglycones and sugars as substrates, various novel PKs were generated *in vivo* (Scheme 7.18) [68]. A glycosyltransferase was recently

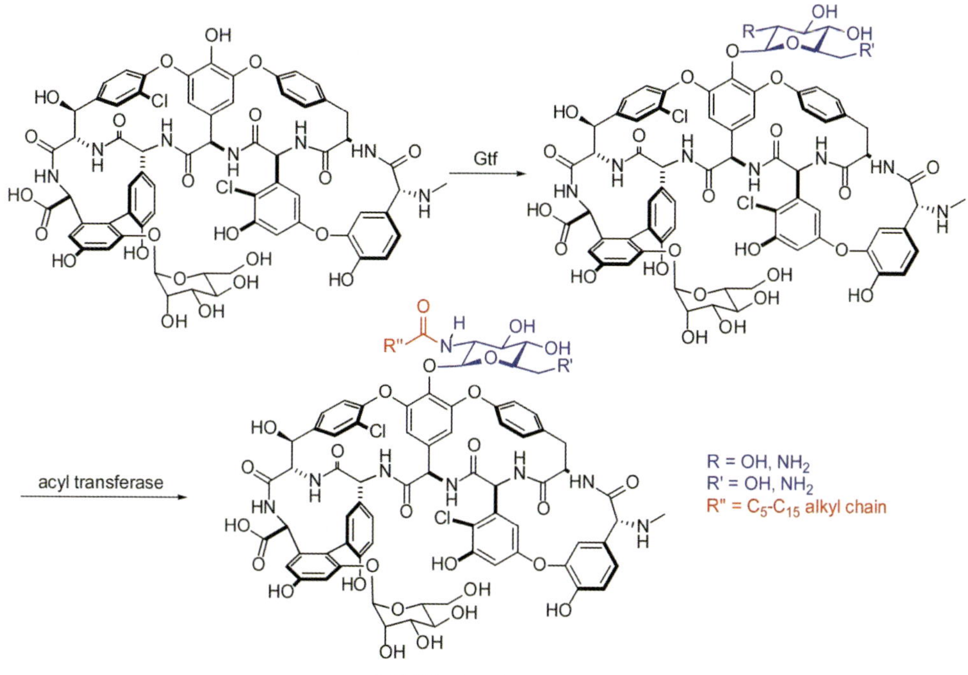

**Scheme 7.16** Enzymatic glycosylation to prepare novel glycopeptides.

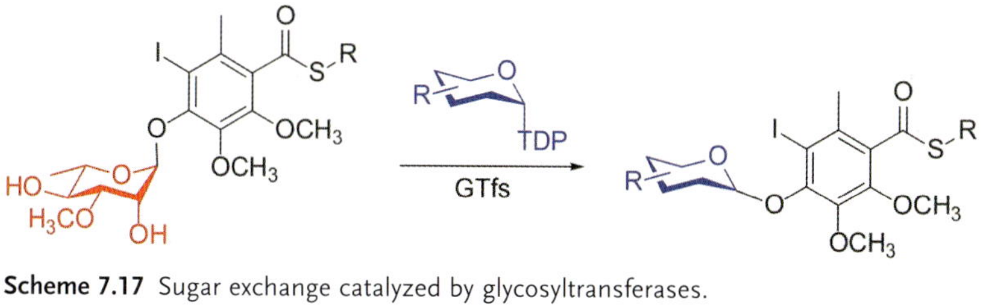

**Scheme 7.17** Sugar exchange catalyzed by glycosyltransferases.

expressed in *E. coli* to synthesize novel glycol-PKs *in vitro* [69]. Similar work has been demonstrated for the synthesis of rebeccamycin analogs from indolocarbazole aglycons *in vivo* [70].

### 7.2.6 Methyltransferases

Methyltransferases are responsible for the methylation a variety of nucleophiles, typically using SAM as the carbon donor. For example, enzymic synthesis of methyl halides through this mechanism has been discovered in cell extracts of *Phellinus promaceus* (a white fungus), *Endocladia muricata* (a marine red algae) and

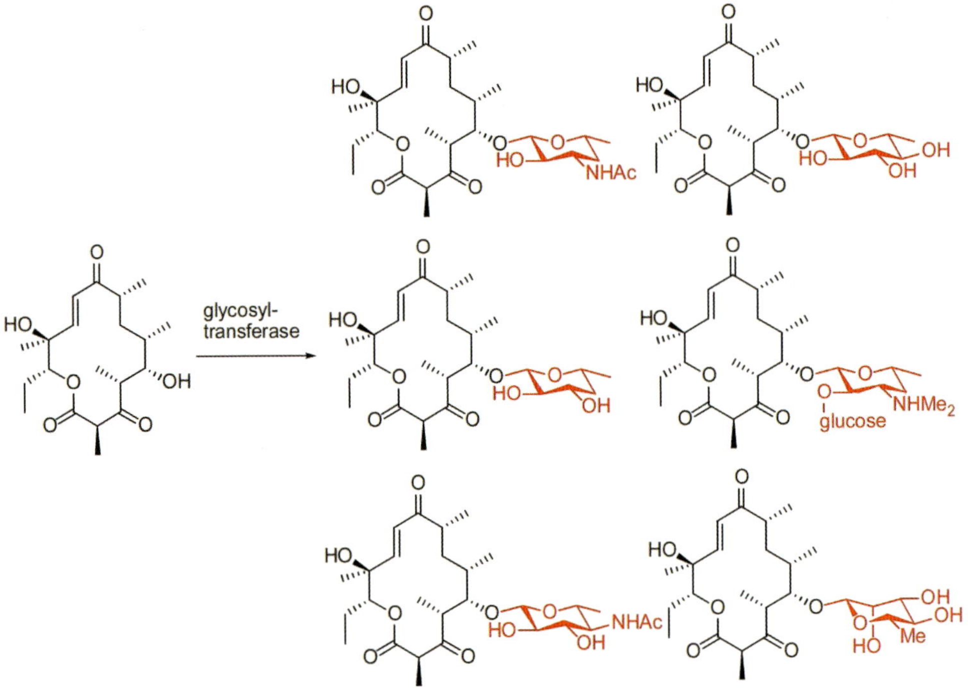

**Scheme 7.18** *In vivo* glycosylation to create novel macrolides.

*Mesembryanthemum crystallium* (an ice plant) [71]. As expected, the reactivity of methyl transfer follows the order of iodide, bromide and chloride, with chloride being the poorest acceptor. In addition to halides, methyltransferases can also catalyze the transfer of a methyl group to other heteroatoms such as nitrogen – a common mechanism in biological amine methylation ranging from natural precuts to DNA and protein. Interestingly, a methyltransferase can be degenerated into a racemase by losing the SAM-binding site [72, 73].

The electrophilic potency of the positively charged SAM allows unusual acceptors. In the biosynthesis of myxothiazol A to myxothiazol Z, a methyltransferase catalyzes the esterification of the corresponding carboxylic acid (Scheme 7.19) [74].

One of the striking features of enzymatic methylation is its exquisite regioselectivity. For example, in the biosynthesis of novobiocin, a methyltransferase NovO catalyzes the methylation of only one of the three phenolic carbons and none of the three hydroxyl groups are methylated (Scheme 7.20). The exquisite regioselectivity is also compatible with substrate promiscuity. For example, the methyltransferase CouO from coumermycin A1 synthase catalyzes the methylation of both the mono- and bis-amides [75, 76].

Methyl transfer for unactivated substrates often takes place under radical mechanism using methylcobalamin. For example, the key step in the biosynthesis of

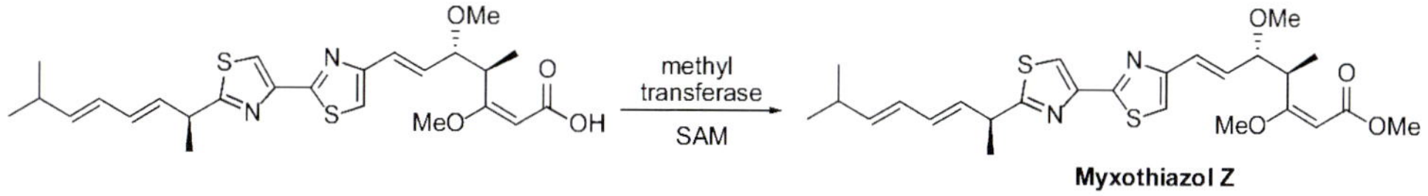

methyl transferase
SAM
Myxothiazol Z

**Scheme 7.19** Esterification catalyzed by methyltransferases.

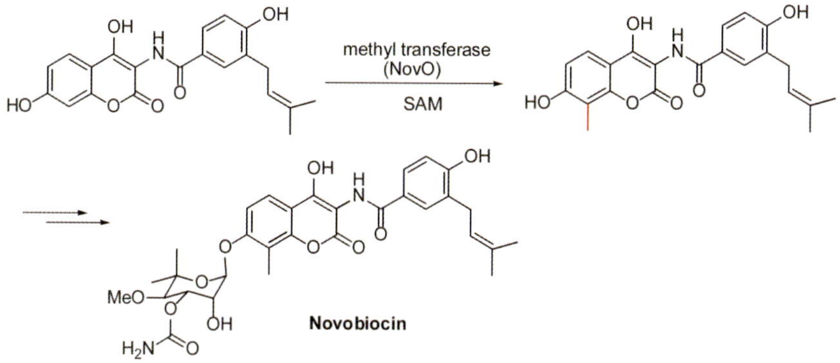

methyl transferase (NovO)
SAM
Novobiocin

**Scheme 7.20** C-methylation in the biosynthesis of novobiocin.

fosfomycin was shown to be methylation of the carbon adjacent to the hydroxyl group catalyzed by a methyltransferase in the presence of methylcobalamin. This remarkable radical methylation distinguishes two prochiral hydrogens [77].

### 7.2.7
### Oxidation

Nature often uses two pathways to produce oxygenated molecules. One is oxidative fictionalization of substrates through enzymatic hydration, monohydroxylation, dihydroxylation or epoxidation. Alternatively, oxygenated compounds can be generated by C—C formation catalyzed by aldolase, transketolase, oxynitrilase and related enzymes, where chiral centers are created simultaneously without an overall change in oxidation states [78].

Oxidative oxygenation is catalyzed by oxygenases, which direct the incorporation of molecular oxygen into substrates. Depending upon whether an oxygenase catalyzes the insertion of one or both atoms of $O_2$ into its substrates, it is called a monooxygenase or dioxygenase. Oxygenases primarily use metal cofactors, which in their low oxidation states can complex with dioxygen, substrate or both. The best understood enzymes using heme cofactors are cytochrome P450s, heme oxygenases and nitric oxide synthases. Other oxygenases utilize an organic cofactor such as dihydroflavin or tetrahydropterin, which can donate an electron to dioxygen generating activated oxygen species, or the copper or a non-heme

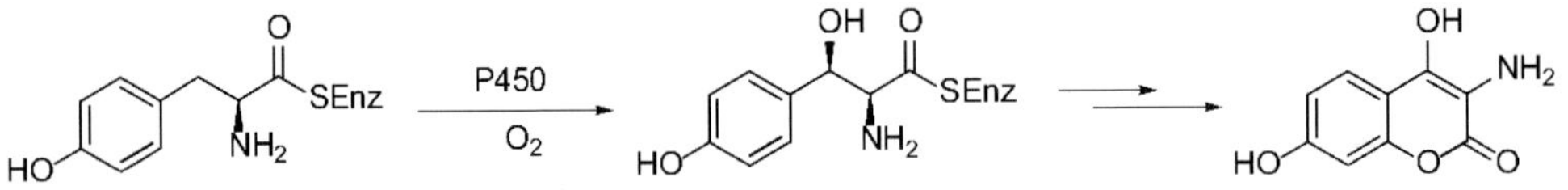

Scheme 7.21 P450 oxygenation of amino acids in coumarin biosynthesis.

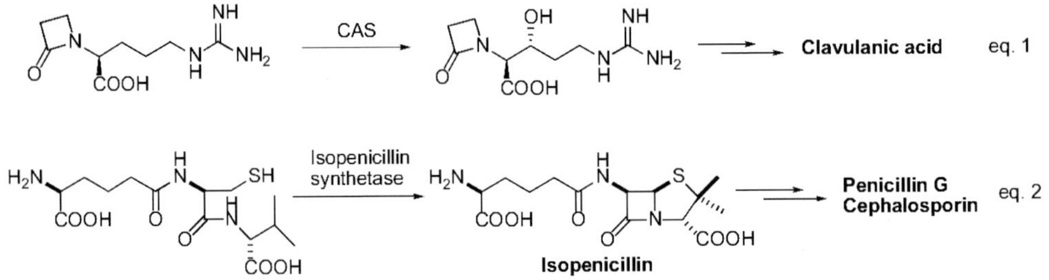

Scheme 7.22 Non-heme iron oxygenase in biosynthesis of clavulanic acid.

iron centers. Cytochrome P450s catalyze a variety of oxidative modifications. For example, the coumarin formation in novobiocin biosynthesis starts with stereospecific β-hydroxylation of an enzyme-bound tyrosine intermediate (Scheme 7.21) [79].

The common motif shared by non-heme iron oxygenases contains an active site, where two histidines and one carboxylate occupy one face of the Fe(II) coordination sphere. These enzymes catalyze a variety of oxidative modification of natural products. For example, in the biosynthesis of clavulanic acid, clavaminic acid synthase demonstrates remarkable versatility by catalyzing hydroxylation, oxidative ring formation and desaturation in the presence of α-ketoglutarate (eq. 1 in Scheme 7.22) [80]. The same theme was seen in the biosynthesis of isopenicillin, the key precursor to penicillin G and cephalosporin, from a linear tripeptide proceeded from a NRPS, where non-heme iron oxygenases catalyze radical cyclization and ring expansion (eq. 2 in Scheme 7.22) [81, 82].

It should be noted that non-heme oxygenases can also degrade aromatics such as biphenyls and naphthalene (Scheme 7.23). A naphthalene dioxygenase consists of a catalytic oxygenase component with a mononuclear iron site, an iron–sulfur flavoprotein reductase and an iron–sulfur ferredoxin transferring electrons from

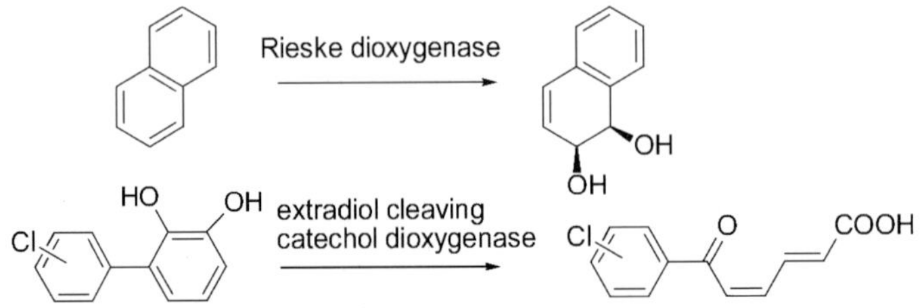

Scheme 7.23 Aromatic dioxygenases.

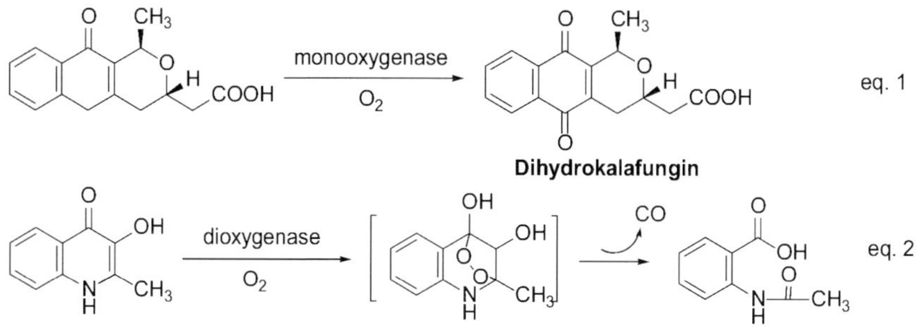

**Scheme 7.24** Cofactorless dioxygenases.

NADP(H) [83, 84]. In addition to dioxygenation, non-heme iron oxygenases can perform monohydroxylation as in the synthesis of L-DOPA from L-tyrosine in mammalian species [85].

While most oxygenases require cofactors, several have been reported to catalyze dioxygenation without apparent reliance on cofactors [86]. For example, an important tailoring step in the synthesis of a number of PKs is the oxidation of aromatic precursors to form corresponding quinine moieties present in tetracenomycin C, actinorhodin, daunorubicin and doxorubicin. In many cases, the oxygenation takes place without cofactors. In the biosynthesis of actinorhodin, the monooxygenation producing dihydrokalafungin is proposed to take place without exogenous cofactors or metal ions (eq. 1 in Scheme 7.24) [87]. Rather, dioxygen and/or substrate activation is likely to be performed in the catalytic site through formation of a protein radical intermediate (e.g. cysteine residual) or direct electron transfer from the deprotonated substrate to molecular oxygen to a form a radical pair. It should be noted that aerobic bacterial degradation of aromatic compounds can also proceed without cofactors, where cleavage of an aromatic ring is often proceeded by formation of catechol derivatives or *p*-dihydroxy aromatic compounds for ring-opening dioxygenation. For example, in the cleavage of *N*-heterocyclic aromatics, the isolated dioxygenase contains no metal or organic cofactors (eq. 1 in Scheme 7.24) [88]. The generality and practicality of these enzymes have not been demonstrated. For synthetic applications, it is very attractive to use these cofactorless oxygenases, where cofactor regeneration can be avoided.

Enzymatic hydroxylation activation has perhaps the highest potential of all enzyme-catalyzed transformations for synthetic applications. Currently, whole-cell processes are used and the outcomes are often unpredictable. The discovery of new oxygenases and efficient hosts for protein expression remain keys to further expanding the synthetic applications of biocatalytic C—H activation [89, 90].

## 7.3
## Conclusions

The use of biological tools in synthetic organic chemistry has only recently attained larger acceptance in the chemical community and significant benefits for pharmaceutical process chemistry have been realized. Enzymes as *in vitro* chemical tools have proven quite useful in process chemical applications, where the selectivity of enzyme transformations is accentuated by the benefits of green chemistry. The introduction of metabolic engineering comes as a natural outgrowth of the use of enzymes in chemistry, bringing these reactions into the cell in order to capitalize on the innate metabolism of living organisms. This move, however, brings with it the complexity of metabolic systems pathways, and as a result requires an understanding of metabolism, molecular biology and microbial cell culture, along with the requisite organic chemistry. As new secondary metabolic pathways become elucidated through genome-sequencing efforts, the application of these tools for synthetic chemistry will continue to thrive.

## References


1 Kleinkauf, H. and von Dohren, H. (1995) *Journal of Antibiotics (Tokyo)*, **48**, 563–7.
2 Schwarzer, D. and Marahiel, M.A. (2001) *Naturwissenschaften*, **88**, 93–101.
3 Cane, D.E. and Walsh, C.T. (1999) *Chemistry and Biology*, **6**, R319–25.
4 Kleinkauf, H. and von Dohren, H. (1996) *European Journal of Biochemistry*, **236**, 335–51.
5 Eppelmann, K., Stachelhaus, T. and Marahiel, M.A. (2002) *Biochemistry*, **41**, 9718–26.
6 Gokhale, R.S., Tsuji, S.Y., Cane, D.E. and Khosla, C. (1999) *Science*, **284**, 482–5.
7 Doekel, S. and Marahiel, M.A. (2001) *Metabolic Engineering*, **3**, 64–77.
8 Mootz, H.D., Schwarzer, D. and Marahiel, M.A. (2002) *ChemBioChem*, **3**, 490–504.
9 Leadlay, P.F., Staunton, J., Oliynyk, M., Bisang, C., Cortes, J., Frost, E., Hughes-Thomas, Z.A., Jones, M.A., Kendrew, S.G., Lester, J.B., Long, P.F., McArthur, H.A., McCormick, E.L., Oliynyk, Z., Stark, C.B. and Wilkinson, C.J. (2001) *Journal of Industrial Microbiology and Biotechnology*, **27**, 360–7.
10 Kwon, H.J., Smith, W.C., Scharon, A.J., Hwang, S.H., Kurth, M.J. and Shen, B. (2002) *Science*, **297**, 1327–30.
11 Du, L., Sanchez, C. and Shen, B. (2001) *Metabolic Engineering*, **3**, 78–95.
12 Crosa, J.H. and Walsh, C.T. (2002) *Microbiology and Molecular Biology Reviews*, **66**, 223–49.
13 Dixon, R.A. (2001) *Nature*, **411**, 843–7.
14 Floss, H.G. and Yu, T.W. (1999) *Current Opinion in Chemical Biology*, **3**, 592–7.
15 Cane, D.E., Walsh, C.T. and Khosla, C. (1998) *Science*, **282**, 63–8.
16 Mootz, H.D. and Marahiel, M.A. (1997) *Current Opinion in Chemical Biology*, **1** (4), 543–51.
17 Weber, T. and Marahiel, M.A. (2001) *Structure (Cambridge)*, **9**, R3–9.
18 Hill, A. (2006) *Natural Product Reports*, **23**, 256–320.
19 Walsh, C.T. (2004) *Science*, **303**, 1805–10.
20 Roy, R.S., Gehring, A.M., Milne, J.C., Belshaw, P.J. and Walsh, C.T. (1999) *Natural Product Reports*, **16**, 249–63.
21 Schneider, T.L., Walsh, C.T. and O'Connor, S.E. (2002) *Journal of the American Chemical Society*, **124**, 11272–3.
22 Roy, R.S., Kelleher, N.L., Milne, J.C. and Walsh, C.T. (1999) *Chemistry and Biology*. **6**, 305–18.
23 Kupke, T. and Götz, F. (1997) *Journal of Biological Chemistry*, **272**, 4759.
24 Jung, G. (1991) *Angewandte Chemie (International Edition in English)*, **30**, 1051.
25 Kim, M.-Y., Vankayalapati, H., Shin-ya, K., Wierzba, K. and Hurley, L.H. (2002) *Journal of the American Chemical Society*, **124**, 2098–9.
26 Hughes-Thomas, Z.A., Stark, C.B.W., Böhm, I.U., Staunton, J. and Leadlay, P.F.

(2003) *Angewandte Chemie (International Edition in English)*, **42**, 4478.

**27** Sun, Y., Zhou, X., Dong, H., Tu, G., Wang, M., Wang, B. and Deng, Z. (2003) *Chemistry and Biology*, **10**, 431–41.

**28** Duerfahrt, T., Eppelmann, K., Muller, R. and Marahiel, M.A. (2004) *Chemistry and Biology*, **11**, 261–71.

**29** You, S.-L., Razavi, H. and Kelly, J.W. (2003) *Angewandte Chemie (International Edition in English)*, **42**, 83–5.

**30** Li, A. and Piel, J. (2002) *Chemistry and Biology*, **9**, 1017.

**31** Schneider, G. (2005) *Current Opinion in Structural Biology*, **15**, 629–36.

**32** Shen, B. (2000) *Topics in Current Chemistry*, **209**, 1–51.

**33** Dorrestein, P. and Begley, T.P. (2005) *Bioorganic Chemistry*, 136–48.

**34** Freeman, J.C., Nayar, T.P., Begley, T.P. and Villafranca, J.J. (1993) *Biochemistry*, **32**, 4826–930.

**35** Trauger, J.W., Kohli, R.M., Mootz, H.D., Marahiel, M.A. and Walsh, C.T. (2000) *Nature*, **407**, 215–18.

**36** Kohli, R.M., Walsh, C.T. and Burkart, M.D. (2002) *Nature*, **418**, 658–61.

**37** Kohli, R.M., Burke, M.D., Tao, J. and Walsh, C.T. (2003) *Journal of the American Chemical Society*, **125**, 7160–1.

**38** Lin, H. and Walsh, C.T. (2004) *Journal of the American Chemical Society*, **126**, 13998–4003.

**39** Sieber, S.A., Tao, J., Walsh, C.T. and Marahiel, M.A. (2004) *Angewandte Chemie (International Edition in English)*, **43**, 493–8.

**40** Lu, H., Tsai, S.-C., Khosla, C. and Cane, D.E. (2002) *Biochemistry*, **41**, 12590–7.

**41** Boddy, C.N., Schneider, T.L., Hotta, K., Walsh, C.T. and Khosla, C. (2003) *Journal of the American Chemical Society*, **125**, 3428–9.

**42** Bruner, S.D., Weber, T., Kohli, R.M., Schwarzer, D., Marahiel, M.A., Walsh, C.T. and Stubbs, M.T. (2002) *Structure*, **10**, 301–10.

**43** Tsai, S.-C., Miercke, L.J.W., Krucinski, J., Gokhale, R., Chen, J.C.-H., Foster, P.G., Cane, D.E., Khosla, C. and Stroud, R.M. (2001) *Proceedings of the National Academy of Sciences of the United States of America*, **98**, 14808–13.

**44** Akey, D.L., Kittendorf, J.D., Giraldes, J.W., Fecik, R.A., Sherman, D.H. and Smith, J.L. (2006) *Nature Chemical Biology*, **2**, 537–42.

**45** Weber, G. and Leitner, E. (1994) *Current Genetics*, **26**, 461–7.

**46** Becker, J.E., Moore, R.E. and Moore, B.S. (2004) *Gene*, **325**, 35–42.

**47** Yarnell, A. (2006) *Chemical and Engineering News*, **84** (21), 12–18.

**48** Harris, C.M., Kannan, R., Kopecka, H. and Harris, T.M. (1985) *Journal of the American Chemical Society*, **107**, 6652–8.

**49** Burd, V.N. and van Pée, K.-H. (2003) *Biochemistry*, **68**, 1132–5.

**50** Dong, C., Flecks, S., Unversucht, S., Haupt, C. and van Pée, K.-H., Naismith, J.H. (2005) *Science*, **309**, 2216–19.

**51** Blasiak, L.G., Vaillancourt, F.H., Walsh, C.T. and Drennan, C.L. (2006) *Nature*, **440**, 368–71.

**52** HÖlzer, M., Burd, W., Reiβig, H.-U. and van Pée, K.-H. (2001) *Advanced Synthesis Catalysis*, **343**, 591–5.

**53** Unversucht, S., Hollmann, F., Schmid, A. and van Pée, K.-H. (2005) *Advanced Synthesis Catalysis*, **347**, 1163–7.

**54** Sánchez, C., Zhu, L., Braña, A.F., Salas, A.P., Rohr, J., Méndez, C. and Salas, J.A. (2005) *Proceedings of the National Academy of Sciences of the United States of America*, **102**, 461–6.

**55** Vaillancourt, F.H., Yeh, E., Vosburg, D.A., O'Connor, S.E. and Walsh, C.T. (2005) *Nature*, **436**, 1191–4.

**56** Galonic, P.D., Vaillancourt, F.H. and Walsh, C.T. (2006) *Journal of the American Chemical Society*, **128**, 3900–1.

**57** Vaillancourt, F.H., Vosburg, D.A. and Walsh, C.T. (2006) *ChemBioChem*, **7**, 748–52.

**58** O'Hagan, D., Schaffrath, C., Cobb, S.L., Hamilton, J.T. and Murphy, C.D. (2002) *Nature*, **416**, 279.

**59** O'Hagan, D. (2006) *Journal of Fluorine Chemistry*, **127**, 1479–83.

**60** Borman, S. (2006) *Chemical and Engineering News*, **36** (84), 13–22.

**61** Nissen, P., Hansen, J.L., Ban, N., Moore, P.B. and Steitz, T.A. (2000) *Science*, **289**, 920–30.

**62** Griffith, B.R., Langenhan, J.M. and Thorson, J.S. (2005) *Current Opinion in Biotechnology*, **16**, 622–30.

**63** Lu, W., Leimkuhler, C., Oberthur, M., Tao, J., Kahne, D. and Walsh, C.T. (2004) *Proceedings of the National Academy of Sciences of the United States of America*, **101**, 4390–5.

**64** Kruger, R.G., Lu, W., Oberthur, M., Tao, J., Kahne, D. and Walsh, C.T. (2005) *Chemistry and Biology*, **12**, 131–40.

**65** Fu, X., Albermann, C., Zhang, C.S., Thorson, J.S. (2005) *Organic Letters*, 1513–15.

**66** Zhang, C., Griffith, B.R., Fu, Q., Albermann, C., Fu, X., Lee, I.-K., Li, L. and Thorson, J.S. (2006) *Science*, **313**, 1291–4.

**67** Blanchard, S. and Thorson, J.S. (2006) *Current Opinion in Chemical Biology*, **10**, 263–71.

**68** Borisova, S., Zhao, L., Melancon, C.E. and Kao, C.-L., Liu, H.-W. (2004) *Journal of the American Chemical Society*, **126**, 6534.

**69** Borisova, S., Zhang, C., Takahashi, H., Zhang, H., Wong, A.W., Thorson, J.S. and Liu, H.-W. (2006) *Angewandte Chemie – International Edition*, **45**, 2748–53.

**70** Mendez, C. and Salas, J.A. (2005) *Methods in Biotechnology*, 131–48.

**71** Wuosmaa, A.M. and Hager, L.P. (1990) *Science*, **249**, 160–2.

**72** Blackburn, G.M., Gamblin, S.J. and Wilson, J.R. (2003) *Helvetica Chimica Acta*, **86**, 4000–6.

**73** Patel, H., Tao, J. and Walsh, C.T. (2003) *Biochemistry*, **42**, 10514–27.

**74** Weinig, S., Hecht, H.-J., Mahmud, T. and Muller, R. (2003) *Chemistry and Biology*, **10**, 939–52.

**75** Pacholec, M., Tao, J. and Walsh, C.T. (2005) *Biochemistry*, **44**, 14969–76.

**76** Tao, J., Hu, S., Pacholec, M. and Walsh, C.T. (2003) *Organic Letters*, **5**, 3233–6.

**77** Woodyer, R.D., Li, G., Zhao, H. and van der Donk, W.A. (2007) *Chemical Communications*, **4**, 359–61.

**78** Fessner, W.-D. and Helaine, V. (2001) *Current Opinion in Biotechnology*, **12**, 574–86.

**79** Chen, H. and Walsh, C. (2001) *Chemistry and Biology*, **8**, 301–12.

**80** Que, L. Jr. (2000) *Nature Structural and Molecular Biology*, **7**, 182–4.

**81** Baldwin, J.E., Byford, M.F., Shiau, C.-Y. and Schofield, C.J. (1997) *Chemical Reviews*, **97**, 2631–49.

**82** Schofield, C., Baldwin, J.E., Byford, M.F., Clifton, I., Hajdu, J., Hensgens, C. and Roach, P. (1997) *Current Opinion in Structural Biology*, **7**, 857–64.

**83** Gibson, D.T. and Parales, R.E. (2000) *Current Opinion in Biotechnology*, **11**, 236–43.

**84** Boyd, D.R., Sharma, N.D. and Allen, C. (2000) *Current Opinion in Biotechnology*, **12**, 564–73.

**85** Flatmark, T. and Stevens, R.C. (1999) *Chemical Reviews*, **99**, 2137–60.

**86** Fetzner, S. (2002) *Applied Microbiology and Biotechnology*, **60**, 243–57.

**87** Kendrew, S.G., Hopwood, D.A. and Marsh, E.N.G. (1997) *Journal of Bacteriology*, **179**, 4305–10.

**88** Bauer, I., Max, N., Fetzner, S. and Lingens, F. (1996) *European Journal of Biochemistry*, **240**, 576–83.

**89** Holland, H.L. and Webber, H.K. (2000) *Current Opinion in Biotechnology*, **11**, 547–53.

**90** Holland, H.L. (1999) *Current Opinion in Chemical Biology*, **3**, 22–7.

# 8
# Modifying the Glycosylation Pattern in Actinomycetes by Combinatorial Biosynthesis

*José A. Salas and Carmen Méndez*

## 8.1
## Bioactive Natural Products in Actinomycetes

Actinomycetes represent an important group of Gram-positive bacteria with interest from the basic and applied points of view. These microorganisms show interesting cell differentiation processes in their life cycle [1]. They are filamentous bacteria forming hyphae that constitute a mycelia. As a response to the lack of some nutritional stimuli they can produce chains of spores, which can remain in a dormant state until appropriate physicochemical conditions are restored. Then, these spores germinate generating a new mycelium. Actinomycetes are also important industrial microorganisms because of their capability to synthesize a number of bioactive natural products, being responsible for the biosynthesis of approximately two-thirds of all natural products produced by microorganisms [2]. Many of these natural products are bioactive compounds with clinical (antibiotics, antifungals, antitumors), industrial (enzymes) or agricultural (herbicides, insecticides) application [3]. Within the actinomycetes, members of the genus *Streptomyces* are the largest contributors to the production of these bioactive natural products.

If we consider bioactive natural products synthesized by actinomycetes, a feature common to many of them is that they are glycosylated compounds [4]. Therapeutically important antibiotics (erythromycin), antifungals (amphotericin B), antiparasitics (avermectins) and anticancer drugs (doxorubicin) contain sugars attached to the aglycone core. These sugars contribute to the structural biodiversity in natural products, and they participate in the interaction between the drug and the cellular target. In many cases, the presence of the sugars is very important for the biological activity of these compounds. The absence of some of these sugars causes a dramatic reduction in activity and, in some cases, the complete loss of biological activity of the drug.

A large number of sugars in natural products belong to the 6-deoxyhexose (6DOH) family, of which more than 70 different variants have been reported in plants, fungi and bacteria (Figure 8.1) [4b, 4c, 5]. These sugars are linked to the

*Multi-Step Enzyme Catalysis: Biotransformations and Chemoenzymatic Synthesis*
Edited by Eduardo Garcia-Junceda

ISBN: 978-3-527-31921-3

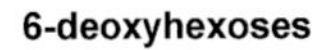

**6-deoxyhexoses**

NDP-D-quinovose NDP-D-fucose NDP-6- deoxy-D-allose NDP-6- deoxy-D-gulose

NDP-L-rhamnose NDP-6-deoxy-L-talose

**2,6-dideoxyhexoses**

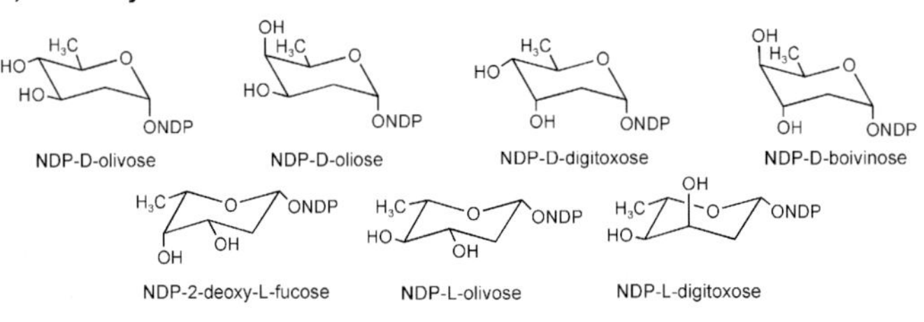


**2,3,6-trideoxyhexoses**

NDP-D-amicetose NDP-D-rhodinose NDP-L-rhodinose NDP-L-amicetose

**branched chain deoxyhexoses**

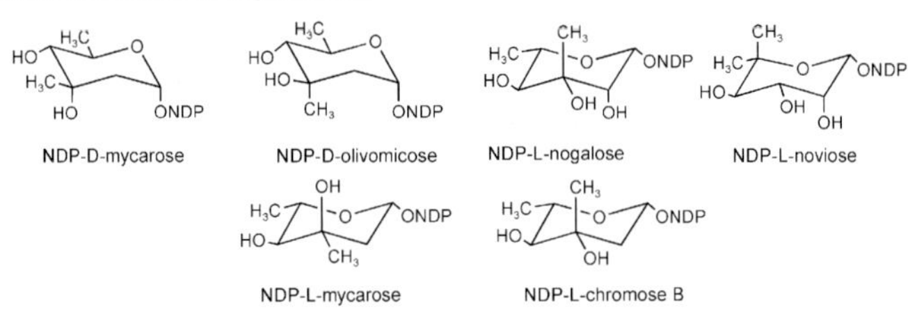


**deoxyaminohexoses**

NDP-D-mycaminose NDP-D-desosamine NDP-D-forosamine NDP-L-daunosamine

NDP-L-rhodosamine NDP-L-ristosamine NDP-L-vancosamine NDP-L-epivancosamine

**pentoses**

NDP-D-xylose NDP-L-lyxose

**Figure 8.1** Chemical structures of some DOHs.

aglycone as saccharides of variable sugar length. They can consist of monosaccharides (doxorubicin), disaccharides (mithramycin, tylosin), trisaccharides (mithramycin, urdamycin A) or larger oligosaccharide chains with six (landomycin), seven (avilamycin) or even very long sugar chains with as many as 17 sugars (saccharomycin). Attachment of the sugars to the aglycone usually occurs in most of the cases through *O*-glycosidic linkages. However, examples exist of *C*- (as in urdamycin) and *N*-glycosidic (as in rebeccamycin) linkages, both occurring less frequently.

## 8.2 Deoxy Sugar Biosynthesis and Gene Clusters

Most 6DOH are synthesized from NDP-activated hexoses (mainly D-glucose) via 4-keto-6-deoxy intermediates [5]. During the early stages of 6DOH biosynthesis, two common enzymatic steps lead to the formation of a 4-keto-6-deoxy intermediate. These reactions are catalyzed by an NDP-D-hexose synthase and an NDP-D-hexose-4,6-dehydratase, respectively. This key intermediate will then undergo further modifications leading to the formation of a variety of different 6DOHs. These modifications include deoxygenations, transaminations, ketoreductions and *C*-, *N*- or *O*-methylations. Additionally, D- and L-isomeric forms of many 6DOHs are also formed through the action of 5- or 3,5-epimerases.

A number of gene clusters involved in the biosynthesis of different 6DOH have been characterized in recent years. The biosynthetic gene cluster for the same 6DOH has been isolated from several organisms (e.g. D-olivose from seven different producers, D-mycosamine from six, D-desosamine and L-rhodinose from five or D-deoxyallose, L-mycarose and L-noviose from three). As a general rule, all genes encoding enzymes required for the biosynthesis of a 6DOH are present within the antibiotic biosynthesis cluster. However, some exceptions exist mainly involving genes encoding the two earliest steps leading to the formation of the central 4-keto-6-deoxy intermediate [4b, 6]. In the case of the mithramycin gene cluster, there are no specific genes for the 2,3-dehydratase and for the 4-ketoreductase of D-mycarose. It has been proposed that other enzymes of the pathway could fulfill these two functions [7]. Furthermore, in the four clusters so far characterized that contain an L-rhamnose moiety (spinosyn, elloramycin, steffimycin and aranciamycin), the genes required for the biosynthesis of this sugar have been identified outside the limits of the gene clusters (Ramos *et al.*, unpublished) [8].

## 8.3 Characterization of Sugar Biosynthesis Enzymes

Sequence analysis of genes in sugar biosynthesis gene clusters has resulted in tentative assignments for putative functions for the encoded proteins in the biosynthesis of the corresponding deoxy sugars. However, a lack of biochemical characterization of these gene products leaves many of these assignments uncertain. *In vitro* assays for testing the activity of these enzymes have been hampered by the unavailability of biosynthetic intermediates. Several academic research groups have addressed the chemical and (chemo)enzymatic synthesis of sugar biosynthesis intermediates in order to *in vitro* characterize enzymes involved in the biosynthesis of NDP-activated deoxy sugars [9]. Enzymatic assays have been carried out for a number of enzymes in order to demonstrate the mechanism and subsequently identify the real substrate of the enzyme. Also, in spite of the difficulties for *in vitro* enzymatic assays, the biosynthesis of two 6DOHs has been fully reconstituted *in vitro*. In the case of L-epivancosamine, the five enzymes involved in its biosynthesis were heterologously produced in *Escherichia coli*, and were used to reconstitute the complete biosynthesis of TDP-L-epivancosamine from TDP-4-keto-6-deoxy-D-glucose in a process involving successively C2 deoxygenation, C3 transamination, C3 methylation, C5 epimerization and C4 ketoreduction [10]. The structure of one of these enzymes, the EvaD C5 epimerase, has been recently reported, and it has been found that fine tuning of a Tyr residue position allows EvaD to distinguish between its natural substrate and other substrates [11]. Recently, the biosynthesis of TDP-L-mycarose has been also reconstituted using as precursors glucose-1-phosphate and thymidine and five enzymes from *Streptomyces fradiae*–a tylosin producer [12].

The first two enzymatic steps are common to the biosynthesis of all 6DOHs and lead to the biosynthesis of the intermediate TDP-4-keto-6-deoxy-D-glucose. These enzymatic reactions have been extensively studied in different sugar biosynthesis pathways, such as tylosin or mithramycin [13].

In the biosynthesis of 2,6-diDOHs (2,6DOHs) and 2,3,6-triDOHs (2,3,6DOHs), the next step is C2 deoxygenation of TDP-4-keto-6-deoxy-D-glucose. This process requires a dehydration step followed by a reduction reaction. Two different products can arise from this deoxygenation process, only differing in the configuration at the C3 hydroxyl group. The C2 deoxygenation process was first *in vitro* characterized in granaticin and oleandomycin biosyntheses [14]. Using a pair of enzymes involved in D-olivose biosynthesis in the granaticin pathway in *Streptomyces violaceoruber* Tü22, and also two other pair of enzymes involved in L-olivose biosynthesis in the oleandomycin pathway in *Streptomyces antibioticus* Tü99, they were able to establish the mode of removal of the 2-OH group. They showed that Gra Orf27 and Tü99 Orf10 act as TDP-4-keto-6-deoxyglucose 2,3-dehydratase, converting TDP-4-keto-6-deoxy-D-glucose into an unstable compound, presumably TDP-3,4-diketo-2,6-dideoxyglucose or its 2,3-enol. Then, this compound is reductively captured by the 3-ketoreductase Gra Orf26 or TÜ99 Orf10 to render TDP-4-keto-2,6-dideoxyglucose, which contains a hydroxyl group at C3 in equatorial configuration. In the absence of this last step, the putative intermediate TDP-3,4-diketo-2,6-dideoxyglucose undergoes facile elimination of TDP giving rise to maltol. Generation of a 2,6DOH with a hydroxyl group at C3 in axial configuration (i.e. TDP-2,6-dideoxy-D-glycero-D-glycero-4-hexulose) also requires the action of a 2,3-dehydratase, but it uses a different 3-ketoreductase. This deoxygenation step has been biochemically characterized using the enzymes TylX3 and TylC1 from *S. fradiae*, which participate in the tylosin pathway [15].

Many DOHs, such as L-daunosamine, L-epivancosamine or L-ristosamine, contain an amino group at C3, which is introduced by an aminotransferase. The substrate for this reaction is the 3-keto sugar intermediate that arises as a consequence of the action of a 2,3-dehydratase. This transamination reaction has been biochemically characterized in the biosynthesis of L-epivancosamine [10]. Using a coupled reaction with EvaB (2,3-dehydratase) and EvaC (aminotransferase), with pyridoxal-5-phosphate (PLP) as a coenzyme and L-glutamate as a cosubstrate, they were able to show conversion of TDP-4-keto-2,6-dideoxyglucose into thymidine-5′-diphospho-3-amino-2,3,6-trideoxy-D-*threo*-hexopyranos-4-ulose.

For the biosynthesis of 2,3,6DOHs such as D-forosamine or L-rhodinose, the C3 deoxygenation of TDP-4-keto-2,6-dideoxyglucose is required. By studying the enzyme SpnQ involved in D-forosamine biosynthesis in the spinosyn pathway in *Saccharopolyspora spinosa*, this deoxygenation step has been characterized [16]. In an *in vitro* assay using reductase pairs (ferredoxin/ferredoxin reductase or flavodoxin/flavodoxin reductase) as electron donors, in the presence of NADPH and pyridoxamine 5′-monophosphate, they found that SpnQ converted TDP-4-keto-2,6-dideoxyglucose into TDP-4-keto-2,3,6-trideoxyglucose. In the biosynthesis of D-forosamine, formation of this compound is followed by a transamination at C4 and dimethylation of the 4-amino group. The protein responsible for the introduction of the amino group is SpnR and it has been characterized by using the reverse reaction [17]. Since the substrate for this reaction (i.e. TDP-4-amino-2,3,4,6-tetradeoxy-D-glucose) was quite unstable, an isostere of it was used in which the TMP group was joined to the hexose core through a stable C-glycosidic phosphonate linker.

Some DOHs are *C*-methylated, mostly at C3 (i.e. L-mycarose, L-chromose B, D-olivomycose, L-vancosamine or L-nogalose) or at other positions such as C5 in the case of L-noviose. The biosynthesis of this type of the so-called methyl branched sugars, requires the involvement of a C-methyltransferase. The first *C*-methyltransferase isolated and characterized was TylC3 involved in the biosynthesis of L-mycarose in the tylosin pathway [18]. It was shown that TylC3 was able to methylate, in a *S*-adenosylmethionine (SAM)-dependent reaction, the expected substrate TDP-2,6-dideoxy-D-glycero-D-glycero-4-hexulose, generating a methylated derivate with a methyl group in axial configuration and indicating that TylC3-catalyzed methylation proceeded with overall inversion of configuration at C3. The biosynthesis of L-epivancosamine by *Amycolaptosis orientalis* also requires a *C*-methylation step, which is carried out by EvaC. This enzymatic step has been characterized and it has been shown that incubation of thymidine-5′-diphospho-3-amino-2,3,6-trideoxy-D-*threo*-hexopyranos-4-ulose as substrate and SAM, in the presence of the *C*-methyltransferase EvaC, produced a compound identified as thymidine-5′-diphospho-3-amino-2,3,6-trideoxy-3-*C*-methyl-D-*erythro*-hexopyranos-4-ulose, which contains an equatorial methyl group at C3 [10].

Many DOHs, such as L-mycarose, L-epivancosamine, L-noviose or L-daunosamine, show an L configuration. Formation of L-DOH requires the action of a 5- or 3,5-epimerase. The epimerase EvaD from the biosynthesis pathway of L-epivancosamine was shown to act as a 5-epimerase on the intermediate thymidine-5′-diphospho-3-amino-2,3,6-trideoxy-D-*threo*-hexopyranos-4-ulose [10]. On the other hand, the epimerization reaction that takes place during the biosynthesis of TDP-L-mycarose in the tylosin pathway has been also characterized [19]. It was shown that the TylK epimerase behaves as a 5-epimerase acting on an already $C^3$-methylated intermediate. Also, they found that TylK showed relaxed substrate flexibility, being able to epimerize C5 of TDP-2,6-dideoxy-D-glycero-D-glycero-4-hexulose. In addition, the epimerization reaction catalyzed by NovW, involved in the biosynthesis of L-noviose of novobiocin in *Streptomyces spheroides* has been also characterized [20]. In this case, by using kinetic studies and deuterium incorporation analysis it was shown that NovW serves as only 3-epimerase and not 3,5-epimerase, using as substrate TDP-4-keto-6-deoxy-D-glucose.

The final step in the biosynthesis of most DOHs is the reduction of the C4 ketone in a stereospecific manner. Ketoreductases responsible for these reactions in the biosynthesis of L-epivancosamine and L-mycarose have been tested in *in vitro* assays, using NADPH as cosubstrate [10, 19].

Some macrolides contain deoxyaminohexoses such as D-desosamine (erythromycin and pikromycin) or D-mycaminose (tylosin). Biosynthesis of these DOHs has some specific peculiarities. The desosamine biosynthetic pathway from the methymycin/pikromycin producer has been *in vitro* characterized. Starting from TDP-4-keto-2,6-dideoxyglucose, the next step is the C4 deoxygenation of this intermediate. This deoxygenation takes place in two stages via an amino sugar intermediate and is carried out by the enzymes DesI (4-aminotransferase) and DesII (deaminase). *In vitro* assays showed that, in the presence of L-glutamate and PLP, purified DesI converts TDP-4-keto-2,6-dideoxyglucose into TDP-4-amino-4,6-dideoxy-D-glucose [21]. Incubation of this intermediate with reduced DesII in the presence of SAM and under anaerobic conditions leads to the production of 3-keto-4,6-dideoxyhexose [22]. This 3-keto intermediate is the substrate for DesV– the aminotransferase that catalyzes the PLP-dependent replacement of the 3-keto group with an amino functionality [23]. The final step consists in the *N,N*-dimethylation of the C3 amino group. This reaction is catalyzed by the SAM-dependent methyltransferase DesVI in a stepwise manner [9b, 24].

Biosynthesis of D-mycaminose has been characterized in the tylosin producer *S. fradiae*. The first step after formation of TDP-4-keto-2,6-dideoxyglucose involves the 3,4-isomerase Tyl1a to render a 3-keto sugar intermediate [25]. This intermediate will serve as substrate for a transamination reaction catalyzed by TylB. The activity of this enzyme was assayed in the reverse reaction, using as substrate TDP-3-amino-3,6-dideoxyglucopyranose together with PLP and α-ketoglutarate [26]. As it occurs in D-desosamine biosynthesis, the last step in the biosynthesis of D-mycaminose is the *N,N*-dimethylation of the 3-amino group, which is catalyzed by TylM1 [27].

Some bioactive compounds contain unusual pentoses in their molecules. This is the case of the antibiotic avilamycin, which contain a L-xylose. Quite recently Bechthold *et al.* have biochemically characterized the genesis of this sugar and they have shown that enzyme AviE2 converted UDP-glucuronic acid to UDP-xylose via decarboxylation, indicating that the pentose residue of avilamycin A is derived from D-glucose and not from D-ribose [28].

The final sugars in some natural products do not exactly match the sugar transferred, because tailoring modification after sugar transfer must occur. Thus, L-cladinose and L-oleandrose are present in the 14-membered macrolides erythromycin A and oleandomycin, respectively, but they are synthesized and transferred to the aglycones as L-mycarose and L-olivose, respectively. These modifications are caused by methyltransferase EryG in the former and OleY in the latter. The methylation reaction carried out by OleY has been characterized *in vitro* [29]. Using purified OleY it was shown that this methyltransferase acts as a dimer, and is able to methylate in a SAM-dependent reaction the C3′ hydroxy group of L-olivosyl-erythronolide B and L-rhamnosyl-erythronolide B. Sugar modification by acylation has been shown to be an interesting peculiarity of some bioactive compounds. Two acyltransferases from the biosynthesis pathway for two glycopeptide antibiotics have been characterized [30]. It was shown that they were able to modify different vancomycin and teicoplanin scaffolds glycosylated with different sugars, and also to use different acyl-CoA donors to create variant lipoglycopeptides.

## 8.4 Strategies for the Generation of Novel Glycosylated Derivatives

With the progress in the knowledge of the biosynthesis of different sugar gene clusters in microorganisms producing bioactive compounds, researchers in the area have envisaged different approaches to use recombinant DNA technology in order to engineer the biosynthetic pathways and to generate novel derivatives. Thus, 'combinatorial biosynthesis' has emerged as a new technology allowing the generation of recombinant strains containing gene combinations for secondary metabolites biosynthesis that have not been previously found in nature [31]. In particular, combinatorial biosynthesis is especially useful for the introduction of chemical modifications in selected compounds that are not easily obtained by chemical means. The use of this technology has made it possible to alter the glycosylation pattern of bioactive compounds [32]. The following strategies have been used to produce novel glycosylated derivatives from bioactive natural products.

### 8.4.1 Gene Inactivation

Specific inactivation of genes coding for sugar biosynthesis in a pathway leading to the formation of a glycosylated bioactive compound can cause the accumulation of a biosynthetic intermediate with a different sugar or a sugar biosynthesis intermediate attached to the aglycone. In this approach, the normal expression of the gene is altered either by creating a mutation or by the insertion of an antibiotic-resistance cassette. This strategy is widely used to generate non-producing mutants in gene clusters in order to demonstrate the involvement of a cloned gene (or DNA fragment) in the biosynthesis of a secondary metabolite.

### 8.4.2
### Gene Expression

The sugars in a bioactive compound can be modified by expression of a few sugar biosynthesis genes from another pathway. In this case, the products of the expressed genes can act on a native sugar biosynthesis intermediate of the host causing its deviation to another sugar pathway–the final result being the formation of a new sugar. The transfer of the new sugar to the corresponding aglycone will now be dependent upon the substrate flexibility of the glycosyltransferase.

### 8.4.3
### Combining Gene Inactivation and Gene Expression

A third approach combines the inactivation of a selected gene with the expression in the corresponding mutant of genes encoding enzymes able to cause specific modifications in the accumulated sugar biosynthesis intermediates. The mutant could therefore accumulate sugar biosynthesis intermediates that will now be substrates for the newly expressed foreign genes, potentially converting them into a final sugar. Again, the possible flexibility of the glycosyltransferase regarding the sugar donor can produce the formation of novel derivatives harboring different sugars.

### 8.4.4
### Endowing a Host with the Capability of Synthesizing Different Sugars

Another alternative is to provide a host with the capability of synthesizing new sugars. This can be achieved either in a producer organism or in a non-producer heterologous host through the expression of genes coding for the enzymatic activities required for the biosynthesis of the sugar. If a non-producer host is used, the aglycone and the glycosyltransferase gene must be provided to the recombinant strain. This strategy has been used successfully in recent years through the generation of a family of plasmids, each directing the biosynthesis of different sugars [32]. In these plasmids, the different genes can be easily removed and replaced by other genes coding for different enzymatic activities, thus providing the capability of synthesizing different sugars.

## 8.5
## Generation of Glycosylated Derivatives of Bioactive Compounds

In recent years a number of glycosylated derivatives of known bioactive natural products have been generated using the different strategies described above. The generation of these bioactive compounds is described in the following sections.

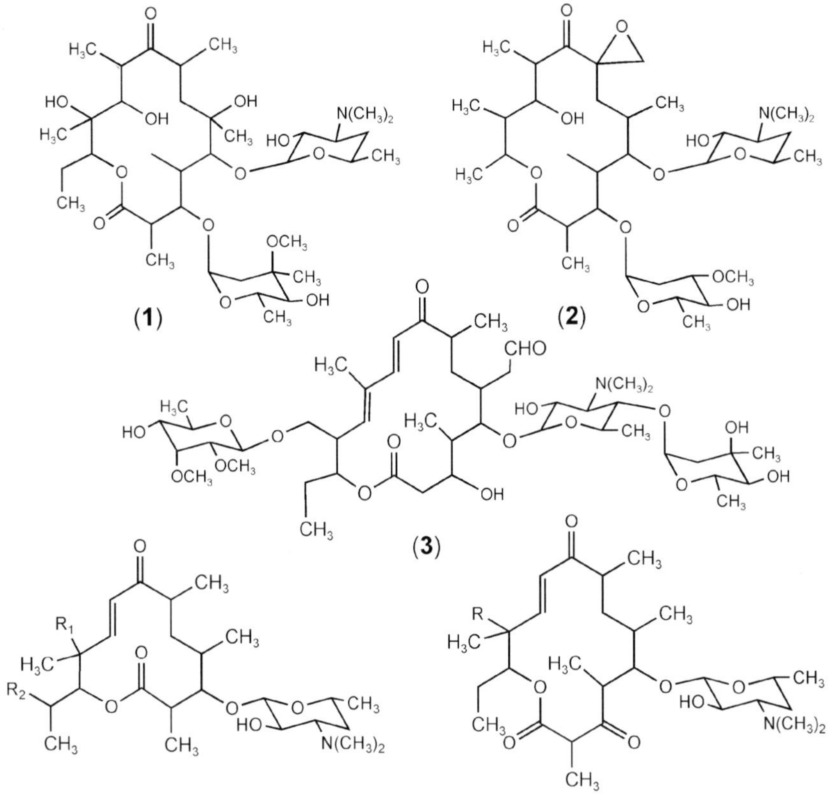

$R_1$= OH $R_2$=H: (**4**)
$R_1$=H $R_2$=OH: (**5**)

R= OH: (**6**)
R=H: (**7**)

**Figure 8.2** Chemical structure of several macrolides.

### 8.5.1
### Macrolides

Macrolides are an important group of antibiotics containing 12- to 18-membered macrocyclic lactone rings to which one or more amino sugars and/or deoxy sugars are attached (Figure 8.2). They are active against Gram-positive bacteria, Gram-negative bacteria such as *Neisseria* and *Haemophilus*, and also against *Mycoplasma*. They have clinical and agricultural applications as antiinfective, inmunosupressive, parasiticidal and insecticidal agents. These antibiotics inhibit protein synthesis by binding to the 50S ribosomal subunit. The biosynthesis gene clusters of several macrolides have been cloned and characterized, and this information has been used to generate novel glycosylated derivatives.

Erythromycin A (Figure 8.2, **1**) is a clinically important antibiotic produced by *Saccharopolispora erythraea*. It consists of a 14-membered macrolactone ring to

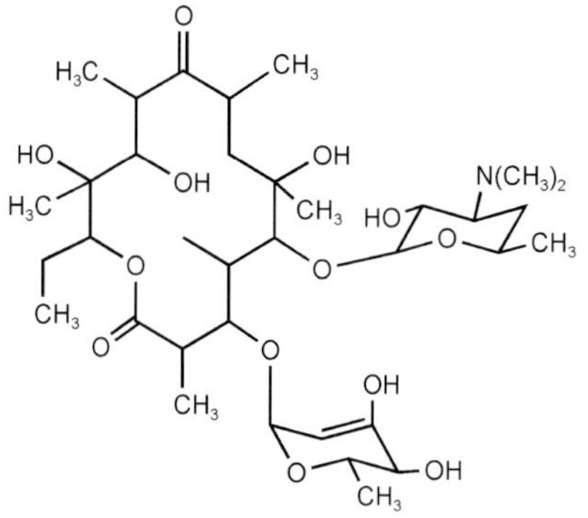

(8)

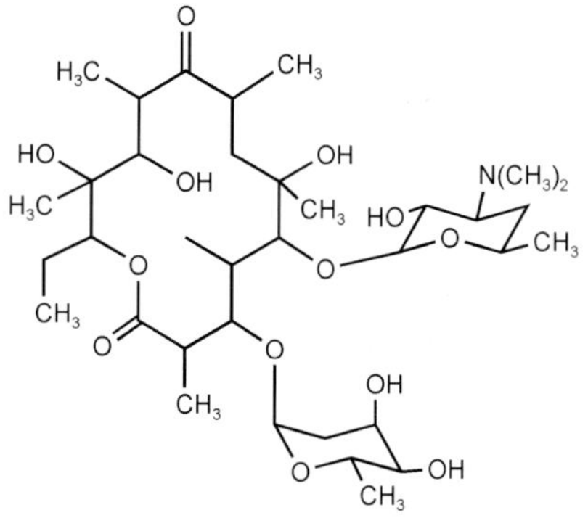

(9)

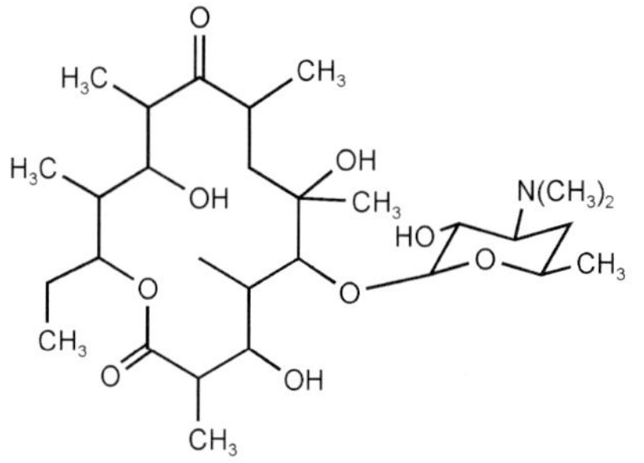

(10)

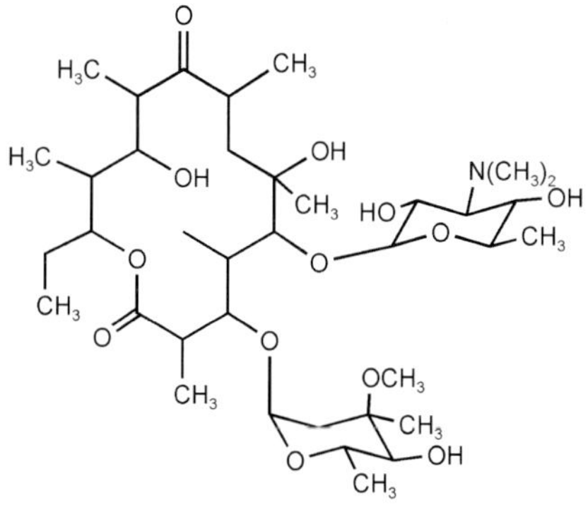

(11)

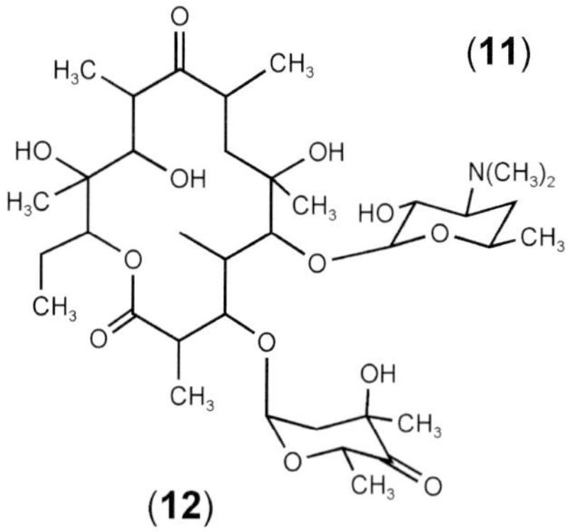

(12)

Figure 8.3 Erythromycin derivatives obtained by gene inactivation.

which two DOHs, L-cladinose and D-desosamine, are attached. The genes for the biosynthesis of erythromycin A have been cloned and sequenced, including the *eryB* and *eryC* genes that code for enzymes involved in the biosynthesis and transfer of L-mycarose and D-desosamine, respectively [33]. Several novel derivatives were isolated as minor compounds, as a consequence of the inactivation of specific *eryB* or *eryC* genes (Figure 8.3). Thus, by inactivating the *eryBII* gene that codes for the enoil reductase responsible for the reduction of the 2,3-double bond in L-mycarose biosynthesis, compounds 3″-C-demethyl-2″,3″-ene-erythromycin C (**8**) and 3″-C-demethyl erythromycin C (**9**) were produced [33d]. On the other hand, inactivation of either the *eryBVI* gene coding for the 2,3-dehydratase or the *eryBV* gene coding for the mycarosyl glycosyltransferase, resulted in the accumulation of desosaminyl erythronolide B (**10**) [33d]. Inactivation of the *eryCIV* gene that codes for the PLP-dependent dehydratase involved in C4 deoxygenation during the biosynthesis of D-desosamine led to the accumulation of 4′-hydroxy erythromycin D (**11**) [33d] and inactivation of *eryBIV* that codes for a 4-ketoreductase involved in L-mycarose biosynthesis resulted in the accumulation of 4″-keto-mycarosyl erythromycin C (**12**) [33d].

Exchanging glycosyltransferase genes from structural related macrolide pathways has also been used to produce a new erythromycin analog. By expressing the OleG2 glycosyltransferase from the oleandomycin pathway into a *Sac. erythraea* mutant, in which the *eryBV* mycarosyl glycosyltransferase had been deleted, a new bioactive erythromycin 3-L-rhamnosyl-6-deoxyerythromycin B (**13**) containing an L-rhamnose moiety instead of L-mycarose was produced (Figure 8.4) [34]. Also, by

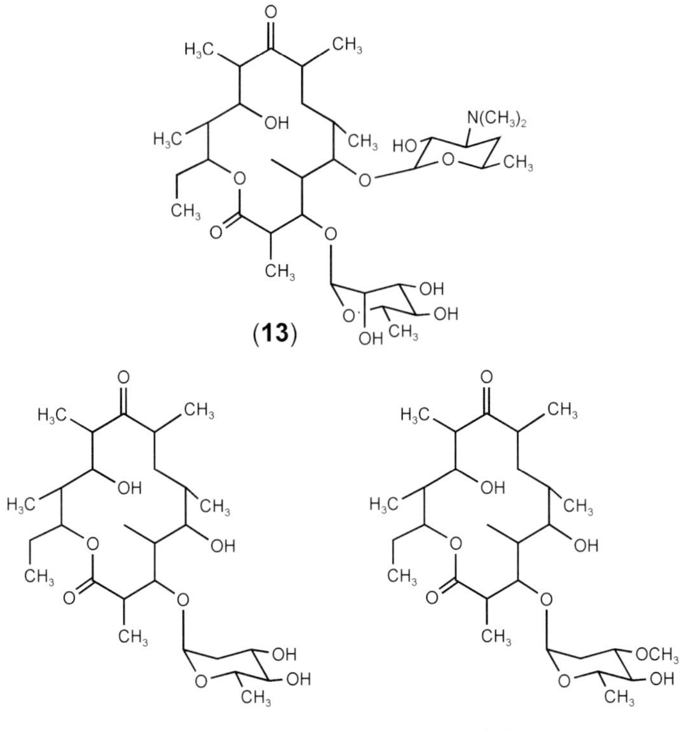

(**13**)

(**14**) (**15**)

**Figure 8.4** Erythromycin derivatives generated by using the OleGII glycosyltransferase of the oleandomycin pathway.

endowing a non-producer host with the capability to synthesize a DOH, new erythronolide B derivatives with a different sugar pattern were produced [35]. A recombinant *Streptomyces albus* strain was generated (named NAG2) by integrating into the chromosome the oleandomycin glycosyltransferase gene *oleG2* responsible for the attachment of L-oleandrose to the oleandomycin aglycone. This strain was transformed with two plasmid constructs containing genes from the oleandomycin gene cluster involved in the biosynthesis of L-oleandrose and able to synthesize L-olivose (plasmid pOLV) or L-oleandrose (plasmid pOLE). Independent biotransformation of the two recombinant strains with erythronolide B (the erythromycin aglycone), resulted in the formation of two novel monoglycosylated derivatives, 3-α-L-olivosyl-erythronolide B (**14**) and 3-α-L-oleandrosyl-erythronolide B (**15**), respectively (Figure 8.4) [35].

Tylosin (Figure 8.2, **3**) is a macrolide produced by *S. fradiae*, which consists of a 16-membered macrolactone ring and three DOHs (D-mycaminose, L-mycarose and D-mycinose). This antibiotic has been used to treat veterinary Gram-positive and *Mycoplasma* infections, and also to promote livestock growth. The biosynthetic gene cluster for this antibiotic has been cloned and characterized, and the genes involved in the biosynthesis of DOHs have been identified [13b, 25, 27, 36]. Two of these genes are involved in the biosynthesis of all three DOHs: *tylA1* and *tylA2* encoding an α-D-glucose-1-phosphate thymidyltransferase and a TDP-D-glucose 4,6-dehydratase, respectively. Other genes are assigned to enzymes involved in D-mycaminose (genes *tyl1a, tylB, tylM1, tylM2* and *tylM3*), L-mycarose (genes *tylX3, tylC1, tylC3, tylK, tylC2* and *tylC4*) and D-mycinose (genes *tylJ, tylD* and *tylN*) biosynthesis. Novel tylosin derivatives were produced in *S. fradiae* by altering its capability of synthesizing sugars. To do that, two genes, *nbmJ* and *nbmK*, involved in C4 deoxygenation during D-desosamine biosynthesis in the narbomycin pathway were imported from the narbomycin producer *Streptomyces narbonensis* and expressed in *S. fradiae*. Expression of these genes caused the appearance of a branched pathway in D-mycaminose biosynthesis leading to the formation of D-desosamine in addition to D-mycaminose. Analysis of the cultures of the recombinant strain showed the formation of two tylosin derivatives both containing desosamine instead of mycaminose, but differing in the presence or absence of the other sugars: 4′-deoxy-5-*O*-mycaminosyl-tylonolide (**16**) and 4′-deoxy-20-dehydro-demycarosyl-tylosin (**17**) (Figure 8.5) [37]. These two compounds had been available before only in small amounts, and produced by chemical synthesis in a time-consuming and expensive process.

On the other hand, several glycosylated derivatives of tylosin were produced in an engineered *Sac. erythraea* strain, by combining gene inactivation and gene expression with a biotransformation procedure. First, a mutant of *Sac. erythraea* was generated by deleting the polyketide synthase (PKS) (*eryAI, eryAII* and *eryAIII*) and the two glycosyltransferase (*eryBV* and *eryCIII*) genes. This mutant still retained the ability to synthesize D-desosamine and L-mycarose. After expressing the tylosin TylM2 glycosyltransferase in this mutant, and feeding with the tylosin aglycone (tylonolide), two new glycosides, 5-*O*-desosaminyl-tylactone (**18**) and 5-*O*-glucosyl-tylactone (**19**), were obtained (Figure 8.6) [38]. The same approach

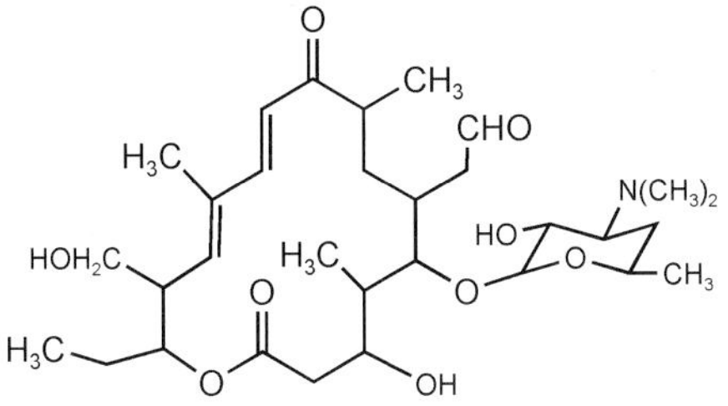

(16)

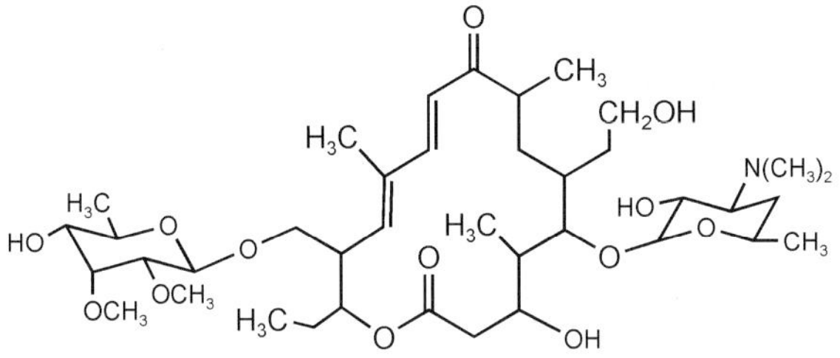

(17)

**Figure 8.5** Tylosin derivatives obtained by expressing sugar genes from *S. narbonensis* into *S. fradiae*.

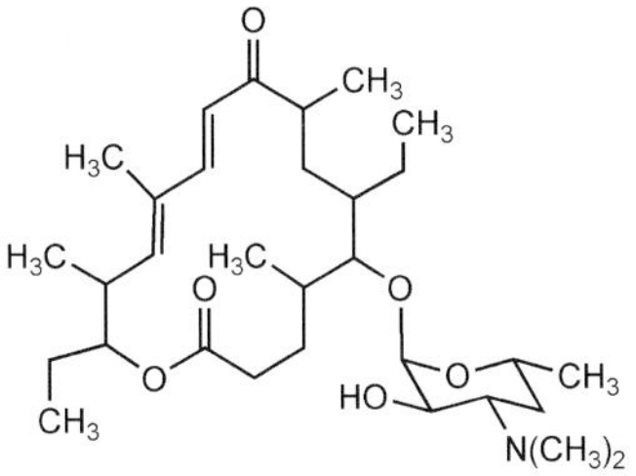

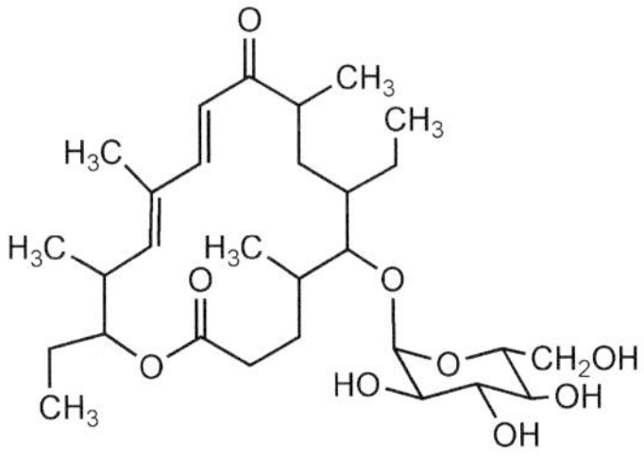

(18) (19)

**Figure 8.6** Tylosin derivatives obtained by combining gene inactivation, gene expression and feeding experiments.

was used to produce glycosylated derivatives of the macrolide bioinsecticide spinosyn. The forosaminyl transferase SpnP of the spinosyn cluster was expressed in the same mutant and novel derivatives were produced: 17-*O*-α-L-mycarosyl-spinosyn A (**20**), 17-*O*-α-L-mycarosyl-spinosyn D (**21**), 17-*O*-β-D-glucosyl-spinosyn A (**22**) and 17-*O*-β-D-glucosyl-spinosyn D (**23**) (Figure 8.7) [39].

Methymycin (**4**) and neomethymycin (**5**) are 12-membered macrolides produced by *Streptomyces venezuelae*–a strain that also produces the 14-membered

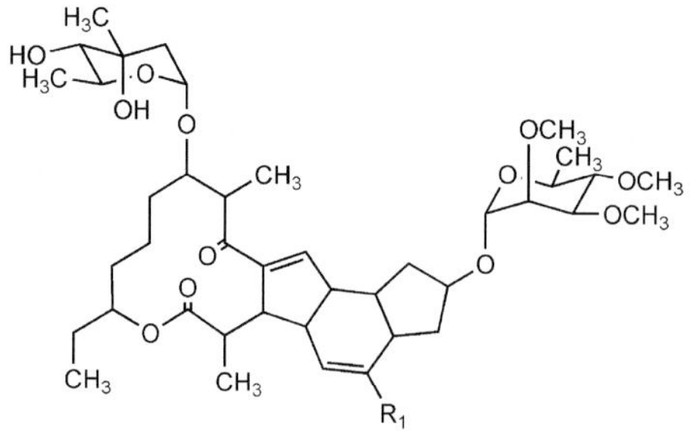

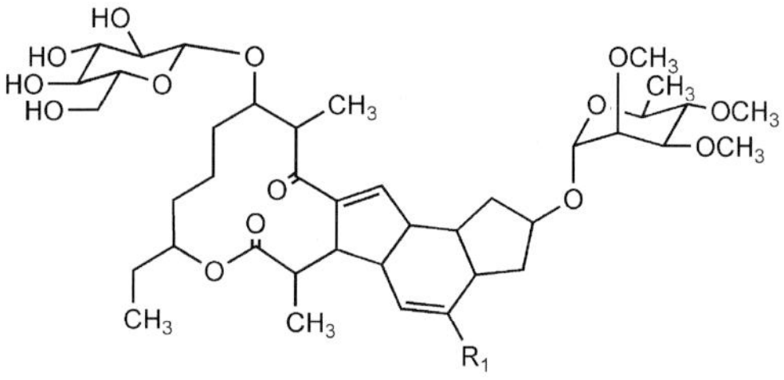

$R_1$=H: (**20**)
$R_1$=$CH_3$: (**21**)

$R_1$=H: (**22**)
$R_1$=$CH_3$: (**23**)

**Figure 8.7** Spinosyn derivatives obtained by combining gene inactivation, gene expression and feeding experiments.

macrolides pikromycin (**6**) and narbomycin (**7**). All these macrolides are monoglycosylated by a D-desosamine. A single biosynthesis gene cluster is responsible for the biosynthesis of both the 12- and 14-membered macrolides [40], and interestingly all the genes responsible for the biosynthesis and transfer of desosamine are contiguous in the chromosome (genes *desI*–*desVIII*). By inactivating specific desosamine biosynthesis genes, several derivatives of these macrolides were produced containing different modified sugars (Figure 8.8). Thus, by inactivating the *desV* gene, which encodes an aminotransferase, two new macrolides containing a 4,6-dideoxy sugar were obtained (**24** and **25**). Formation of these compounds was explained by the accumulation in the mutant of a 3-keto-4,6-dideoxy sugar, followed by a stereospecific reduction [23]. Similarly, by deleting the *desVI* gene, which encodes an *N*-methyltransferase, two new macrolides containing an *N*-acetylated amino sugar were obtained (**26** and **27**). Apparently this mutant accumulates *N*-demethyl desosamine, which after being transferred to the aglycone will suffer postsynthetic acetylation [41]. On the other hand, a new analog of methymycin and neomethymycin containing the sugar residue D-quinovose instead of D-desosamine was generated by deleting the *desI* gene, which codes for a dehydratase responsible for the C4 deoxygenation in the biosynthesis of D-desosamine (**28**). In this mutant, a 4-keto-6-deoxyhexose intermediate is accumulated and reduced by a pathway-independent D-hexulose reductase of *S. venezuelae*, thus resulting in the formation of D-quinovose [42]. Finally, deletion of *desII*, which is involved in the C4 deoxygenation process, caused the production of two macrolides containing an *N*-acetylated 4-amino sugar (**29** and **30**) [21].

New analogs of methymycin/pikromycin were obtained by taking advantage of targeted gene deletion combined with expression of sugar biosynthesis genes from other pathways in order to reconstitute hybrid sugar pathways. Thus, a *desI* mutant was used as host for expressing the *strM* and *strL* genes recruited from the streptomycin gene cluster. The resultant recombinant strain produced four novel macrolide derivatives all of them containing an L-rhamnose moiety as sugar component

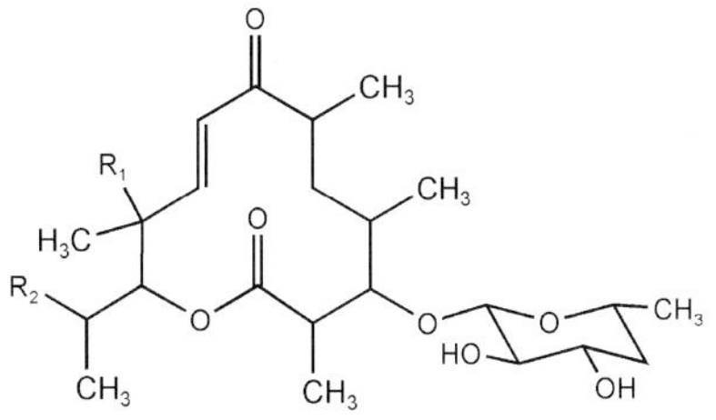

$R_1$=OH $R_2$=H (**24**)
$R_1$=H $R_2$=OH (**25**)

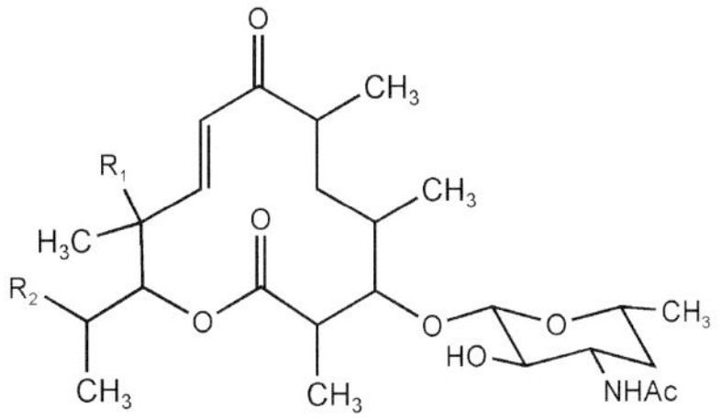

$R_1$=OH $R_2$=H (**26**)
$R_1$=H $R_2$=OH (**27**)

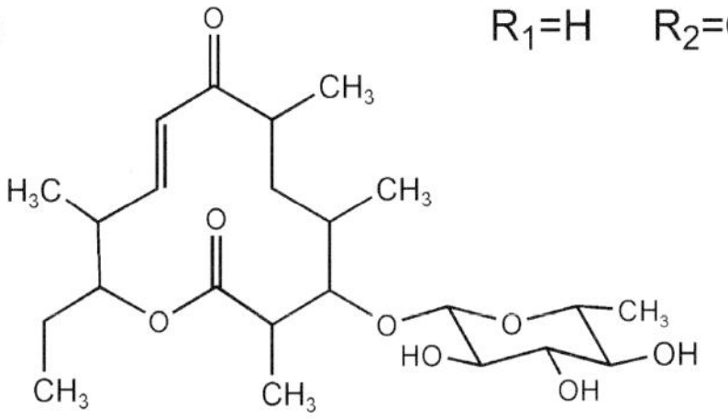

(**28**)

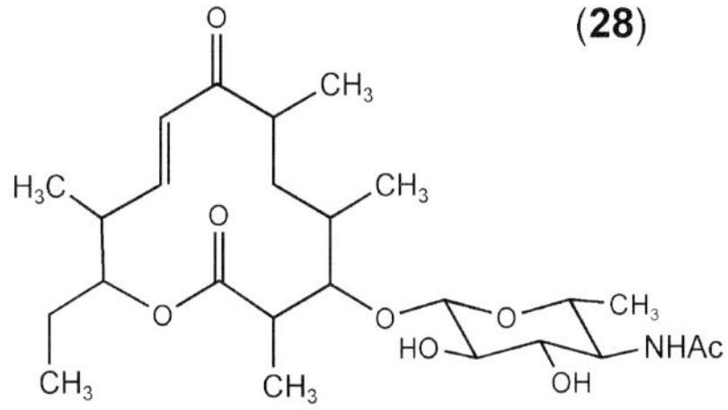

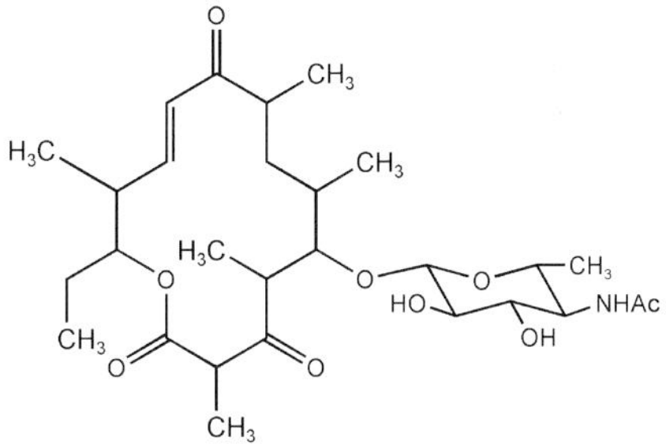

(**29**)

(**30**)

**Figure 8.8** Methymycin/pikromycin derivatives generated by gene inactivation.

(Figure 8.9, **31–34**). Formation of these compounds resulted of the action of the 3,5-epimerase StrM followed by the reductase StrL on 4-keto-6-deoxy-D-glucose, which is the compound accumulated by the *desI* mutant [43]. In a similar way, a mutant of *S. venezuelae* was generated by deleting the entire D-desosamine cluster and this mutant was used as host for the heterologous expression of genes involved in the biosynthesis of an intermediate sugar (TDP-4-keto-6-deoxy-D-glucose) or an exogenous sugar (TDP-D-olivose) from the oleandomycin and urdamycin pathways. In this way new macrolide derivatives containing quinovose or olivose were generated (Figure 8.10, **35–38**) [44].

The DesVII desosaminyl transferase naturally transfers D-desosamine to both 12- and 14-membered macrolactones, which is an example of inherent flexibility towards the acceptor aglycone. This unusual property of DesVII has been exploited to generate novel macrolide derivatives that differ in size of the macrolactone ring and in their oxygenation/reduction state. A *Streptomyces lividans* strain in which the genes involved in the biosynthesis and transfer of D-desosamine were

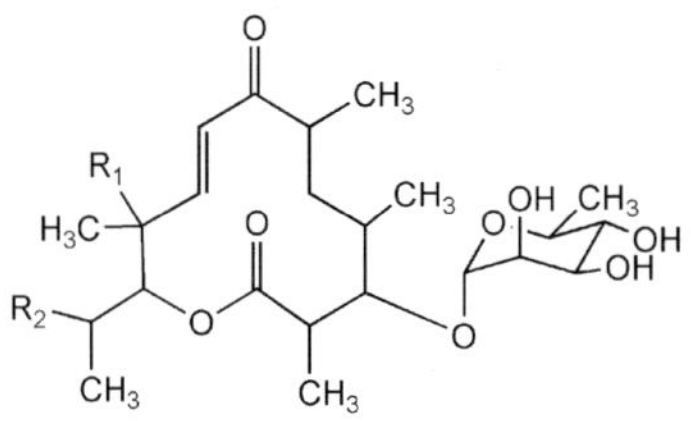

$R_1$=OH $R_2$=H (**31**)
$R_1$=H $R_2$=OH (**32**)
$R_1$=H $R_2$=H (**33**)

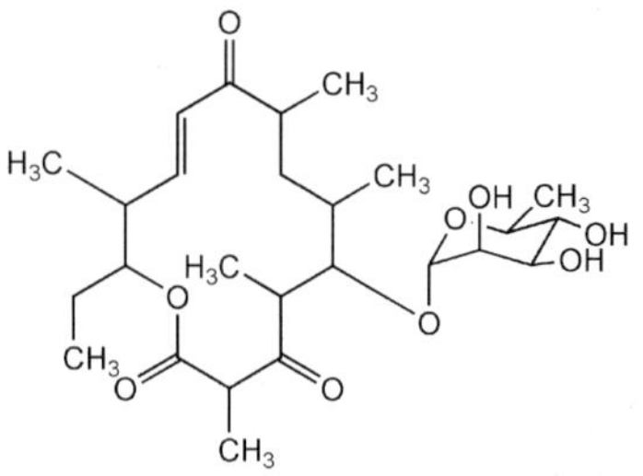

(**34**)

**Figure 8.9** Methymycin/pikromycin derivatives obtained by combining gene inactivation and gene expression.

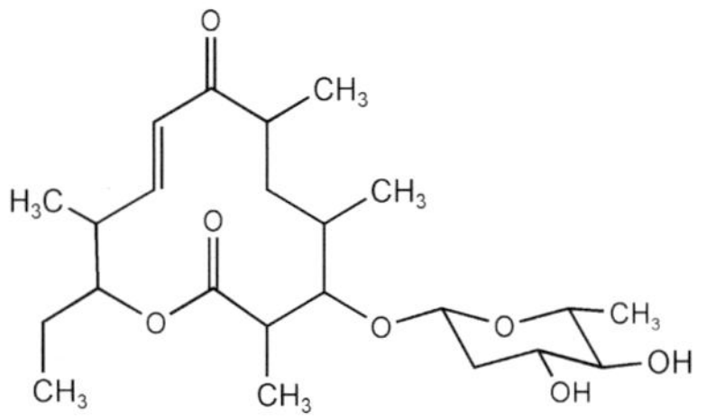

(**35**) 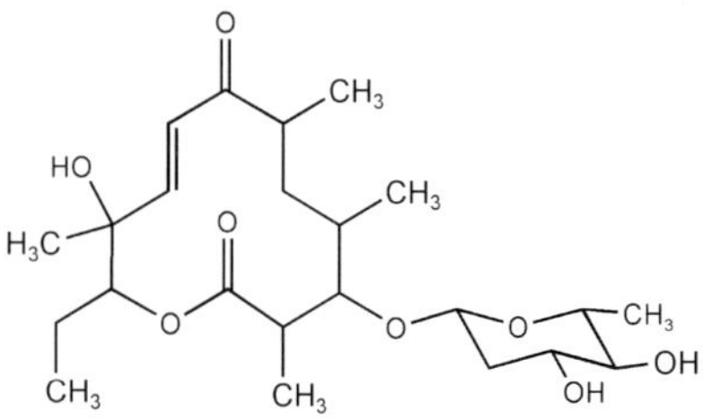 (**36**)

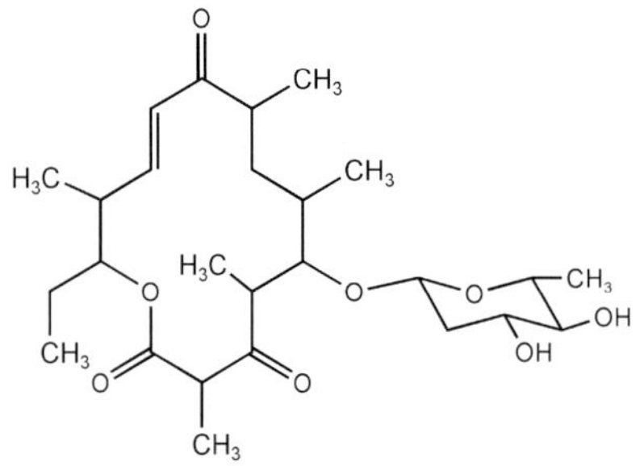

(**37**) 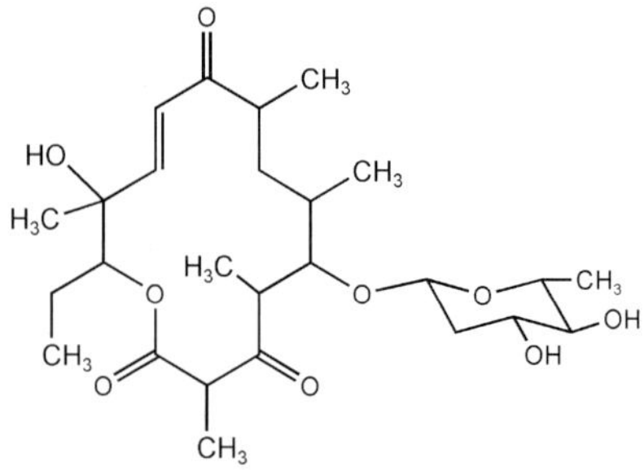 (**38**)

**Figure 8.10** Methymycin/pikromycin derivatives obtained by combining gene inactivation and expression of sugar plasmids.

integrated into its chromosome was transformed with a library of expression plasmids encoding genetically modified PKS [45]. More than 20 different 14-membered desosaminylated macrolides were produced by this recombinant strain based on the versatility of the DesVII desosaminyltransferase to recognize different acceptor substrates. In a similar way, hybrid PikA, TylG and DEBS (deoxyerythronolide B synthase) modular PKS systems engineered at the beginning or end of the PKS complex in *S. venezuelae* resulted in the production of multiple 12- and 14-membered biologically active macrolides [46].

### 8.5.2 Aureolic Acid Group

Mithramycin (**39**) and chromomycin $A_3$ (**40**) (Figure 8.11) are two antitumor drugs of the aureolic acid family produced by *Streptomyces argillaceus* and *Streptomyces griseus*, respectively [47]. Structurally, they consist of a tricyclic chromophore (aglycone) with two aliphatic side-chains attached at C3 and C7. They are glycosylated by a disaccharide and a trisaccharide chains. However, mithramycin (**39**) and chromomycin $A_3$ (**40**) show different glycosylation patterns. While chromomycin $A_3$ (**40**) contains a trisaccharide of D-olivose, D-olivose and 4-*O*-acetyl-L-chromose B, and a disaccharide of 4-*O*-acetyl-D-oliose and 4-*O*-methyl-D-oliose, mithramycin contains a trisaccharide of D-olivose, D-oliose and D-mycarose, and a disaccharide of D-olivose, attached at positions 2 and 6 of the aglycone, respectively (Figure 8.11). These compounds inhibit growth and multiplication of several tumor cell lines. These antitumor properties are ascribed to their inhibitory effects on replication and transcription processes during macromolecular biosynthesis by interacting, in the presence of $Mg^{2+}$, with GC-rich nucleotide sequences located in the minor groove of DNA. They were also found to stimulate K562 cell erythroid

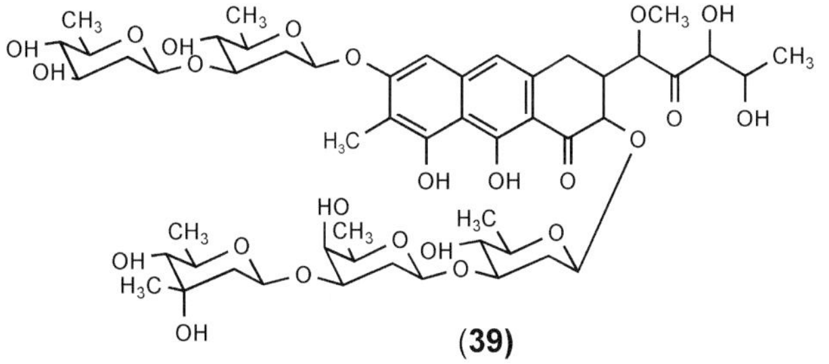

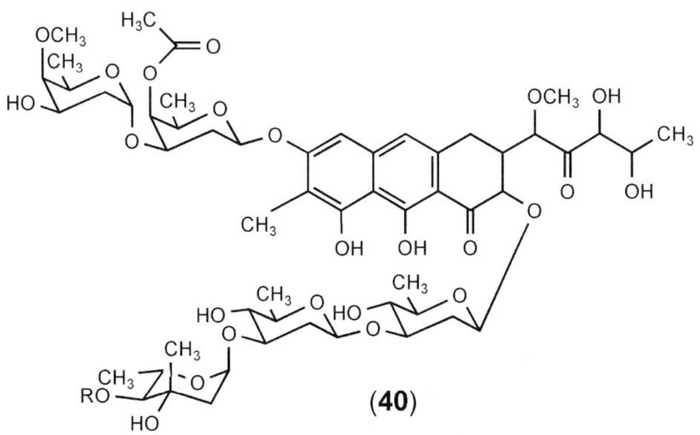

**Figure 8.11** Chemical structures of selected aureolic acid compounds.

differentiation, to be potent inhibitors of neuronal apoptosis and to have antiviral activity against human immunodeficiency virus type 1 [48]. The mithramycin and chromomycin biosynthetic gene clusters have been cloned and sequenced, and genes involved in DOH biosynthesis and transfer have been identified. In the case of mithramycin, three genes (*mtmD, mtmE* and *mtmV*) have been involved in the first three common biosynthetic steps of all DOH. In addition, *mtmU* was specifically involved in D-oliose biosynthesis, *mtmC* in D-mycarose and in D-olivose, and *mtmTIII* in D-olivose [7, 13a]. Four glycosyltransferase genes (*mtmGI–mtmGIV*) have been identified and assigned to specific glycosylation steps [49]. In the chromomycin pathway, four genes (*cmmD, cmmE, cmmV* and *cmmW*) are involved in the biosynthesis of all three DOHs. In addition, *cmmUI* participates in the biosynthesis of D-olivose, *cmmUII* in D-oliose, *cmmC, cmmF* and *cmmUIII* in L-chromose B, and *cmmMIII* and *cmmA* in tailoring modification of deoxy sugars [50]. Also, four glycosyltransefase genes (*cmmGI–cmmGIV*) have been identified and assigned to different glycosylation steps [51].

Several derivatives of mithramycin were produced containing different modified sugars by inactivating specific sugar biosynthesis genes [7]. Thus, by inactivating the *mtmC* gene, which codes for a *C*-methyltransferase involved in D-mycarose biosynthesis, three novel compounds were obtained: 4A-ketopremithramycin A2 (**41**), 4A-keto-9-demethylpremithramycin A2 (**42**) and 4C-keto-demycarosylmithramycin (**43**) (Figure 8.12). Interestingly, all three compounds lack the D-mycarose moiety and possess an unexpected 4-keto sugar moiety instead of D-olivose as the first sugar of the trisaccharide chain. While the absence of the D-mycarose moiety in all three new compounds was in agreement with the function for MtmC as the $C^3$-methyltransferase necessary for the formation of the D-mycarose unit, the appearance of a hydrated keto group at the 4-position of the first sugar of the lower saccharide chain (normally a D-olivose) was unexpected, indicating that a sugar 4-ketoreductase was also affected. It was speculated that the modification of the spatial structure of the protein cluster by inactivation of *mtmC* might be the underlying cause by which the 4-ketoreductase was rendered inactive. In a similar way, inactivation of *mtmTIII* led to the production of a new compound, 4E-ketomithramycin (Figure 8.12, **44**), which revealed that the corresponding enzyme is involved in the 4-ketoreduction during D-mycarose biosynthesis. Also, inactivation of *mtmOIV* gene, which codes for an oxygenase catalyzing the Baeyer–Villiger oxidative cleavage of the fourth ring of the tetracyclic intermediate premithramycin B, led to the production of two new compounds, premithramycin B and 3A-deolivosylpremithramycin B (**45**) [49c]. The latter compound showed a different glycosylation pattern than mithramycin, lacking one olivosyl residue of the disaccharide chain (Figure 8.12).

Inactivation of the different glycosyltransferase genes also led to the production of novel compounds. Thus, mutations in *mtmGI* or *mtmGII* lead to the production of four novel compounds, premithramycin A1 (**46**), premithramycin A2 (**47**), premithramycin A3 (**48**), all of them with a tetracyclic aglycone and differing in the number of sugars attached to the aglycone, and premithramycin A4 (**49**), which shows the fourth ring opened and the trisaccharide chain (Figure 8.13) [49a, 9b].

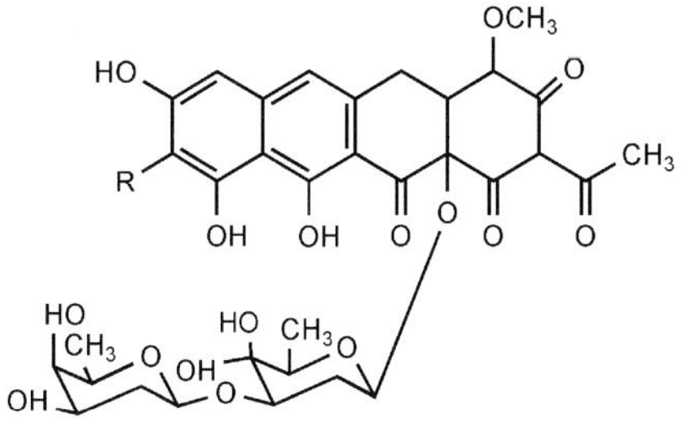

R=$CH_3$: (**41**)
R=H: (**42**)

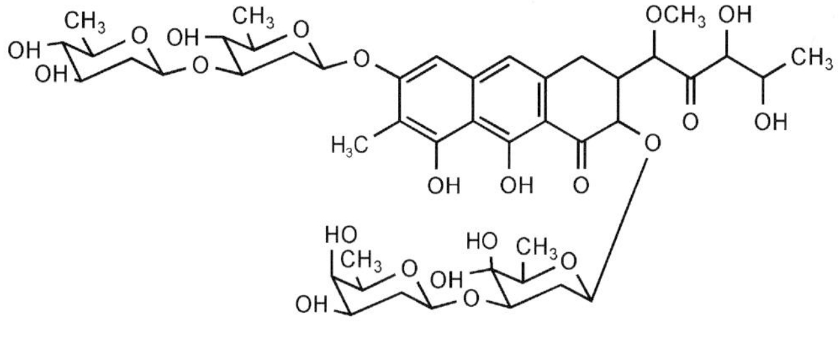

(**43**)

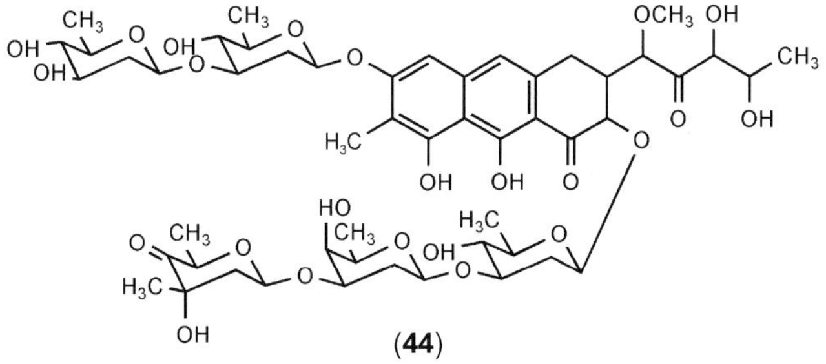

(**44**)

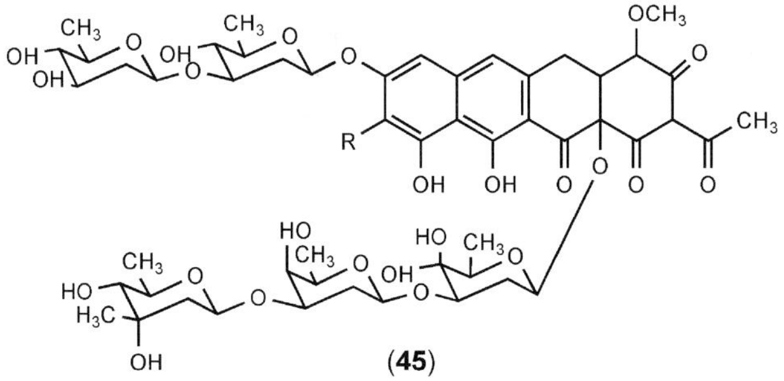

(**45**)

**Figure 8.12** Mithramycin derivatives obtained by inactivation of sugar genes.

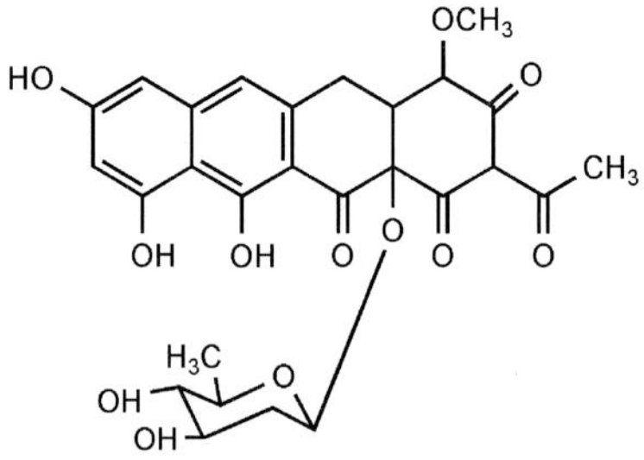

(46)

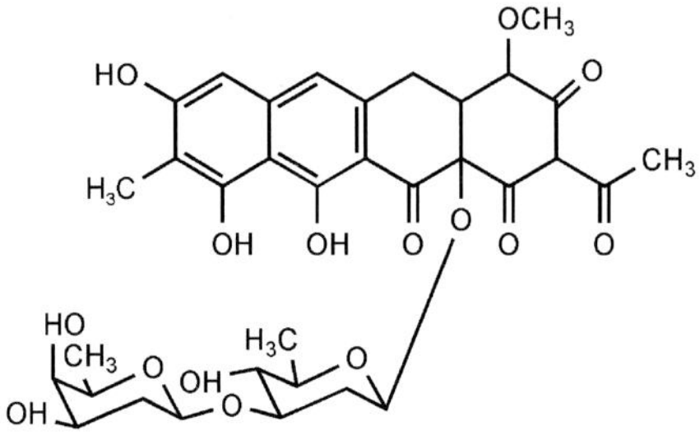

(47)

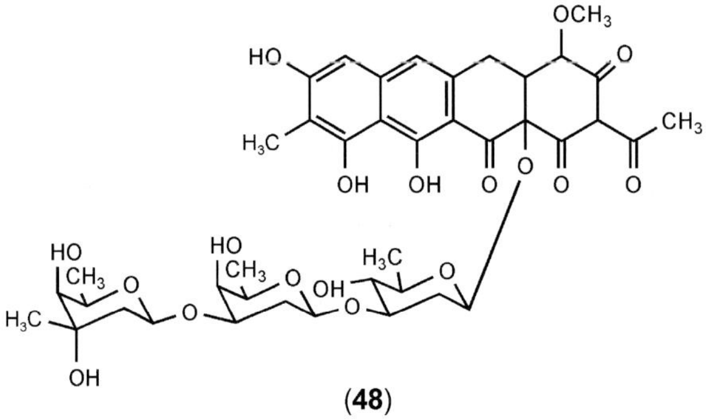

(48)

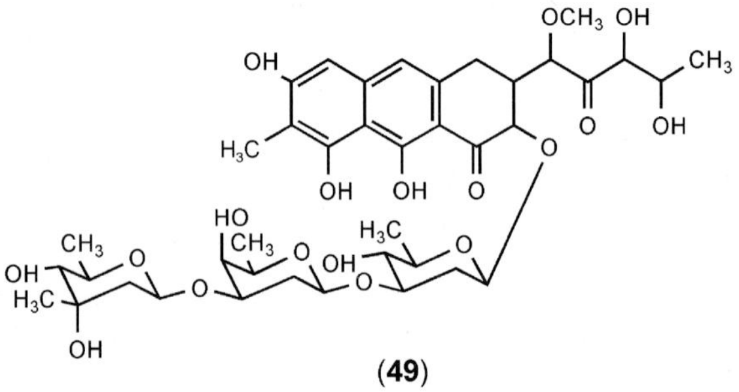

(49)

**Figure 8.13** Mithramycin derivatives obtained by inactivation of glycosyltransferase genes.

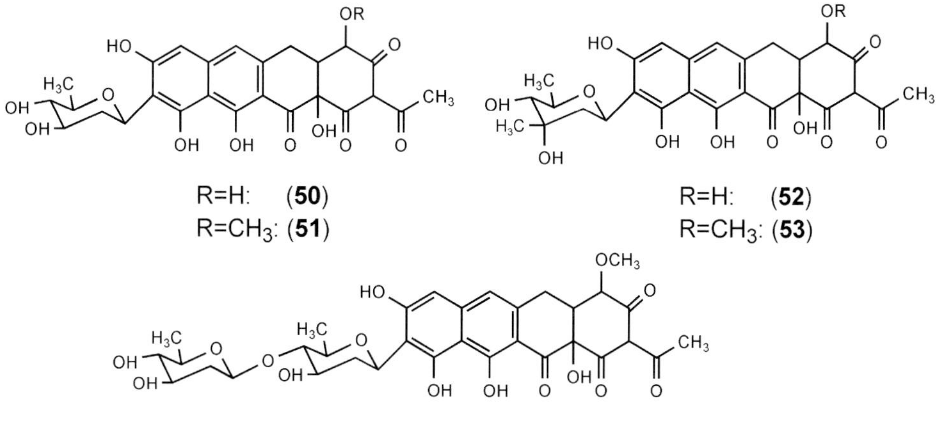

R=H: (**50**)
R=$CH_3$: (**51**)

R=H: (**52**)
R=$CH_3$: (**53**)

(**54**)

**Figure 8.14** Premithramycin derivatives obtained by combining gene inactivation and gene expression.

Mutants affected in *mtmGIII* also produced premithramycin A1 (**46**) [49a]. Inactivation of *mtmGIV* conducted to the formation of premithramycinone, a non-glycosylated tetracyclic compound [49a, 52]. This mutant and another lacking all glycosyltransferase and methyltransferase genes were used as hosts to express glycosyltransferase *urdGT2*. This glycosyltransferase is involved in the biosynthesis of the angucycline urdamycin A, attaching a D-olivose moiety to the urdamycin aglycone through a *C*-glycosidic linkage [53]. The expression of the *urdGT2* gene caused the formation of novel *C*-glycosylated premithramycin-type molecules containing D-olivose or D-mycarose sugar moieties: 9-*C*-olivosyl-4-demethyl-premithramycinone (**50**), 9-*C*-olivosylpremithramycinone (**51**), 9-*C*-mycarosyl-4-demethylpremithramycinone (**52**) and 9-*C*-mycarosylpremithramycinone (**53**) (Figure 8.14). Moreover, by coexpressing the glycosyltranferase LanGT1 from the landomycin pathway, the sugar chain was enlarged by the incorporation of another D-olivose moiety to the saccharide chain. Thus, the final compound, 9-*C*-olivo-1-4-olivosyl-premithramycinone (**54**) resulted from the use of genes from three different biosynthesis pathways (Figure 8.14) [54].

In the case of chromomycin, several novel glycosylated compounds were generated by inactivating specific glycosyltransferases [51]. Thus, by inactivating *cmmGII*, two new tetracyclic compounds lacking the disaccharide chain: prechromomycin A2 (**55**) and prechromomycin A3 (**56**) were obtained (Figure 8.15). On the other hand, inactivation of *cmmGI* conducted to the production of three tricyclic compounds lacking the first sugar of the disaccharide chain: prechromomycin A4 (**57**), 3A-*O*-acetyl-prechromomycin A4 (**58**) and 4A-*O*-deacetyl-3A-*O*-acetyl-prechromomycin A4 (**59**) (Figure 8.15). Three sugars in chromomycin $A_3$ are further modified as late steps of the biosynthesis. One of these modifications is a methylation step catalyzed by the CmmMIII *O*-methyltransferase. Interestingly,

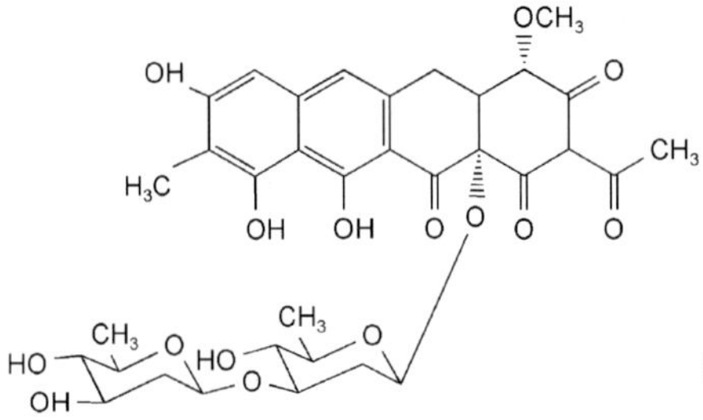

(55)

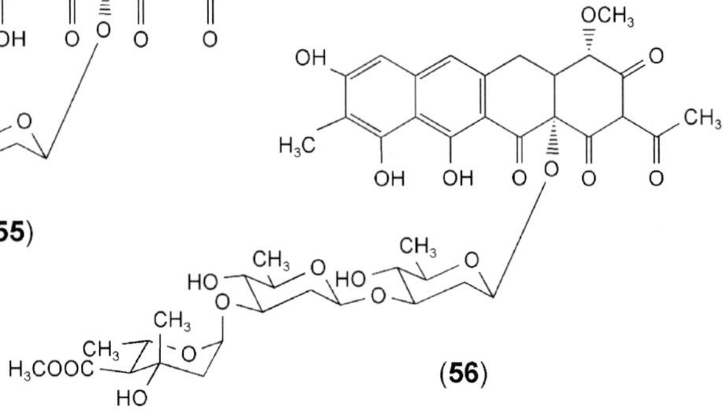

(56)

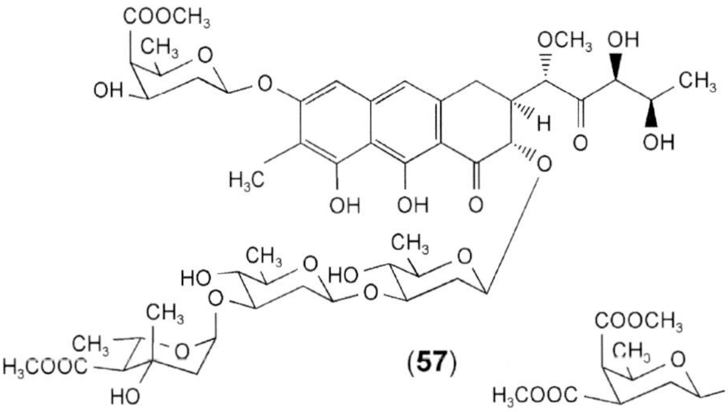

(57)

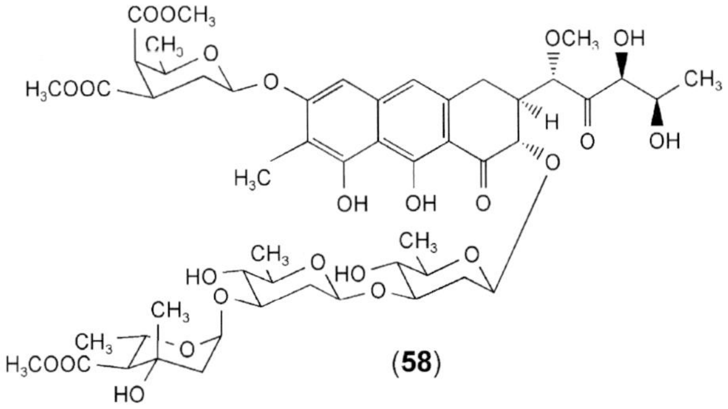

(58)

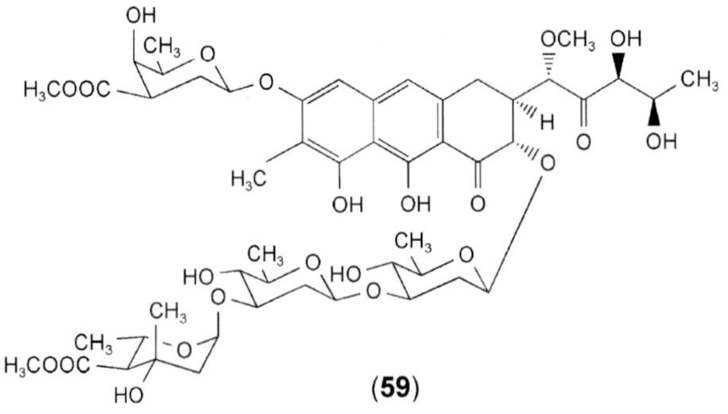

(59)

**Figure 8.15** Chromomycin derivatives obtained by inactivation of glycosyltransferase genes.

the other two modifications are acylations but only one gene *cmmA*, coding for an acyltransferase is responsible for both acylation events. By inactivation of the methyltransferase *cmmMIII*, the novel compounds 4B-*O*-demethyl-4B-*O*-acetyl-4A-*O*-deacetylchromomycin $A_3$ (**60**), 4B-*O*-demethyl-chromomycin $A_3$ (**61**) and 4B-*O*-demethyl-chromomycin $A_2$ (**62**) were produced. All these compounds lacked the methyl group in D-oliose and differ in the number, position and type of acyl group present in D-oliose and L-chromose B (Figure 8.16). Similarly, by inactivating the acetyltransferase *cmmA*, compound 4A,4E-*O*-dideacetyl-chromomycin $A_3$ (**63**) was generated, which lacked the two acetyl groups normally present in sugars A and E of chromomycin $A_3$ (Figure 8.16) [50b]. CmmA is a membrane-associated protein. Recent experiments demonstrate that this enzyme can use various acyl group donors (malonyl-, isobutyril- or propionyl-CoA) to modify the sugars (B. García *et al.*, unpublished) and, consequently, it could be an important genetic tool for modifying sugars in bioactive natural products and therefore contributing to increase structural diversity.

### 8.5.3 Angucyclines

Angucyclines constitute a group of compounds chemically characterized by a tetracyclic angular benz[*a*]anthraquinone ring system. Some of them are glycosylated, such as urdamycin A (**64**) or landomycins A (**65**) and E (**66**), but others are not, such as tetrangomycin (**67**) or oviedomycin (**68**) (Figure 8.17). A broad range of biological activities has been ascribed to members of this family of compounds, including antitumor, antibacterial, platelet aggregation inhibition, vincristrine activity potentiation, endothelin receptor antagonism, inducible nitric oxide synthase inhibition, enzymatic inhibitory effects on hydroxylases and protection against HCl-induced gastric lesions.

Urdamycin A is an angucycline produced by *S. fradiae*, which contains a *C*-glycosidically linked D-olivose and three *O*-glycosidically linked deoxy sugars: two L-rhodinoses and one D-olivoses [55]. The biosynthetic gene cluster for this antibiotic has been cloned and characterized and genes involved in the biosynthesis of all three sugars (*urdG, urdH, urdS* and *urdT*), in D-olivose (*urdR*) and L-rhodinose (*urdQ, urdZ1* and *urdZ3*) have been identified, as well as those coding for the different glycosyltransferases (*urdGT1a, urdGT1b, urdGT1c* and *urdGT2*) [53a, 56].

Inactivation of sugar biosynthesis genes produced glycosylated derivatives of urdamycin containing modified sugars. Thus, inactivation of *urdR*, which catalyzes the final step in TDP-D-olivose biosynthesis, led to the formation of the novel urdamycin derivatives urdamycin M (**69**), urdamycin R (**70**) and urdamycin S (**71**) possessing either L- or D-rhodinose attached at the 9-position instead of D-olivose (Figure 8.18) [56a, 57]. Formation of the new DOH D-rhodinose was explained because the *urdR*-minus mutant is unable to proceed with D-olivose biosynthesis accumulating the 4-keto intermediate, which can be deoxygenated at C3 by the 3-dehydratase UrdQ and apparently further reduced at the C4 keto group by UrdZ3 generating D-rhodinose.

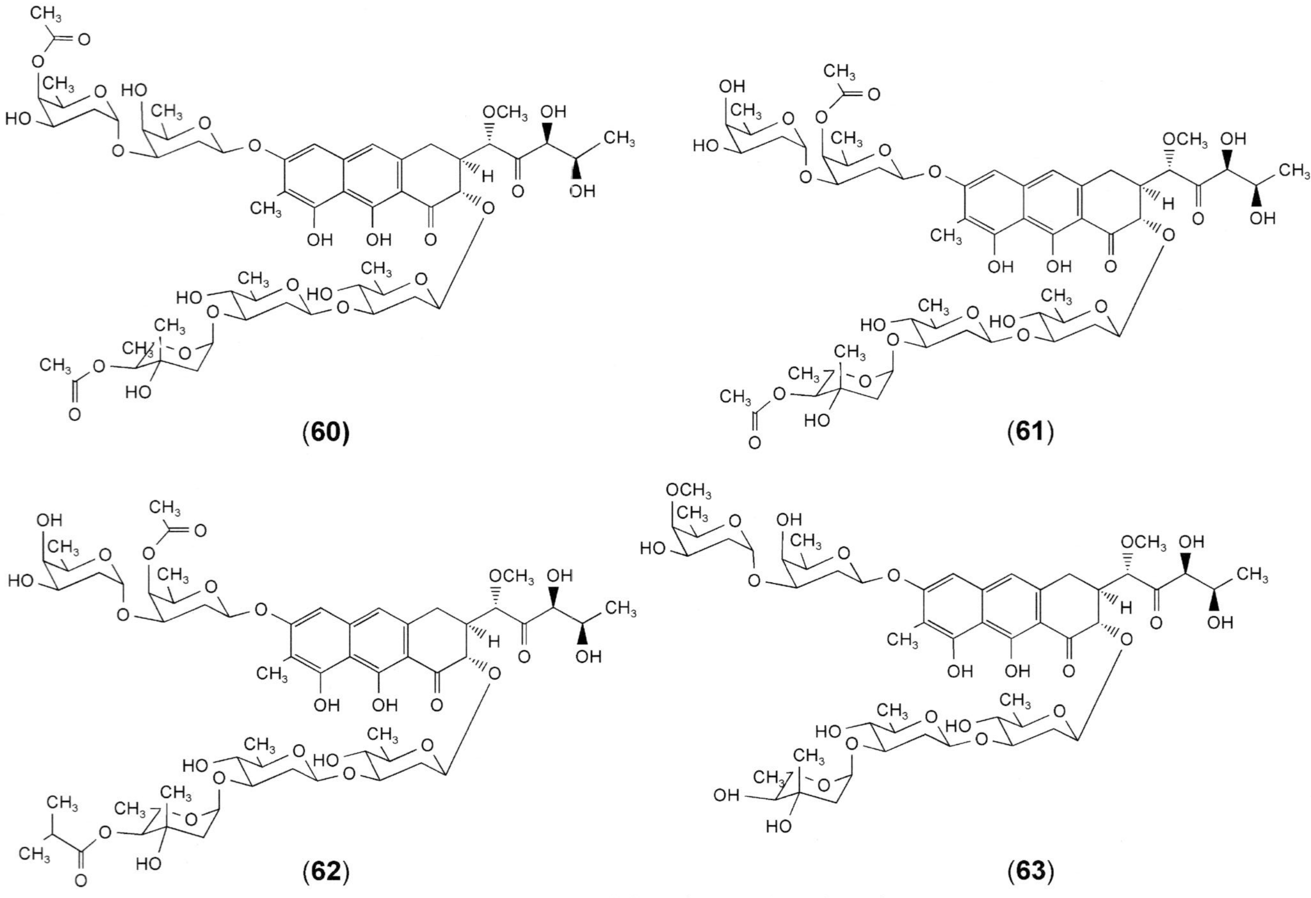


**Figure 8.16** Chromomycin derivatives generated by inactivating methyltransferase and acetyltransferase genes.

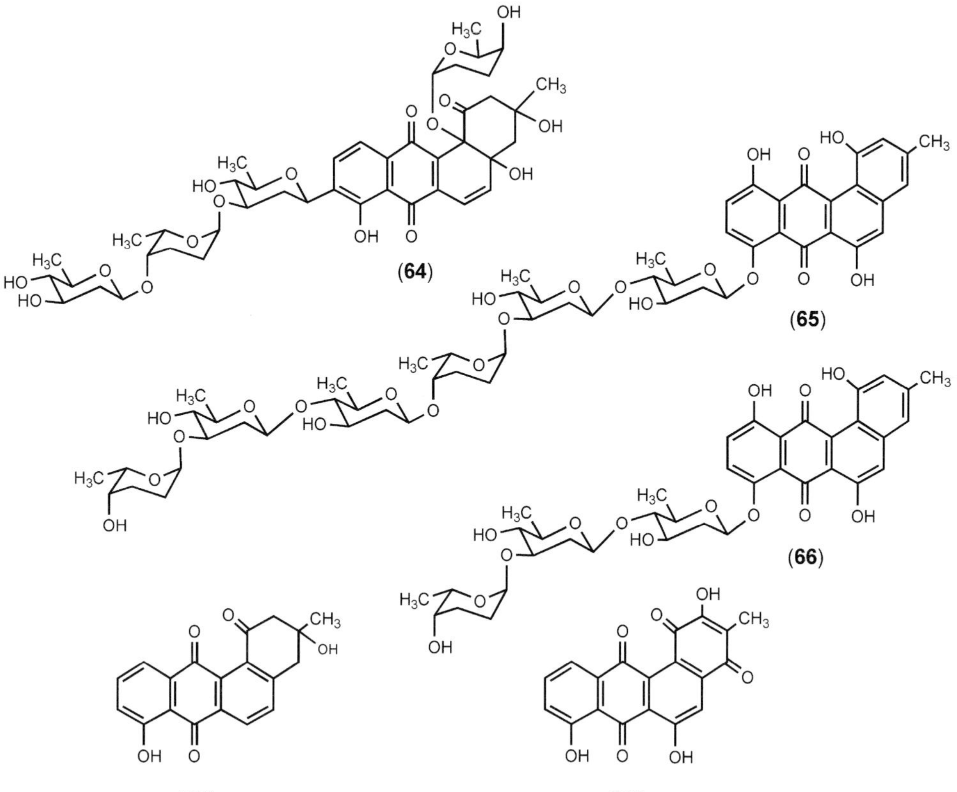

(64) (65) (66) (67) (68)

**Figure 8.17** Chemical structures of selected angucyclines.

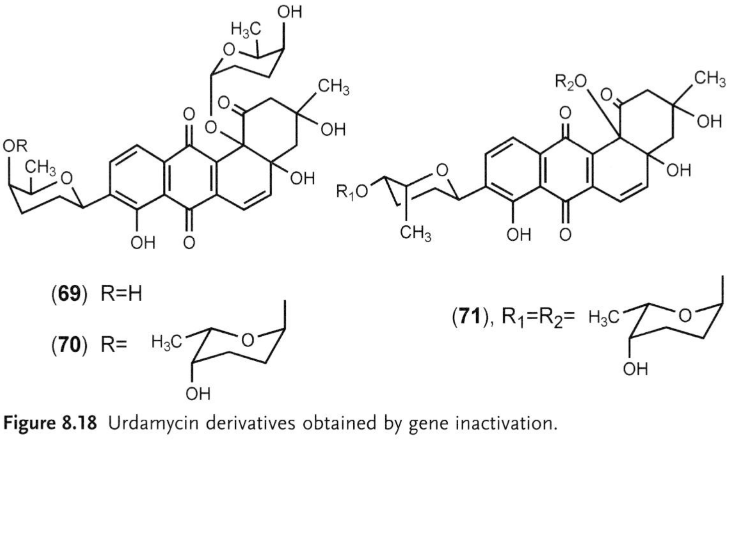

(69) R=H

(70) R=

(71), $R_1$=$R_2$=

**Figure 8.18** Urdamycin derivatives obtained by gene inactivation.

Expression of glycosyltransferase genes into mutants affected in one or several glycosyltransferase genes was used as a strategy to produced new urdamycin derivatives. By expressing the UrdGT1c glycosyltransferase into two different mutants affected in one or several glycosyltransferase genes some new glycosylated urdamycin derivatives were produced. The expression of UrdGT1c in mutant Ax, in which three glycosyltransferase genes had been deleted, resulted in the formation of the new compound urdamycin N (**72**), a urdamycin B analog with a L-rhodinose instead of a D-olivose as the terminal sugar of the *C*-glycosidically bound trisaccharide chain (Figure 8.19). Also, expression of *urdGT1c* in mutant 16-14, which is defective for both *urdGT1b* and *urdGT1c* glycosyltransferases, but carries instead a chromosomal hybrid thereof, resulted in the production of the new compound urdamycin O (**73**), which is an urdamycin A analog with a terminal α-L-rhodinose instead of a β-D-olivose in the trisaccharide chain (Figure 8.19) [56a].

UrdGT1b and UrdGT1c show different specificities for both nucleotide sugar and acceptor substrate, but they share 91% identical amino acids. A region comprising 10 amino acids is potentially involved in substrate specificity. By combining codons corresponding to this region from both glycosyltransferase genes, several hybrid genes were generated, some of them showed novel specificities. One of these hybrids was responsible for the biosynthesis of the novel urdamycin P (**74**) that carries a branched saccharide side-chain (Figure 8.19) [58].

The landomycins are a subgroup of the large family of angucycline compounds characterized by possessing a unique phenylglycoside moiety in their structures.

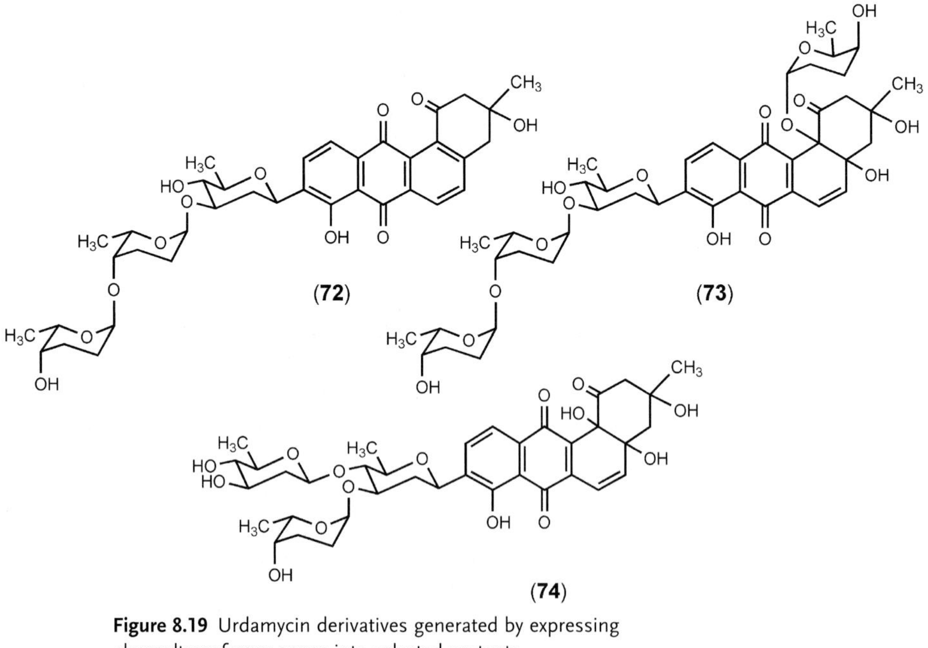

(**72**) (**73**) (**74**)

**Figure 8.19** Urdamycin derivatives generated by expressing glycosyltransferase genes into selected mutants.

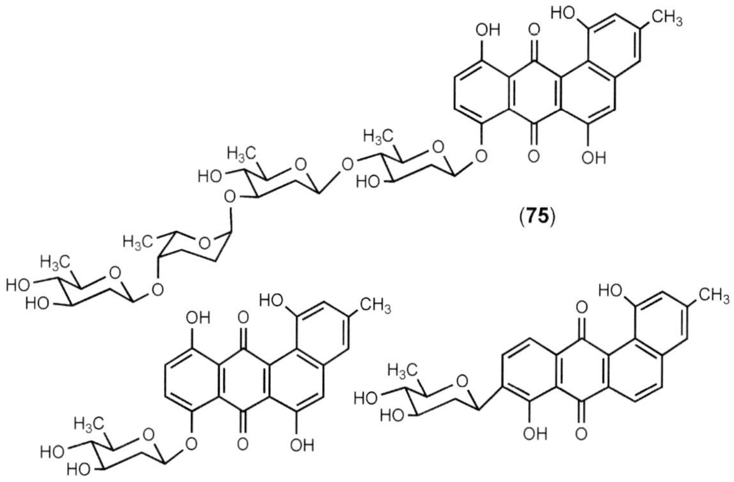

(75)

(76) (77)

**Figure 8.20** Landomycin derivatives generated by expressing glycosyltransferase genes into *S. cyanogenus*.

All share the same aglycone and differ in the saccharide chain. They show diverse biological activities, such as antitumor, antibacterial and enzyme inhibitory activity. The landomycin A (**65**), produced by *Streptomyces cyanogenus* S136 [59], and the landomycin E (**66**), produced by *Streptomyces globisporus* 1912 [60], gene clusters have been cloned and sequenced. The clusters are very similar, both at the level of gene organization and nucleic acid sequence.

By expressing the glycosyltransferase LanGT3 in *S. cyanogenus* two new landomycins were generated, landomycin J (**75**) and landomycin I (**76**), containing a tetrasaccharide and a monosaccharide, respectively (Figure 8.20). The occurrence of these new landomycins was explained by sugar donor substrate depletion of LanGT1, which slows down the second glycosylation step catalyzed by this enzyme and simultaneously accelerates the fourth glycosylation step catalyzed by LanGT3 [61]. Also, a novel *C*-glycosylated angucycline, 9-*C*-D-olivosyl-tetrangulol (Figure 8.20, **77**), was produced by heterologous expression of glycosyltransferase UrdGT2 from *S. fradiae* Tu2717 in a mutant of *S. cyanogenus* affected in glycosyltransferase lanGT2 [62].

On the other hand, two new landomycin derivatives, landomycin F (**78**) and landomycin H (**79**) (Figure 8.21), were produced in a *S. globisporus* mutant strain by inactivating *lndGT4*. A third new compound (landomycin G, **80**) (Figure 8.21) was generated after complementation of this mutant with the corresponding glycosyltransferase gene. All three new compounds lacked the 11-OH group [63]. Moreover, by combining genes from the landomycin E cluster of *S. globisporus* and the glycosyltransferase gene *urdGT2* from the urdamycin cluster, three novel prejadomycin analogs were produced (Figure 8.21) [64]. To do that, the *S. globisporus* ΔlndE mutant was used, which is affected in an oxygenation step of landomycin

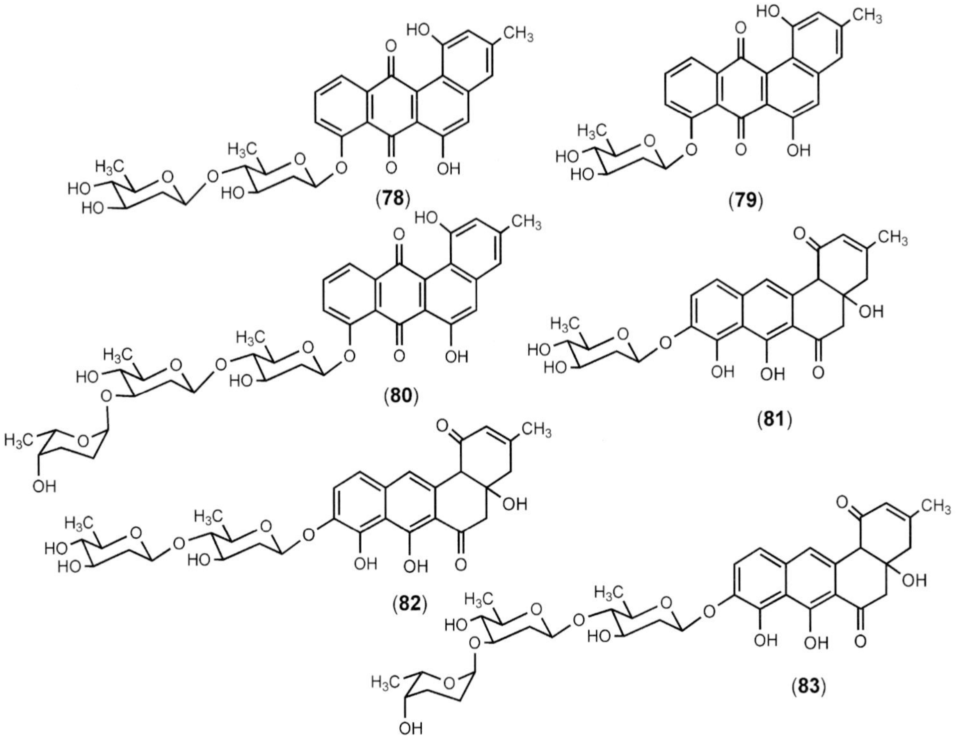

**Figure 8.21** Landomycin derivatives obtained by expressing glycosyltransferase genes into *S. globisporus* mutants.

E and accumulates prejadomycin (2,3-dehydro-UWM6). Expression in this mutant of UrdGT2 caused the production of 9-*C*-β-D-olivosylprejadomycin (**81**), 9-*C*-β-D-olivosyl-1,4-β-D-olivosylprejadomycin (**82**) and 9-*C*-α-L-rhodinosyl-1,3-β-D-olivosyl-1,4-β-D-olivosylprejadomycin (**83**) (Figure 8.21), which differ in the *C*-glycosydically bound moieties attached at C9. The sugar residue and oligosaccharide moieties are the same as those previously found in landomycins H, D and E, but they are attached *C*-glycosidically at C9 instead of *O*-glycosidically at C8-O, a position shift that was expected from using UrdGT2. The glycosyltransferases responsible for the elongation to the landomycin E trisaccharide, LndGT4 and LndGT1, could attach their sugars to the corresponding intermediates, although the aglycone was structurally quite different and the first sugar unit was *C*-attached and at different position.

### 8.5.4 Anthracyclines

Anthracyclines (Figure 8.22) are an important group of antitumor compounds that are characterized by a basic structure of 7,8,9,10-tetrahydro-5,12-naphthacene

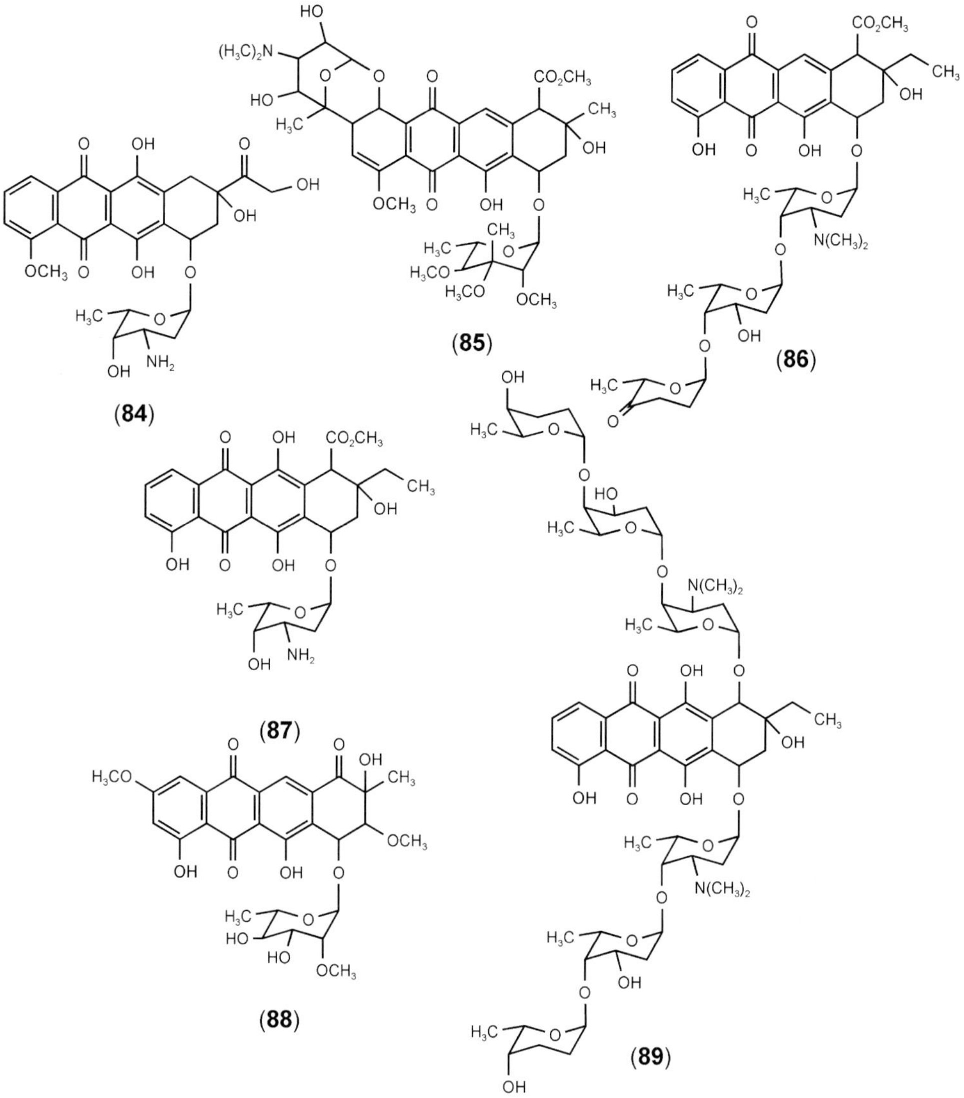

(84) (85) (86) (87) (88) (89)

**Figure 8.22** Chemical structures of several anthracyclines.

quinone. Most anthracyclines are glycosylated by one or more sugar moieties, which form part of one or two saccharide chains. These sugars are generally L-daunosamine in daunorubicin/doxorubicin and carminomycin groups, and L-rhodosamine, 2-deoxy-L-fucose and L-cinerulose in the aclacinomycin, pyrromycin and rhodomycin groups. Nogalamine and L-nogalose are characteristic of the nogalamycin group. Anthracyclines as doxorubicin are known for their complex mechanism of cytotoxicity involving inhibition of enzymes such as DNA topoisomerase II, RNA polymerase, cytochrome *c* oxidase, intercalation into DNA, chelation of iron and generation of reactive oxygen species, and finally induction of apoptosis. The daunorubicin (**84**) gene cluster has been cloned, sequenced and characterized from *Streptomyces peucetius* and genes involved in L-daunosamine, the sugar component of daunorubicin, have been identified [65]. Also, the nogalamycin (**85**), aclacinomycin (**86**), rhodomycin (**87**), steffimycin (**88**) or cosmomycin (**89**) gene clusters have been cloned and characterized from *Streptomyces nogalater, S. galilaeus, Streptomyces purpurascens, Streptomyces steffisburgensis* and *Streptomyces olindensis*, respectively [8a, 66].

By inactivating the glycosyltransferase gene *cosG*, in the cosmomycin producer *S. olindensis*, a novel compound was produced. This compound showed the characteristic cosmomycin aglycone, but containing L-rhodinose residues attached to C2 and C17 instead of L-rhodosamine (Figure 8.23, **90**) [66a]. On the other hand, different novel anthracycline derivatives have been produced by expressing genes from an anthracycline producer in different streptomycetes. Thus, when plasmid

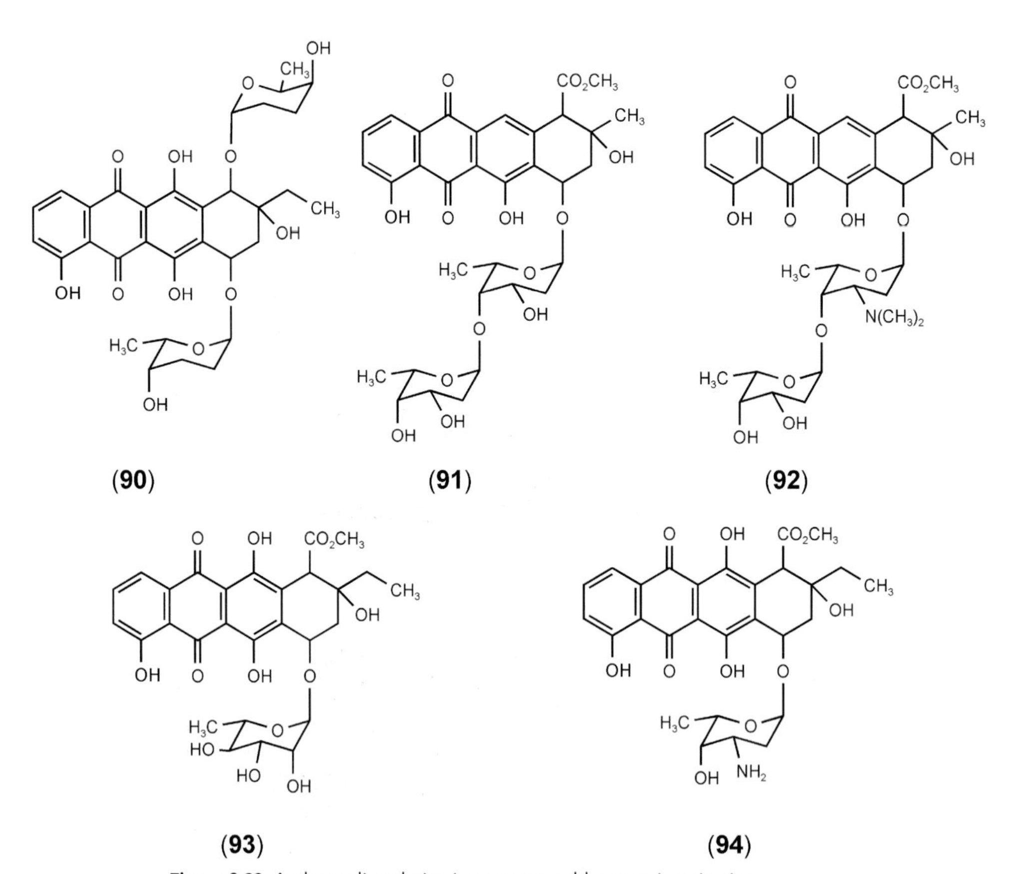


**Figure 8.23** Anthracycline derivatives generated by gene inactivation or by combining gene inactivation and gene expression.

pSY15 containing several nogalamycin genes (including the PKS genes) from *S. nogalater* was introduced into the *S. galilaeus* mutant H028 (this mutant is blocked at an early stage of aclacinomycin biosynthesis before the first stable intermediate is formed) several hybrid compounds were formed, including several glycosylated auramycinone derivatives containing different combinations of sugars (2-deoxy-L-fucose, L-rhodosamine, L-rhodinose) (Figure 8.23, **91** and **92**) [67]. Moreover, by expression of DNA fragments from *S. purpurascens* (rhodomycins producer) in *S. galilaeus* (aclacinomycin producer), several novel compounds were produced. The structures of the aglycone of two of these compounds were 10-demethoxycarbonylaklavinone and 11-deoxy-β-rhodomycinone indicating that they were obtained as a consequence of the cooperative action of genes from *S. galilaeus* and *S. purpurascens*. Three 11-deoxy-β-rhodomycinone derivatives with different saccharide chains were isolated and found to posses cytotoxicity activity against L1210 mouse leukemia cells [68]. Also, some hybrids compounds, L-rhamnosyl-ε-rhodomycinone (**93**) and L-daunosaminyl-ε-rhodomycinone (**94**) (Figure 8.23), were produced by expressing several genes involved in glycosylation in *S. galilaeus* (aclacinomycin producer) into a daunomycin non-producing mutant of *Streptomyces peucetius* var. *cesius* (strain M18), obtained by chemical mutagenesis and affected in daunosamine biosynthesis [66c].

An outstanding and ingenious strategy was designed for the production of the antitumor drugs epirubicin (4′-epidoxorubicin) and 4′-epidaunorubicin. These are important cancer chemotherapeutic drugs, which are usually produced by low-yielding semisynthetic processes in which some chemical modification steps follow the fermentative production of doxorubicin and daunorubicin by *S. peucetius*. A shorter and more economic biosynthetic pathway was developed as an alternative to the semisynthetic procedure, by combining gene inactivation and gene expression [69]. A mutant was generated in the doxorubicin producer *S. peucetius* by inactivating the *dnmV* gene that encodes a 4-ketoreductase involved in the biosynthesis of L-daunosamine. Then, the *eryBIV* or *avrE* genes, both coding for 4-ketoreductases with different stereospecificity at C4 from that of the daunosamine ketoreductase, were expressed in this mutant. In this way, complementation of the *dnmV*-minus mutant with either of these genes changed the configuration of the 4-hydroxyl group, generating 4-epi-L-daunosamine and leading to the production of epirubicin.

Elloramycin (Figure 8.24, **95**) is an anthracycline-like compound produced by *Streptomyces olivaceus* Tü 2353. It belongs to a small group of naphthacenequinones characterized by the highly hydroxylated cyclohexenone moiety. The elloramycin gene cluster has been identified in cos16F4–a cosmid isolated from a *S. olivaceus* genomic library [70]. Further characterization of this cosmid showed that it contained all the genes necessary for the biosynthesis of elloramycin with the exception of genes involved in rhamnose biosynthesis, which are located outside the cluster (Ramos *et al.*, personal communication) [71]. Expressing cosmid 16F4 into different hosts generated new elloramycin compounds. When cos16F4 was expressed into a mutant of the urdamycin producer *S. fradiae* Tü2717, which lacks the PKS genes, it produced two new glycosylated compounds:

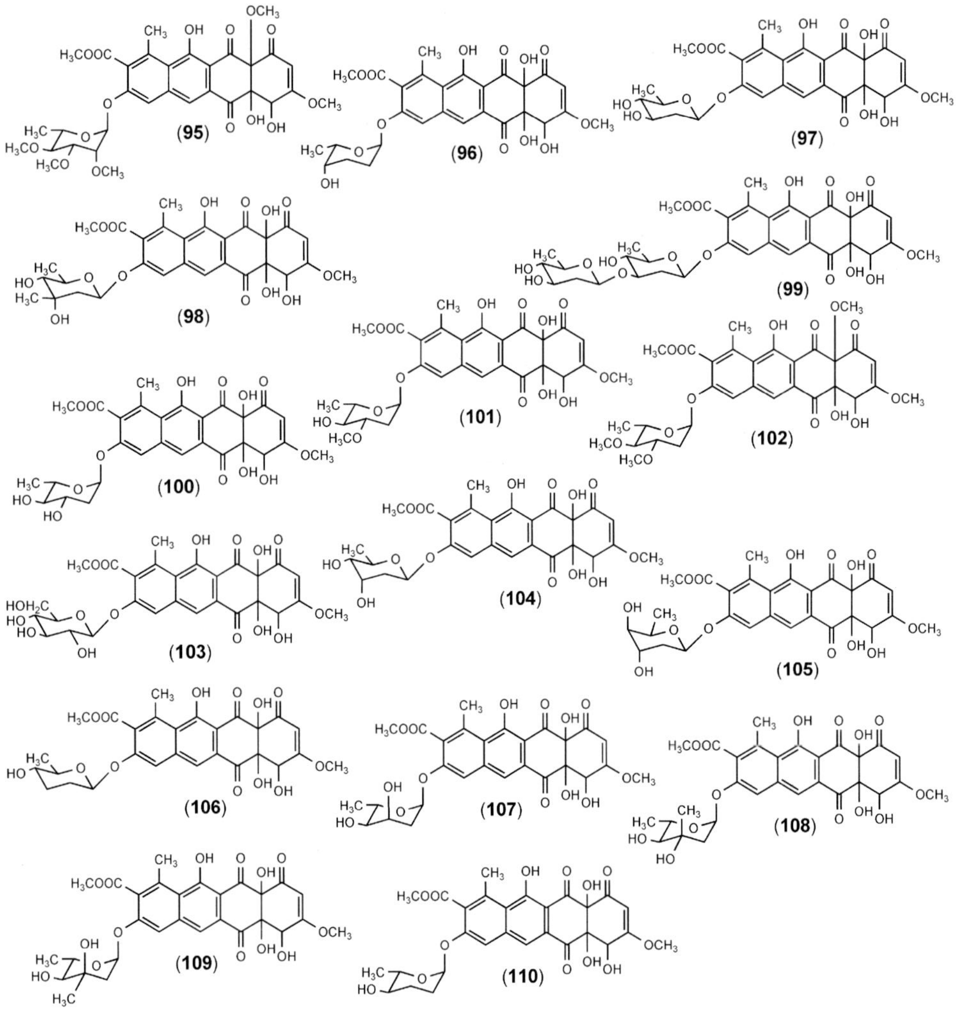

Figure 8.24 Chemical structure of elloramycin and derivatives obtained by combinatorial biosynthesis.

8-demethyl-8-α-L-rhodinosyl-tetracenomycin C (**96**) and 8-demethyl-8-β-D-olivosyl-tetracenomycin C (**97**) (Figure 8.24) [72]. Similarly, when cos16F4 was expressed into the producer of the antitumor mithramycin *S. argillaceus* ATCC 12956 three glycosylated compounds were formed: 8-demethyl-8-β-D-olivosyl-tetracenomycin C (**97**) (the same compound as obtained in the *S. fradiae* mutant), 8-demethyl-8-β-D-mycarosyl-tetracenomycin C (**98**) and 8-demethyl-β-D-diolivosyl-3′-1″-D-β-tetracenomycin C (**99**) (Figure 8.24) [73]. In addition, novel glycosylated elloramycin derivatives have been produced when cosmid 16F4 was used in combination with genes involved in the biosynthesis of L-oleandrose from the oleandomycin producer *S. antibioticus* [35]. In this way, the novel compounds L-olivosyl-tetracenomycin C (**100**), L-oleandrosyl-tetracenomycin C (**101**) and 2′-demethoxy-elloramycin (**102**) were produced (Figure 8.24) [74]. The new elloramycin compounds produced differed in the sugar moieties. Formation of all these glycosylated compounds raised the question about the origin of the gene encoding the glycosyltransferase responsible for transferring so many sugars. After using different experimental approaches, it was concluded that the elloramycin glycosyltransferase ElmGT was responsible for the formation of the glycosylated compounds, showing high substrate flexibility [71a]. The capability of ElmGT to transfer different deoxy sugars, therefore producing novel compounds, was broaden by using different sugar cassette plasmids, which direct the biosynthesis of various deoxy sugars. In this way a number different elloramycins were produced containing different sugars [D-glucose (**103**), D-digitoxose (**104**), D-boivinose (**105**), D-amicetose (**106**), L-digitoxose (**107**), L-mycarose (**108**), L-chromose B (**109**) and L-amicetose (**110**)], and including both D- and L-isomer forms of some of them (Figure 8.24) [50c, 75].

### 8.5.5 Indolocarbazoles

The indolocarbazole alkaloids and the related bisindolylmaleimides constitute an important class of natural products, which have been isolated from actinomycetes, cyanobacteria, slime molds and marine invertebrates [76]. They display a wide range of biological activities, including antibacterial, antifungal, antiviral, hypotensive, antitumor or neuroprotective properties. The antitumor and neuroprotective activities of indolocarbazoles are the result of one, or several, of the following mechanisms: (i) inhibition of different protein kinases, (ii) inhibition of DNA topoisomerases or (iii) direct DNA intercalation. Several indolocarbazoles or derivatives have entered clinical trials for the treatment of diverse types of cancer, Parkinson's disease or diabetic retinopathy [77]. Rebeccamycin (**111**) and staurosporine (**112**) are the two most representative members of the family (Figure 8.25). They inhibit DNA topoisomerase I and protein kinases, respectively. The gene clusters

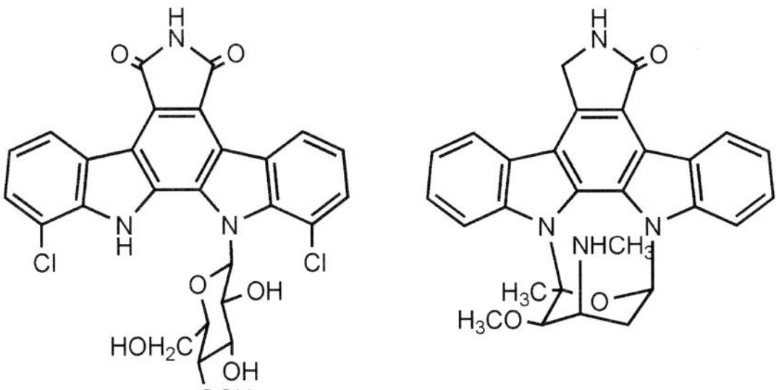

**Figure 8.25** Chemical structure of selected indolocarbazoles.

for these indolocarbazoles have been cloned and characterized by different laboratories [78]. Novel glycosylated derivatives of staurosporine have been produced by developing a two-plasmid system for staurosporine biosynthesis in the non-producer heterologous host *S. albus* [79]. One of these plasmids, the 'aglycone plasmid' was an integrative plasmid, and contained all genes necessary for the biosynthesis of the staurosporine aglycone together with the StaG glycosyltransferase and the StaN P450 monooxygenase, both enzymes required for the establishment of the double C—N linkage in staurosporine. The second plasmid, 'the sugar plasmid', was a replicative plasmid and harbored the genes involved in the biosynthesis of the L-ristosamine moiety. This double-recombinant strain produced staurosporine in good yields. When the 'sugar plasmid' was substituted by several plasmids directing the biosynthesis of different DOHs [32], novel glycosylated derivatives of staurosporine were produced containing either L-rhamnose (**113** and **116**), L-digitoxose (**114** and **117**), L-olivose (**115** and **118**) or D-olivose (**119**) attached to the aglycone (Figure 8.26), either through a single (**113–115**) or a

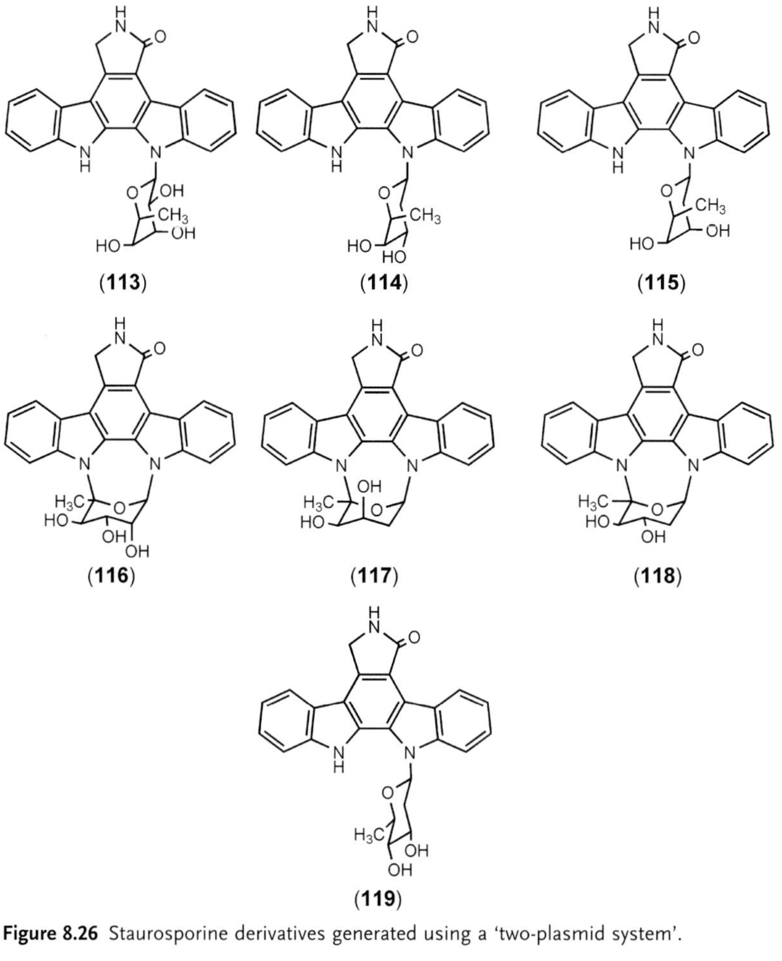

(**113**) (**114**) (**115**)

(**116**) (**117**) (**118**)

(**119**)

**Figure 8.26** Staurosporine derivatives generated using a 'two-plasmid system'.

double (**116–118**) C—N linkage, with the exception of D-olivose (**119**) that was only incorporated through a single C—N bond [79].

### 8.5.6
### Aminocoumarins

This group of antibiotics are potent inhibitors of bacterial gyrases. Structurally, these antibiotics are characterized for containing a 3-amino-4,7-dihydroxy-coumarin core that is attached to the sugar noviose, an unusual branched DOH with a 5,5-gem-dimethyl structure. The clorobiocin (Figure 8.27, **120**) gene cluster has been cloned and characterized, and genes involved in L-noviose biosynthesis have been identified [80]. By inactivating the *cloU*, which codes for a C-methyltransferase involved in L-noviose biosynthesis, a mutant in the clorobiocin gene cluster was created that produces two novel derivatives, novclobiocin 122 (**121**) and novclobiocin 123 (**122**), that contained a L-rhamnosyl moiety instead of the usual L-noviose sugar in clorobiocin (Figure 8.27) [81]. However, the yields of

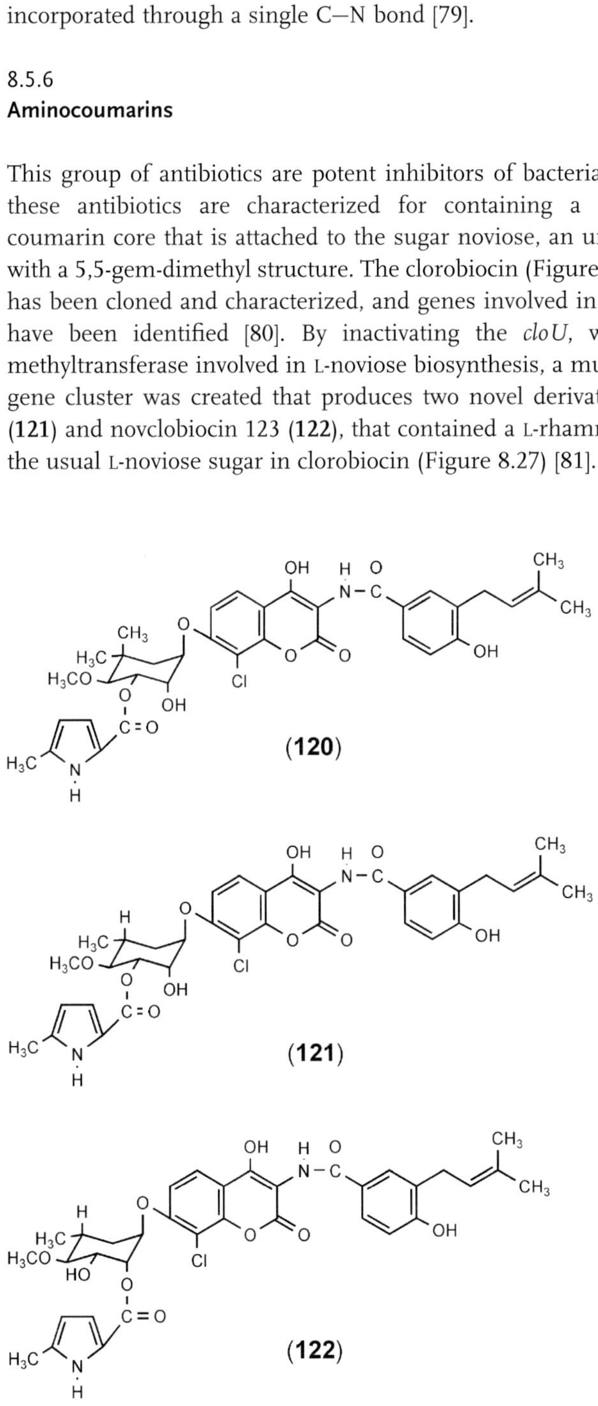

(**120**)

(**121**)

(**122**)

**Figure 8.27** Chemical structure of clorobiocin and derivatives obtained by gene inactivation.

the novel derivatives were quite low due to the ineffective formation of L-rhamnose in this mutant because of the relatively low processing of the accumulated biosynthesis intermediate by the CloS noviose 4-ketoreductase. This bottleneck was overcome by expressing in this mutant the *oleU* gene also coding for a 4-ketoreductase, but more efficient in recognizing the 4-keto intermediate substrate [82].

## Acknowledgments

The authors wish to thank all of the people in their laboratories, particularly those involved in research on sugar biosynthesis and glycosylation. Research at the authors' laboratory has been supported by grants from the Spanish Ministry of Education and Science (BMC2002-03599 and BIO2005-04115 to C.M; BMC2003-00478 and BFU2006-00404 to JAS), and Red Temática de Investigación Cooperativa de Centros de Cáncer (Ministry of Health, Spain; ISCIII-RETIC RD06/00200026).

## References


1 (a) Chater, K.F. (2001) *Current Opinion in Microbiology*, **4**, 667.
(b) Chater, K.F. (2006) *FEMS Microbiology Reviews*, **30**, 651.

2 Demain, A.L. (1999) *Applied Microbiology and Biotechnology*, **52**, 455.

3 (a) Adrio, J.L. and Demain, A.L. (2006) *FEMS Microbiology Reviews*, **30**, 187.
(b) Behal, V. (2000) *Advances in Applied Microbiology*, **47**, 113.
(c) Bibb, M.J. (2005) *Current Opinion in Microbiology*, **8**, 208.

4 (a) Thorson, J.S., Hosted, T., Jiang, J., Biggins, J.B. and Ahleret, J. (2001) *Current Organic Chemistry*, **5**, 139.
(b) Trefzer, A., Salas, J.A. and Bechthold, A. (1999) *Natural Product Reports*, **16**, 283.
(c) Weymouth-Wilson, A.C. (1997) *Natural Product Reports*, **14**, 99.

5 (a) Liu, H.W. and Thorson, J.S. (1994) *Annual Review of Microbiology*, **48**, 223.
(b) Piepersberg, W. (1994) *Critical Reviews in Biotechnology*, **14**, 251.

6 Waldron, C., Matsushima, P., Rosteck, P.R. Jr, Broughton, M.C., Turner, J., Madduri, K., Crawford, K.P., Merlo, D.J. and Baltz, R.H. (2001) *Chemistry and Biology*, **8**, 487.

7 (a) González, A., Remsing, L.L., Lombó, F., Fernandez, M.J., Prado, L., Braña, A.F., Kunzel, E., Rohr, J., Mendez, C. and Salas, J.A. (2001) *Molecular and General Genetics*, **264**, 827.
(b) Remsing, L.L., Garcia-Bernardo, J., Gonzalez, A., Kunzel, E., Rix, U., Braña, A. F., Bearden, D.W., Mendez, C., Salas, J.A. and Rohr, J. (2002) *Journal of the American Chemical Society*, **124**, 1606.

8 (a) Gullón, S., Olano, C., Abdelfattah, M.S., Braña, A.F., Rohr, J., Mendez, C. and Salas, J.A. (2006) *Applied and Environmental Microbiology*, **72**, 4172.
(b) Luzhetskyy, A., Mayer, A., Hoffmann, J., Pelzer, S., Holzenkamper, M., Schmitt, B., Wohlert, S.E., Vente, A. and Bechthold, A. (2007) *ChemBioChem*, **8**, 599.
(c) Madduri, K., Waldron, C. and Merlo, D.J. (2001) *Journal of Bacteriology*, **183**, 5632.

9 (a) Amann, S., Drager, G., Rupprath, C., Kirschning, A. and Elling, L. (2001) *Carbohydrate Research*, **335**, 23.
(b) Chen, H., Yamase, H., Murakami, K., Chang, C.-W., Zhao, L., Zhao, Z. and Liu, H.-W. (2002) *Biochemistry*, **41**, 9165.

10 Chen, H., Thomas, M.G., Hubbard, B.K., Losey, H.C., Walsh, C.T. and Burkart,

M.D. (2000) *Proceedings of the National Academy of Sciences of the United States of America*, **97**, 11942.

**11** Merkel, A.B., Major, L.L., Errey, J.C., Burkart, M.D., Field, R.A., Walsh, C.T. and Naismith, J.H. (2004) *Journal of Biological Chemistry*, **279**, 32684.

**12** Takahashi, H., Liu, Y.N. and Liu, H.W. (2006) *Journal of the American Chemical Society*, **128**, 1432.

**13** (a) Lombo, F., Siems, K., Braña, A.F., Mendez, C., Bindseil, K. and Salas, J.A. (1997) *Journal of Bacteriology*, **179**, 3354. (b) Merson-Davies, L.A. and Cundliffe, E. (1994) *Molecular Microbiology*, **13**, 349.

**14** Draeger, G., Park, S.H. and Floss, H.G. (1999) *Journal of the American Chemical Society*, **121**, 2611.

**15** Chen, H., Agnihotri, G., Guo, Z., Que, N.L.S., Chen, X.H. and Liu, H.W. (1999) *Journal of the American Chemical Society*, **121**, 8124.

**16** Hong, L., Zhao, Z. and Liu, H.W. (2006) *Journal of the American Chemical Society*, **128**, 14262.

**17** Zhao, Z., Hong, L. and Liu, H.W. (2005) *Journal of the American Chemical Society*, **127**, 7692.

**18** Chen, H., Zhao, Z., Hallis, T.M., Guo, Z. and Liu, H.W. (2001) *Angewandte Chemie (International Edition in English)*, **40**, 607.

**19** Takahashi, H., Liu, Y.N., Chen, H. and Liu, H.W. (2005) *Journal of the American Chemical Society*, **127**, 9340.

**20** Tello, M., Jakimowicz, P., Errey, J.C., Freel Meyers, C.L., Walsh, C.T., Buttner, M.J., Lawson, D.M. and Field, R.A. (2006) *Chemical Communications*, 1079.

**21** Zhao, L., Borisova, S., Yeung, S.M. and Liu, H.W. (2001) *Journal of the American Chemical Society*, **123**, 7909.

**22** Szu, P.H., He, X., Zhao, L. and Liu, H.W. (2005) *Angewandte Chemie (International Edition in English)*, **44**, 6742.

**23** Zhao, L., Que, N.L.S., Xue, Y., Sherman, D.H. and Liu, H.W. (1998) *Journal of the American Chemical Society*, **120**, 12159.

**24** Chang, C.W., Zhao, L., Yamase, H. and Liu, H.W. (2000) *Angewandte Chemie (International Edition in English)*, **39**, 2160.

**25** Melancon, C.E., Yu, W.L. and Liu, H.W. (2005) *Journal of the American Chemical Society*, **127**, 12240.

**26** Chen, H., Yeung, S.M., Que, N.L.S., Miller, T., Schmidt, R.R. and Liu, H.W. (1999) *Journal of the American Chemical Society*, **121**, 7166.

**27** Chen, H., Guo, Z. and Liu, H.W. (1998) *Journal of the American Chemical Society*, **120**, 9951.

**28** Hofmann, C., Boll, R., Heitmann, B., Hauser, G., Dürr, C., Frerich, A., Weitnauer, G., Glaser, S.J. and Bechthold, A. (2005) *Chemistry and Biology*, **12**, 1137.

**29** Rodríguez, L., Rodríguez, D., Olano, C., Braña, A.F., Méndez, C. and Salas, J.A. (2001) *Journal of Bacteriology*, **183**, 5358.

**30** Kruger, R.G., Lu, W., Oberthür, M., Tao, J., Kahne, D. and Walsh, C.T. (2005) *Chemistry and Biology*, **12**, 131.

**31** (a) Baltz, R. (2006) *SIM News*, **56**, 148. (b) Khosla, C. and Tosí, C.J. (1995) *Chemistry and Biology*, **2**, 355. (c) Khosla, C. and Zawada, R.J. (1996) *Trends in Biotechnology*, **14**, 335.

**32** Salas, J.A. and Méndez, C. (2007) *Trends in Microbiology*, **15**, 219.

**33** (a) Gaisser, S., Bohm, G.A., Cortés, J. and Leadlay, P.F. (1997) *Molecular and General Genetics*, **256**, 239. (b) Gaisser, S., Bohm, G.A., Doumith, M., Raynal, M.C., Dhillon, N., Cortes, J. and Leadlay, P.F. (1998) *Molecular and General Genetics*, **258**, 78. (c) Haydock, S.F., Dowson, J.A., Dhillon, N., Roberts, G.A., Cortes, J. and Leadlay, P.F. (1991) *Molecular and General Genetics*, **230**, 120. (d) Salah-Bey, K., Doumith, M., Michel, J.M., Haydock, S., Cortés, J., Leadlay, P.F. and Raynal, M.C. (1998) *Molecular and General Genetics*, **257**, 542. (e) Summers, R.G., Donadio, S., Staver, M.J., Wendt-Pienkowski, E., Hutchinson, C.R. and Katz, L. (1997) *Microbiology*, **143**, 3251.

**34** Doumith, M., Legrand, R., Lang, C., Salas, J.A. and Raynal, M.C. (1999) *Molecular Microbiology*, **34**, 1039.

**35** Aguirrezabalaga, I., Olano, C., Allende, N., Rodriguez, L., Braña, A.F., Méndez, C. and Salas, J.A. (2000) *Antimicrobial Agents and Chemotherapy*, **44**, 1266.

**36** (a) Butler, A.R., Bate, N. and Cundliffe, E. (1999) *Chemistry and Biology*, **6**, 287.
(b) Fish, S.A. and Cundliffe, E. (1997) *Microbiology*, **143**, 3871.
(c) Fouces, R., Mellado, E., Diez, B. and Barredo, J.L. (1999) *Microbiology*, **145**, 855.
(d) Gandecha, A.R., Large, S.L. and Cundliffe, E. (1997) *Gene*, **184**, 197.

**37** Butler, A.R., Bate, N., Kiehl, D.E., Kirst, H.A. and Cundliffe, E. (2002) *Nature Biotechnology*, **20**, 713.

**38** Gaisser, S., Reather, J., Wirtz, G., Kellenberger, L., Staunton, J. and Leadlay, P.F. (2000) *Molecular Microbiology*, **36**, 391.

**39** Gaisser, S., Martin, C.J., Wilkinson, B., Sheridan, R.M., Lill, R.E., Weston, A.J., Ready, S.J., Waldron, C., Crouse, G.D., Leadlay, P.F. and Staunton, J. (2002) *Chemical Communications*, **21**, 618.

**40** Xue, Y., Zhao, L., Liu, H.W. and Sherman, D.H. (1998) *Proceedings of the National Academy of Sciences of the United States of America*, **95**, 12111.

**41** Zhao, L., Sherman, D.H. and Liu, H.W. (1998) *Journal of the American Chemical Society*, **120**, 10256.

**42** Borisova, S.A., Zhao, L., Sherman, D.H. and Liu, H.W. (1999) *Organic Letters*, **15**, 133.

**43** Yamase, H., Zhao, L. and Liu, H.W. (2000) *Journal of the American Chemical Society*, **122**, 12397.

**44** Hong, J.S., Park, S.H., Choi, C.Y., Sohng, J.K. and Yoon, Y.J. (2004) *FEMS Microbiology Letters*, **238**, 391.

**45** Tang, L. and McDaniel, R. (2001) *Chemistry and Biology*, **8**, 547.

**46** Yoon, Y.J., Beck, B.J., Kim, B.S., Kang, H.Y., Reynolds, K.A. and Sherman, D.H. (2002) *Chemistry and Biology*, **9**, 203.

**47** Rohr, J., Mendez, C. and Salas, J.A. (1999) *Bioorganic Chemistry*, **27**, 41.

**48** Lombo, F., Menendez, N., Salas, J.A. and Mendez, C. (2006) *Applied Microbiology and Biotechnology*, **73**, 1.

**49** (a) Blanco, G., Fernández, E., Fernandez, M.J., Braña, A.F., Weissbach, U., Kunzel, E., Rohr, J., Mendez, C. and Salas, J.A. (2000) *Molecular and General Genetics*, **262**, 991.
(b) Fernandez, E., Weissbach, U., Sanchez Reillo, C., Braña, A.F., Mendez, C., Rohr, J. and Salas, J.A. (1998) *Journal of Bacteriology*, **180**, 4929.
(c) Nur-e-Alam, M., Mendez, C., Salas, J.A. and Rohr, J. (2005) *ChemBioChem*, **6**, 632.

**50** (a) Menéndez, N., Nur-e-Alam, M., Braña, A.F., Rohr, J., Salas, J.A. and Mendez, C. (2004) *Chemistry and Biology*, **11**, 21.
(b) Menendez, N., Nur-E-Alam, M., Braña, A.F., Rohr, J., Salas, J.A. and Mendez, C. (2004) *Molecular Microbiology*, **53**, 903.
(c) Perez, M., Lombo, F., Baig, I., Braña, A.F., Rohr, J., Salas, J.A. and Mendez, C. (2006) *Applied and Environmental Microbiology*, **72**, 6644.

**51** Menendez, N., Nur-e-Alam, M., Fischer, C., Braña, A.F., Salas, J.A., Rohr, J. and Mendez, C. (2006) *Applied and Environmental Microbiology*, **72**, 167.

**52** Rohr, J., Weibbach, U., Beninga, C., Künzel, E., Siems, K., Bindseil, K., Prado, L., Lombó, F., Braña, A.F., Méndez, C. and Salas, J.A. (1998) *Chemical Communications*, 437.

**53** (a) Faust, B., Hoffmeister, D., Weitnauer, G., Westrich, L., Haag, S., Schneider, P., Decker, H., Kunzel, E., Rohr, J. and Bechthold, A. (2000) *Microbiology*, **146**, 147.
(b) Kunzel, E., Faust, B., Oelkers, C., Weissbach, U., Bearden, D.W., Weitnauer, G., Westrich, L., Bechthold, A. and Rohr, J. (1999) *Journal of the American Chemical Society*, **121**, 11058.

**54** Trefzer, A., Blanco, G., Remsing, L., Kunzel, E., Rix, U., Lipata, F., Braña, A.F., Mendez, C., Rohr, J., Bechthold, A. and Salas, J.A. (2002) *Journal of the American Chemical Society*, **124**, 6056.

**55** Rohr, J. (1989) *Journal of Antibiotics*, **42**, 1482.

**56** (a) Hoffmeister, D., Ichinose, K., Domann, S., Faust, B., Trefzer, A., Drager, G., Kirschning, A., Fischer, C., Kunzel, E., Bearden, D.W., Rohr, J. and Bechthold, A. (2000) *Chemistry and Biology*, **7**, 821.
(b) Trefzer, A., Hoffmeister, D., Kunzel, E., Stockert, S., Weitnauer, G., Westrich, L., Rix, U., FuchserK, J., Bindseil, U., Rohr, J. and Bechthold, A. (2000) *Chemistry and Biology*, **7**, 133.

**57** Hoffmeister, D., Drager, G., Ichinose, K., Rohr, J. and Bechthold, A. (2003) *Journal of the American Chemical Society*, **125**, 4678.

**58** Hoffmeister, D., Wilkinson, B., Foster, G., Sidebottom, P.J., Ichinose, K. and Bechthold, A. (2002) *Chemistry and Biology*, **9**, 287.

**59** Westrich, L., Domann, S., Faust, B., Bedford, D., Hopwood, D.A. and Bechthold, A. (1999) *FEMS Microbiology Letters*, **170**, 381.

**60** Fedorenko, V., Basiliya, L., Pankevych, K., Dubitska, L., Ostash, B., Luzhetskyy, A., Gromyko, O. and Krugel, H. (2000) *Bulletin of the Institute of Agricultural Microbiology (Ukraine)*, **8**, 27.

**61** Zhu, L., Luzhetskyy, A., Luzhetska, M., Mattingly, C., Adams, V., Bechthold, A. and Rohr, J. (2007) *ChemBioChem*, **8**, 83.

**62** Luzhetskyy, A., Taguchi, T., Fedoryshyn, M., Durr, C., Wohlert, S.E., Novikov, V. and Bechthold, A. (2005) *ChemBiochem*, **6**, 1406.

**63** Ostash, B., Rix, U., Rix, L.L., Liu, T., Lombo, F., Luzhetskyy, A., Gromyko, O., Wang, C., Braña, A.F., Mendez, C., Salas, J.A., Fedorenko, V. and Rohr, J. (2004) *Chemistry and Biology*, **11**, 547.

**64** Baig, I., Kharel, M., Kobylyanskyy, A., Zhu, L., Rebets, Y., Ostash, B., Luzhetskyy, A., Bechthold, A., Fedorenko, V.A. and Rohr, J. (2006) *Angewandte Chemie (International Edition in English)*, **45**, 7842.

**65** Hutchinson, C.R. (1997) *Chemical Reviews*, **97**, 2525.

**66** (a) Garrido, L.M., Lombo, F., Baig, I., Nur-E-Alam, M., Furlan, R.L., Borda, C. C., Braña, A.F., Mendez, C., Salas, J.A., Rohr, J. and Padilla, G. (2006) *Applied Microbiology and Biotechnology*, **73**, 122.
(b) Niemi, J. and Mantsala, P. (1995) *Journal of Bacteriology*, **177**, 2942.
(c) Raty, K., Kunnari, T., Hakala, J., Mantsala, P. and Ylihonko, K. (2000) *Molecular and General Genetics*, **264**, 164.
(d) Torkkell, S., Kunnari, T., Palmu, K., Mantsala, P., Hakala, J. and Ylihonko, K. (2001) *Molecular Genetics and Genomics*, **266**, 276.

**67** (a) Kunnari, T., Ylihonko, K., Klika, K.D., Mäntsälä, P. and Hakala, J. (2000) *Journal of Organic Chemistry*, **65**, 2851.
(b) Ylihonko, K., Hakala, J., Kunari, T. and Mäntsälä, P. (1996) *Microbiology*, **142**, 1965.

**68** Niemi, J., Ylihonko, K., Hakala, J., Parssinen, R., Kopio, A. and Mäntsälä, P. (1994) *Microbiology*, **140**, 1351.

**69** Madduri, K., Kennedy, J., Rivola, G., Inventi-Solari, A., Filippini, S., Zanuso, G., Colombo, A.L., Gewain, K.M., Occi, J.L., MacNeil, D.J. and Hutchinson, C.R. (1998) *Nature Biotechnology*, **16**, 69.

**70** Decker, H., Rohr, J., Motamedi, H., Zähner, H. and Hutchinson, C.R. (1995) *Gene*, **166**, 121.

**71** (a) Blanco, G., Patallo, E., Braña, A.F., Trefzer, A., Becthold, A., Rohr, J., Méndez, C. and Salas, J.A. (2001) *Chemistry and Biology*, **75**, 1.
(b) Patallo, E., Blanco, G., Fischer, C., Braña, A.F., Rohr, J., Méndez, C. and Salas, J.A. (2001) *Journal of Biological Chemistry*, **276**, 18765.

**72** Decker, H., Haag, S., Udvarnoki, G. and Rohr, J. (1995) *Angewandte Chemie (International Edition in English)*, **34**, 1107.

**73** Wohlert, S.E., Blanco, G., Lombó, F., Fernández, E., Reich, S., Udvarnoki, G., Braña, A.F., Mendez, C., Decker, H., Salas, J.A. and Rohr, J. (1998) *Journal of the American Chemical Society*, **120**, 10596.

**74** Rodriguez, L., Oelkers, C., Aguirrezabalaga, I., Braña, A.F., Rohr, J., Méndez, C. and Salas, J.A. (2000) *Journal of Molecular Microbiology and Biotechnology*, **2**, 271.

**75** (a) Fischer, C., Rodriguez, L., Patallo, E.P., Lipata, F., Braña, A.F., Mendez, C., Salas, J.A. and Rohr, J. (2002) *Journal of Natural Products*, **65**, 1685.
(b) Lombó, F., Gibson, M., Greenwell, L., Braña, A.F., Rohr, J. and Salas, J.A. (2004) *Chemistry and Biology*, **11**, 1709.
(c) Pérez, M., Lombó, F., Zhu, L., Gibson, M., Braña, A.F., Rohr, J., Salas, J.A. and Méndez, C. (2005) *Chemical Communications*, 1604.
(d) Rodriguez, L., Aguirrezabalaga, I., Allende, N., Braña, A.F., Mendez, C. and Salas, J.A. (2002) *Chemistry and Biology*, **9**, 721.

**76** Sánchez, C., Méndez, C. and Salas, J.A. (2006) *Natural Product Reports*, **23**, 1007.

**77** Butler, M.S. (2005) *Natural Product Reports*, **22**, 162.

**78** (a) Hyun, C.G., Bililign, T., Liao, J. and Thorson, J.S. (2003) *ChemBioChem*, **4**, 114.

(b) Onaka, H., Taniguchi, S., Igarashi, Y. and Furumai, T. (2002) *Journal of Antibiotics*, **55**, 1063.
(c) Onaka, H., Taniguchi, S., Igarashi, Y. and Furumai, T. (2003) *Bioscience, Biotechnology, and Biochemistry*, **67**, 127.
(d) Sánchez, C., Butovich, I.A., Braña, A.F., Rohr, J., Méndez, C. and Salas, J.A. (2002) *Chemistry and Biology*, **9**, 519.

**79** Salas, A.P., Zhu, L., Sánchez, C., Braña, A.F., Rohr, J., Méndez, C. and Salas, J.A. (2005) *Molecular Microbiology*, **58**, 17.

**80** Pojer, F., Li, S.-M. and Heide, L. (2002) *Microbiology*, **148**, 3901.

**81** Freitag, A., Li, S.M. and Heide, L. (2006) *Microbiology*, **152**, 2433.

**82** Freitag, A., Mendez, C., Salas, J.A., Kammerer, B., Li, S.M. and Heide, L. (2006) *Metabolic Engineering*, **8**, 653.

# 9
# Microbial Production of DNA Building Blocks

*Jun Ogawa, Nobuyuki Horinouchi, and Sakayu Shimizu*

## 9.1
## Introduction

There will be a need for 2′-deoxyribonucleoside in the near future due to increasing demand in new medical and biotechnology fields. 2′-Deoxyribonucleoside is a building block of promising antisense drugs for cancer therapy. For some recently developed antiviral reagents, such as azidothymidine for the treatment of human immunodeficiency virus infections, 2′-deoxyribonucleoside is a synthesis intermediate. 2′-Deoxyribonucleoside is also a precursor of an indispensable material used for widespread polymerase chain reaction applications – 2′-deoxyribonucleoside triphosphate. The current 2′-deoxyribonucleoside sources include hydrolyzed herring and salmon sperm DNA, which are not suitable sources for sudden high demands. Microbial/enzymatic processes could possibly remove this bottleneck of raw material supply [1].

Processes for the production of various amino acids [2], fatty acids [3], and nucleotides such as 5′-inosinic acid [4] and 5′-guanylic acid [5] using microorganisms or enzymes have been established. However, the production of 2′-deoxyribonucleoside, which is one of the most important biological compounds and a building block of DNA, has not been established. DNA fermentation from sugar, *n*-paraffin [6] and acetic acid [7] by microorganisms was reported. However, the difficulty in controlling the amount of DNA in the cell caused low accumulation of DNA.

We focused on the reactions involved in 2′-deoxyribonucleoside degradation for the synthesis of 2′-deoxyribonucleoside. All reactions in 2′-deoxyribonucleoside degradation are reversible, so the biochemical retrosynthesis of 2′-deoxyribonucleosides from their metabolites is possible. Some of the metabolites such as triosephosphates can be obtained from cheap sugar materials through the glycolytic pathway. Therefore, we proposed the multi-step enzymatic process presented in Scheme 9.1, with glucose, acetaldehyde and a nucleobase as the starting materials. The enzymes involved in the process are: (i) glycolytic enzymes that generate D-glyceraldehyde-3-phosphate (G3P) from glucose, (ii) deoxyriboaldolase

*Multi-Step Enzyme Catalysis: Biotransformations and Chemoenzymatic Synthesis*
Edited by Eduardo Garcia-Junceda

ISBN: 978-3-527-31921-3

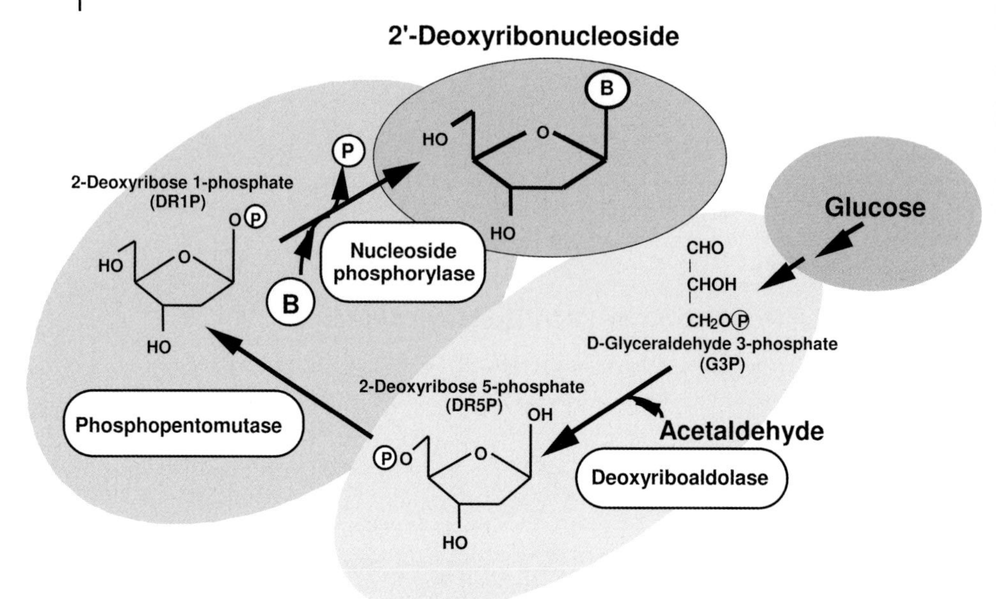


**Scheme 9.1** Multi-step enzymatic process for the production of 2′-deoxyribonucleoside from glucose, acetaldehyde and a nucleobase through the reverse reactions of 2′-deoxyribonucleoside degradation.

that catalyzes the condensation of acetaldehyde and G3P to yield 2-deoxyribose-5-phosphate (DR5P), (iii) phosphopentomutase that catalyzes the intermolecular transfer of phosphate from DR5P to 2-deoxyribose-1-phosphate (DR1P), and (iv) nucleoside phosphorylase that catalyzes nucleobase transfer to the pentosyl moiety to generate 2′-deoxyribonucleosides. In this chapter, we describes the development of this multi-step enzymatic process starting from the screening of a key enzyme, deoxyriboaldolase, followed by metabolic analysis of deoxyriboaldolase-expressing *Escherichia coli*, coupling of yeast glycolytic fermentation with reverse reactions of 2′-deoxyribonucleoside degradation and optimization of reaction conditions.

## 9.2
## Screening of Acetaldehyde-Tolerant Deoxyriboaldolase and Its Application for DR5P Synthesis

The difficulty in the chemical synthesis of 2′-deoxyribonucleoside lies in the generation of 2-deoxyribosyl groups. The chemical syntheses of 2-deoxyribosyl groups and further to 2′-deoxyribonucleosides involve complex protection and deprotection steps [8–11]. It is likely that the introduction of biochemical reactions with high selectivity will solve this problem. In general metabolism, there are two reactions concerning with 2-deoxyriobosyl groups. One is reduction of ribonucleotide to 2′-deoxyribonuleotide catalyzed by ribonucleotide reductase in biosynthesis, and the other is DR5P cleavage to G3P and acetaldehyde catalyzed by deoxyriboaldolase in the degradation of 2′-deoxyribonucleoside. The former reaction is under complex regulation and is difficult to handle for practical purposes [12, 13]. We focused on the latter reaction for the generation of the 2-deoxyribose frame (Scheme 9.2).

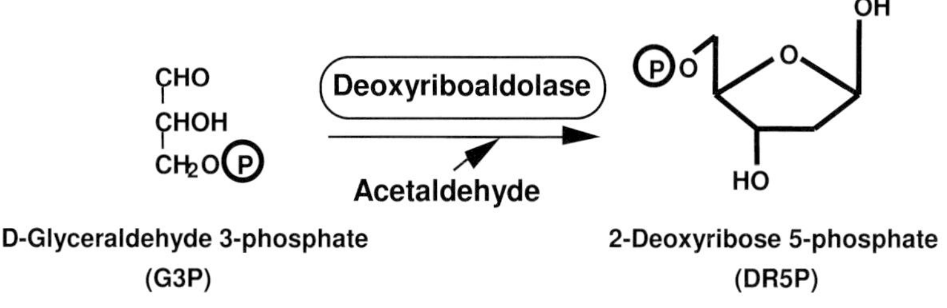


**Scheme 9.2** DR5P synthesis by the reverse deoxyriboaldolase reaction.

Screening for suitable catalysts for the production of DR5P through the reverse deoxyriboaldolase reaction and evaluation of DR5P production from acetaldehyde and G3P using the potent strain were carried out. The screening reactions were done with a high acetaldehyde concentration to reveal acetaldehyde-tolerant enzymes, because a high concentration of acetaldehyde is required to push the deoxyriboaldolase reaction in the direction of DR5P synthesis. Among about 200 strains tested, *Klebsiella pneumoniae* B-4-4 was selected as an acetaldehyde-tolerant deoxyriboaldolase producer for further investigation. Under the optimum conditions, 70.8 mM DR5P was produced from 200 mM acetaldehyde and 87.5 mM G3P in 5 h with a molar yield of 80.9% [14]. The deoxyriboaldolase of this strain is tolerant to high concentrations of acetaldehyde up to 300 mM.

## 9.3 Construction of Deoxyriboaldolase-Overexpressing *E. coli* and Metabolic Analysis of the *E. coli* Transformants for DR5P Production from Glucose and Acetaldehyde

The gene encoding a deoxyriboaldolase was cloned from the chromosomal DNA of *K. pneumoniae* B-4-4. It contains an open reading frame consisting of 780 nucleotides corresponding to 259 amino acid residues. The predicted amino acid sequence exhibited 94.6% homology with that of deoxyriboaldolase from *E. coli* [15]. The deoxyriboaldolase of *K. pneumoniae* was expressed in recombinant *E. coli* cells. Two deoxyriboaldolase-expressing *E. coli* transformants, JM109/pTS8 and 10B5/pTS8, were constructed. The specific activity of deoxyriboaldolase in the cell extracts of these *E. coli* transformants reached as high as 2.5 U/mg, which was 3

**Table 9.1** Evaluation of glycolysis intermediates as substrates for DR5P production by deoxyriboaldolase-expressing *E. coli* without and with ATP.

| Substrate | DR5P production (mM) | | | | | |
|---|---|---|---|---|---|---|
| | *K. pneumoniae* B-4-4 | | JM109/pTS8 | | 10B5/pTS8 | |
| | – | +ATP | – | +ATP | – | +ATP |
| G3P | 68.8 | 52.1 | 36.7 | 40.3 | 59.2 | 53.4 |
| Glyceraldehyde | ND | ND | ND | 0.3 | ND | 0.5 |
| DHAP | 64.5 | 49.5 | 53.7 | 45.2 | 48.9 | 44.1 |
| Dihydroxyacetone | ND | ND | ND | ND | ND | 0.6 |
| Glucose | ND | 3.9 | ND | 3.4 | 0.3 | 6.3 |
| Glucose-6-phosphate | ND | 8.5 | ND | 9.5 | 3.9 | 10.0 |
| Glucose-1,6-diphosphate | ND | ND | ND | 0.4 | ND | 0.5 |
| Fructose | ND | ND | ND | 0.6 | ND | 1.1 |
| Fructose-6-phosphate | ND | 0.6 | 0.6 | 0.5 | ND | 0.8 |
| FDP | 36.1 | 31.4 | 33.0 | 39.7 | 38.9 | 41.3 |

The reactions were carried out with various glycolysis and glycerol metabolism intermediates (100 mM) without (–) or with ATP (+ATP; 10 mM) under the conditions described in the text. ND, not detected.

times higher than that in the *K. pneumoniae* cell extract. Strain 10B5/pTS8 was superior for DR5P production because of a defect in alkaline phosphatase activity decomposing DR5P to 2-deoxyribose [15].

To determine whether the glycolytic pathway could provide G3P, various compounds involved in the glycolytic pathway were examined as precursors of G3P in the reactions with washed cells of *E. coli* transformants and *K. pneumoniae* B-4-4 (Table 9.1). Dihydroxyacetone phosphate (DHAP) and fructose-1,6-diphosphate (FDP) were utilized well for G3P generation by all strains. Since several phosphorylated compounds were found to be precursors of G3P, screening of substrates that are alternatives to G3P for DR5P production was carried out in the presence of a phosphate (energy) donor (ATP). Xylene and polyoxyethylenelaurylamine (PL) were added to the reaction mixtures to enhance ATP permeability into the cells. As shown in Table 9.1, glucose, glucose-6-phosphate and fructose-6-phosphate were used as substrates by all strains in the presence of ATP. These results suggested that the activities of glycolytic enzymes for G3P generation in these strains are high enough, but that the energy required for glucose transformation is insufficient. Other than ATP, ADP and AMP apparently provided the energy (phosphate) for G3P generation from glucose. Under the optimal conditions, 100 mM DR5P was produced from 900 mM glucose, 200 mM acetaldehyde and 100 mM ATP by strain 10B5/pTS8 [15].

## 9.4
## Efficient Production of DR5P from Glucose and Acetaldehyde by Coupling of the Alcoholic Fermentation System of Baker's Yeast and Deoxyriboaldolase-Expressing *E. coli*

For the production of DR5P from glucose and acetaldehyde by using strain 10B5/pTS8, the supply of energy (ATP) is required. The sugar-fermentative system of baker's yeast had been used as the ATP donor in several bioprocesses [16, 17]. The temporary accumulation of FDP in the course of ATP regeneration was reported [18–20]. Based on this, we designed a novel metabolic and enzymatic DR5P production process consisting of FDP synthesis through glucose fermentation and DR5P synthesis through deoxyriboaldolase-catalyzed aldol condensation with glucose and acetaldehyde as starting materials (Scheme 9.3). In this process, toluene-treated baker's yeast generates FDP from glucose and inorganic phosphate, and then the strain 10B5/pTS8 converts the FDP into DR5P via G3P.

It is important that almost all the inorganic phosphate is consumed in the first step (FDP production from glucose) because the second step (deoxyriboaldolase reaction) is inhibited by inorganic phosphate. When the FDP production was

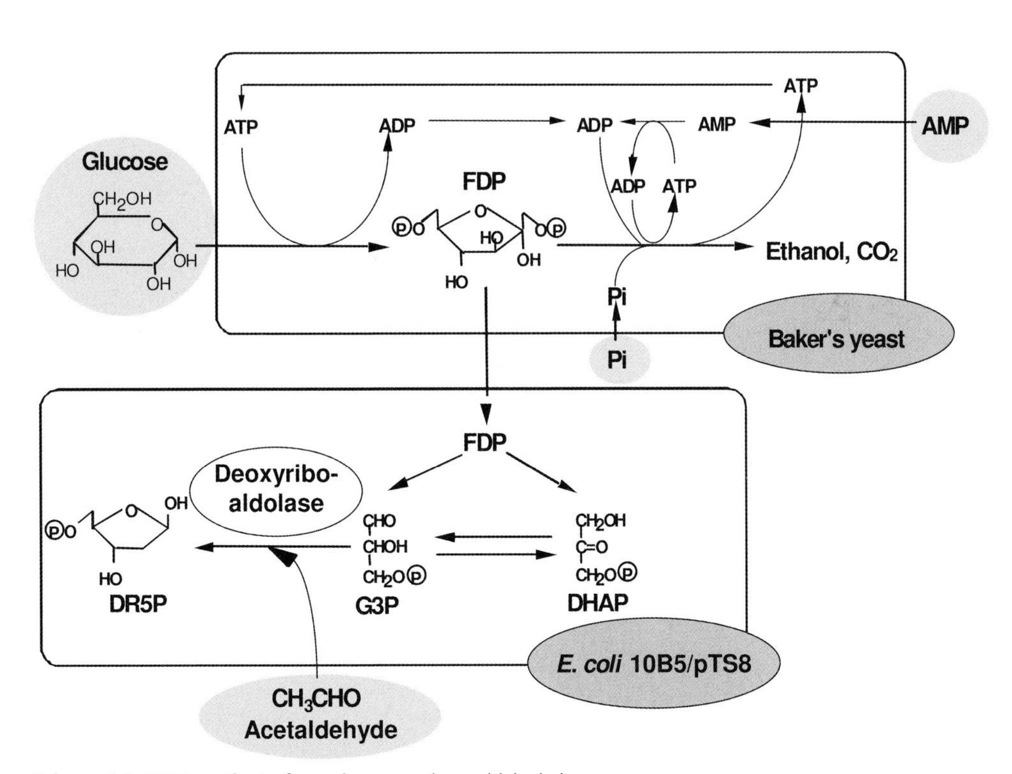


**Scheme 9.3** DR5P synthesis from glucose and acetaldehyde by baker's yeast and deoxyriboaldolase-expressing *E. coli*.

carried out with 750 mM inorganic phosphate, almost all the inorganic phosphate was converted into FDP, so the subsequent DR5P production proceeded well. Addition of acetaldehyde was effective on FDP production. In the presence of a downstream intermediate of alcoholic fermentation, acetaldehyde, FDP degradation might be depressed, resulting in an increase in FDP accumulation. A standing reaction with an almost fully filled Erlenmeyer flask, to maintain a low soluble oxygen concentration, was effective. Under the optimized conditions with toluene-treated yeast cells, 356 mM (121 g/l) FDP was produced from 1111 mM glucose and 750 mM potassium phosphate buffer (pH 6.4) with a catalytic amount of AMP in 5 h. The molar yields of FDP as to glucose and inorganic phosphate were 32 and 95.9%, respectively, and the apparent AMP turnover was 47.5 [21].

For optimization of DR5P production from the enzymatically prepared FDP and acetaldehyde with deoxyriboaldolase-expressing *E. coli* 10B5/pTS8, surfactants, xylene and PL were added for improvement of the permeation of phosphorylated compounds [22]. Under the preparative reaction conditions at 28 °C with 178 mM enzymatically prepared FDP and 400 mM acetaldehyde as the substrates, 246 mM DR5P (52.6 g/l) was produced in 2 h. The molar yields of DR5P as to FDP and acetaldehyde were 69.1 and 61.5%, respectively. The molar yield as to glucose through the total two-step reaction was 22.1% [21]. It is noteworthy that the coupling of the glycolytic pathway of baker's yeast to DR5P synthesis greatly reduced the amount of endogenous energy source required.

## 9.5
## Biochemical Retrosynthesis of 2′-Deoxyribonucleosides from Glucose Acetaldehyde and a Nucleobase: Three-Step Multi-Enzyme-Catalyzed Synthesis

As mentioned above, DR5P was successfully produced through the two-step multi-enzyme-catalyzed process: (Step 1) FDP production from glucose, a high concentration of inorganic phosphate and a catalytic amount of AMP by toluene treated-baker's yeast, and (Step 2) DR5P production from acetaldehyde and enzymatically prepared FDP via G3P by deoxyriboaldolase-expressing *E. coli* 10B5/pTS8 (Scheme 9.3). The DR5P enzymatically produced from glucose and acetaldehyde was further converted to 2′-deoxyribonucleosides through the reverse reactions of 2′-deoxyribonucleoside degradation. All reactions in 2′-deoxyribonucleoside degradation are reversible, so the biochemical retrosynthesis of 2′-deoxyribonucleosides from their metabolites is possible. DR5P was further converted to 2′-deoxyribonucleosides through phosphopentomutase-catalyzing isomerization of DR5P to DR1P and nucleoside phosphorylase-catalyzing nucleobase transfer to the pentosyl moiety to generate 2′-deoxyribonucleosides (Step 3) (Scheme 9.4). *E. coli* transformants expressing phosphopentomutase (*E. coli* BL21/pTS17) and commercial nucleoside phosphorylase were used for the third step. Adenine was the best nucleobase for the Step 3 reaction; however, the adenosine deaminase activity of *E. coli* BL21/pTS17 transformed the 2′-deoxyadenosine produced to 2′-deoxyinosine. From 12.3 mM enzymatically prepared DR5P and 50 mM adenine, 9.9 mM

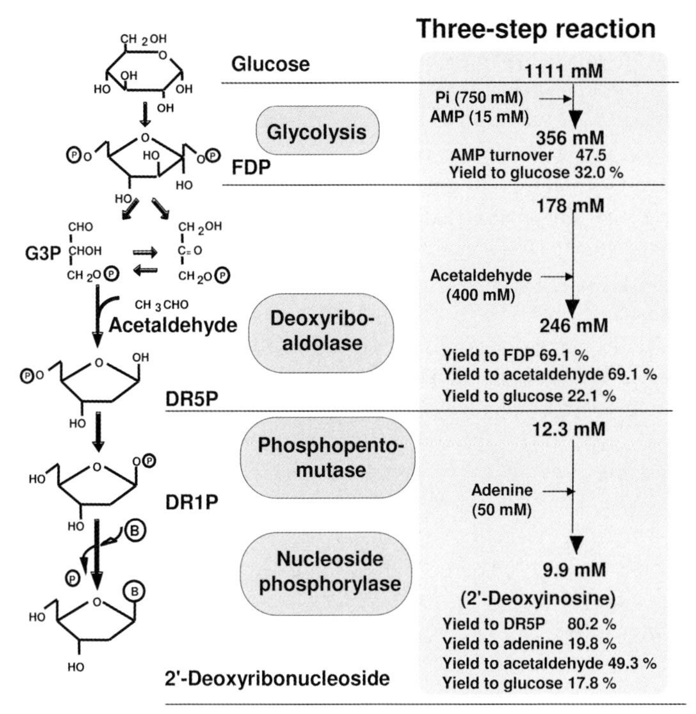


**Scheme 9.4** Biochemical retrosynthesis of 2′-deoxyribonucleosides from glucose, acetaldehyde and a nucleobase (adenine) through the glycolytic pathway and the reverse reactions of 2′-deoxyribonucleoside degradation. P, phosphate; B, adenine.

2′-deoxyinosine was produced [23]. The molar yield of 2′-deoxyinosine as to DR5P obtained here (80.2%) was much higher than the 14–60% reported by Ouwerkerk *et al.* [24]. The yield as to glucose throughout the whole reaction process was 17.7%. The results obtained here are summarized in Scheme 9.4.

The molar yield of 2′-deoxyinosine as to glucose obtained here was also comparable to the 7–18% on *de novo* fermentative production of a ribonucleoside (inosine) [25–27] and much higher than the 0.2–2.0% on DNA fermentation with glucose [6, 28]. The amount of 2′-deoxyribonucleoside produced (approximately 2.5 g/l) was much higher than that (0.22 g/l) on preparative production of plasmid DNA with recombinant *E. coli* strains [29]. The above results indicate the possibility of biochemical retrosynthesis of 2′-deoxyribonucleosides from their metabolites through the reverse reactions of degradation. This was the first report of 2′-deoxyribonucleoside production via DR5P from such simple materials such as glucose, acetaldehyde and nucleobase through the multi-step enzymatic process without expensive energy sources such as ATP.

## 9.6
## One-Pot Multi-Step Enzymatic Synthesis of 2′-Deoxyribonucleoside from Glucose, Acetaldehyde and a Nucleobase

We established a means of multi-enzyme-catalyzed three-step synthesis of 2′-deoxyribonucleoside by coupling the glycolytic pathway and the reverse reactions of 2′-deoxyribonucleoside degradation [23]. The 2′-deoxyribonucleoside production amounted to about 10 mM. However, it was difficult to further increase the 2′-deoxyribonucleoside production since phosphate and the phosphorylated glycolysis intermediates such as FDP, G3P and DHAP generated through the Steps 1 and 2 severely inhibited Step 3 [23]. Then, we investigated a means of one-pot enzymatic synthesis of 2′-deoxyribonucleoside in which the concentrations of phosphate and phosphorylated glycolysis intermediates could be controlled more easily, and attained much higher production of 2′-deoxyribonucleoside. In this one-pot system, phosphate generated by nucleoside phosphorylase-catalyzed base transfer is well reused for ATP regeneration by the yeast glycolytic pathway (Scheme 9.5).

The reaction mixture was comprised of 600 mM glucose, 333 mM acetaldehyde, 100 mM adenine, 26 mM $MgSO_4 \cdot 7H_2O$, 33 mM potassium phosphate buffer (pH 7.0), 1.3 mM $MnCl_2 \cdot 4H_2O$, 0.13 mM glucose-1,6-diphosphate, 0.53% (v/v) PL,

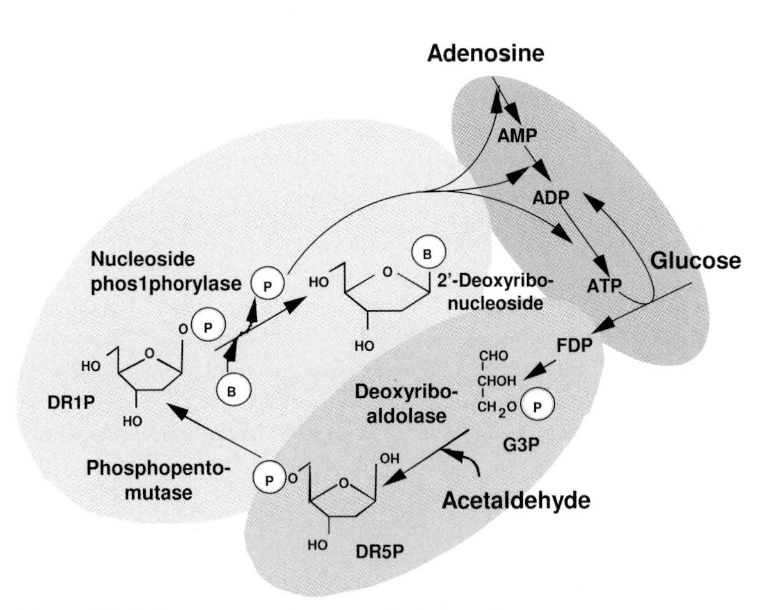


**Scheme 9.5** Multi-step enzymatic process for 2′-deoxyribonucleoside production from glucose, acetaldehyde and a nucleobase through glycolysis, reverse reactions of 2′-deoxyribonucleoside degradation and ATP regeneration by the yeast glycolytic pathway recycling the phosphate generated by nucleoside phosphorylase.

1.3% (v/v) xylene, 10 mM adenosine, 4% (w/v) acetone-dried cells, 6.6% (w/v) wet cells [approximately 0.7% (w/v) dry cells] of *E. coli* 10B5/pTS8, 26.6% (w/v) wet cells [approximately 2.7% (w/v) dry cell] of *E. coli* BL21/pTS17 and 5.3 U/ml commercial purine nucleoside phosphorylase. After 24 h reaction at 33 °C with shaking (120 r.p.m.), the 2′-deoxyribonucleoside (2′-deoxyinosine) production had reached 33.3 mM. This was almost 3.4 times higher than the production with the three-step process. The energy carrier (adenosine) turnover through the whole process was 3.3. The yields as to glucose, acetaldehyde and adenine were 2.8, 10 and 33.3%, respectively [30].

The one-pot process showed different features from those of the three-step process. Adenosine could be used as an energy carrier in the one-pot process, while AMP was required for the three-step process. High concentrations of phosphate (750 mM) and anaerobic conditions were required for the first step (FDP production through glycolysis) of the three-step process according to the Harden–Young equation [18, 19]. On the other hand, 2′-deoxyribonucleoside production with the one-pot process proceeded well with a low concentration of phosphate (33 mM), which could prevent inhibition of phosphopentomutase- and nucleoside phosphorylase-catalyzed reactions by phosphorylated glycolysis intermediates and phosphate itself, resulting in an increase in 2′-deoxyribonucleoside production. Furthermore, 2′-deoxyribonucleoside production with the one-pot process increased with shaking, while a static condition (anaerobic condition) was required for the first step of the three-step process. Adenine, which is one of the substrates, exhibits low solubility. Accordingly, the agitation efficiency might contribute to the reactivity [30].

## 9.7 Improvement of the One-Pot Multi-Step Enzymatic Process for Practical Production of 2′-Deoxyribonucleoside from Glucose, Acetaldehyde and a Nucleobase

The one-pot multi-step enzymatic process described above seemed still impractical on an industrial scale due to the low amount of 2′-deoxyribonucleoside accumulation and the low molar yield to added nucleobase (yield to adenine was 33.3%). In addition, it required three kinds of catalyst cells (baker's yeast, deoxyriboaldolase- or phosphopentomutase-expressing *E. coli* and commercial nucleoside phosphorylase). It is difficult and complicated to operate the multi-catalysts. If the molar yield of 2′-deoxyribonucleoside to the most expensive material, nucleobase, was improved and the number of catalysts could be reduced, a practical enzymatic process could be developed.

We constructed deoxyriboaldolase/phosphopentomutase coexpressing *E. coli*, strain BL21/pACDR-pTS17, and optimized the one-pot 2′-deoxyribonucleoside production from glucose, acetaldehyde and adenine using acetone-dried yeast, *E. coli* BL21/pACDR-pTS17 and commercial nucleoside phosphorylase. It was found that maintaining a low adenine concentration in the reaction mixture was effective for improvement of the yield of 2′-deoxyribonucleoside to adenine. As for the

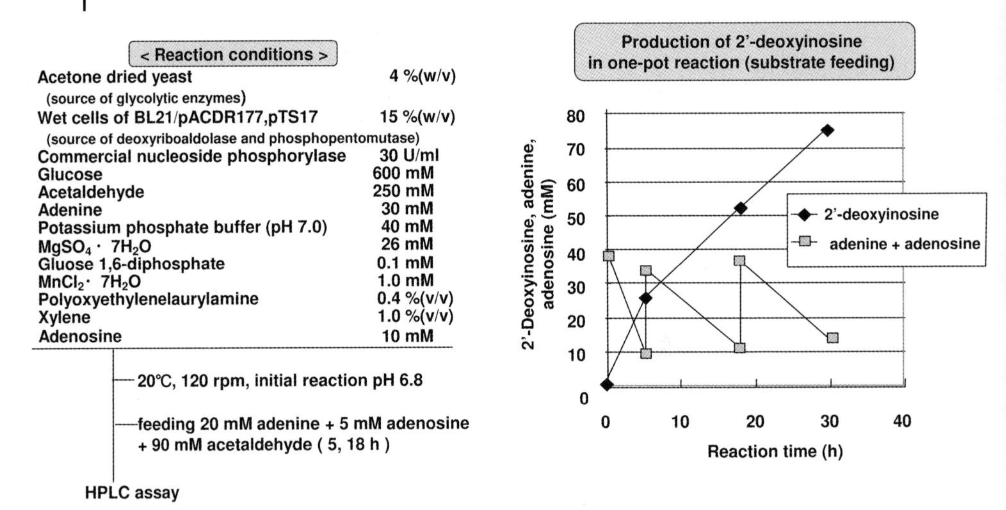


**Figure 9.1** 2′-Deoxyribonucleoside production from glucose, acetaldehyde and a nucleobase in a one-pot (substrate-feeding) reaction.

endurance of catalysts, acetone-dried yeast could not keep its activity more than 20 h. Thus, by feeding a little amount of adenine, adenosine and acetone-dried yeast into the reaction mixture, the amount of 2′-deoxyribonucleoside production was greatly increased to 75 mM (2′-deoxyinosine) and the yield of 2′-deoxyribonucleoside to added base was considerably improved (83.3%) (Figure 9.1). Once a 2′-deoxyribonucleoside was produced, the other 2′-deoxyribonucleosides could be synthesized through base-exchange reactions catalyzed by various nucleoside phosphorylases [31, 32].

## 9.8 Conclusions

Above all, we could demonstrate the practical production of 2′-deoxyribonucleoside from glucose, acetaldehyde and a nucleobase by multi-step enzymatic processes. In the processes, especially in the one-pot process, successive reactions (i.e. glycolysis for FDP generation, reverse reactions of 2′-deoxyribonucleoside degradation and ATP regeneration by the yeast glycolytic pathway) recycling the phosphate generated by nucleoside phosphorylase reaction might be well coupled. The use of economical substrates (glucose, acetaldehyde and a nucleobase), the high amount of product accumulation and the high yield to the starting material made the multi-step enzymatic process suitable for industrial application (Scheme 9.6).

A further enlargement of utilization of DNA is anticipated in the future. The nature of DNA (i.e. repetitious structure and stability) is expected to be used as a

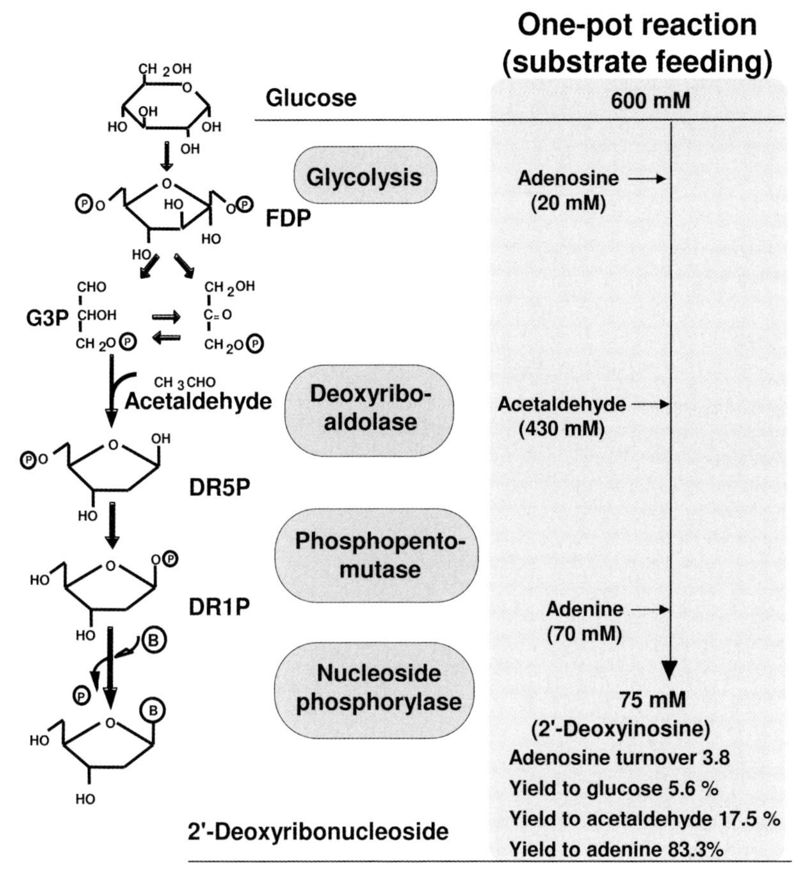


**Scheme 9.6** Summary of 2′-deoxyribonucleoside production from glucose, acetaldehyde and a nucleobase in a one-pot (substrate-feeding) reaction.

commodity and in electronic materials. The demand for DNA will expand not only the field of biotechnology, but also the chemical and electronic industries. Therefore, the efficient supply of a building block of DNA (i.e. 2′-deoxyribonucleoside) will be necessary. We hope our multi-step enzymatic production process of 2′-deoxyribonucleoside and its synthesis intermediates will contribute to the development of the life sciences and various industries through the practical supply of 2′-deoxyribonucleosides.

## Acknowledgments

This work was partially supported by the Industrial Technology Research Grant Program (02A07001a to J.O.) of the New Energy and Industrial Technology Development Organization, Japan, and Grants-in-Aid for Scientific Research (16688004 to J.O.) and COE for Microbial-Process Development Pioneering Future Production Systems (to S.S.) from the Ministry of Education, Science, Sports, and Culture, Japan.

## References

1 Schmid, A., Dordick, J.S., Hauer, B., Kiener, A., Wubbolts, M. and Witholt, B. (2001) *Nature*, **409**, 258–68.
2 Leuchtenberger, W., Huthmacher, K. and Drauz, K. (2005) *Applied Microbiology and Biotechnology*, **69**, 1–8.
3 Ogawa, J., Sakuradani, E. and Shimizu, S. (2002) *Lipid Biotechnology* (eds T.M. Kuo and H.W. Gardner), Mercel Dekker, New York, pp. 563–74.
4 Mori, H., Iida, A., Fujio, T. and Teshiba, S. (1997) *Applied Microbiology and Biotechnology*, **48**, 693–8.
5 Fujio, T., Nishi, T., Ito, S. and Maruyama, A. (1997) *Bioscience, Biotechnology, and Biochemistry*, **61**, 840–5.
6 Tomita, F. and Suzuki, S. (1972) *Agricultural and Biological Chemistry*, **36**, 133–40.
7 Hara, T. and Ueda, S. (1981) *Agricultural and Biological Chemistry*, **45**, 2457–61.
8 Aoyama, H. (1987) *Bulletin of the Chemical Society of Japan*, **60**, 2073–7.
9 Hoffer, M. (1960) *Chemische Berichte*, **93**, 2777–81.
10 Kawakami, H., Matsushita, M., Naoi, Y., Itoh, K. and Yoshikoshi, H. (1989) *Chemistry Letters*, 235–8.
11 Park, M. and Rizzo, C.J. (1996) *Journal of Organic Chemistry*, **61**, 6092–3.
12 Brunella, A. and Ghisalba, O. (2000) *Journal of Molecular Catalysis B: Enzymatic*, **10**, 215–22.
13 Brunella, A., Abrantes, M. and Ghisalba, O. (2000) *Bioscience, Biotechnology, and Biochemistry*, **64**, 1836–41.
14 Ogawa, J., Saito, K., Sakai, T., Horinouchi, N., Kawano, T., Matsumoto, S., Sasaki, M., Mikami, Y. and Shimizu, S. (2003) *Bioscience, Biotechnology, and Biochemistry*, **67**, 933–6.
15 Horinouchi, N., Ogawa, J., Sakai, T., Kawano, T., Matsumoto, S., Sasaki, M., Mikami, Y. and Shimizu, S. (2003) *Applied and Environmental Microbiology*, **69**, 3791–7.
16 Tochikura, T., Kuwahara, M., Yagi, S., Okamoto, H., Tominaga, Y., Kano, T. and Ogata, K. (1967) *Journal of Fermentation Technology*, **45**, 511–29.
17 Yamamoto, K., Kawai, H. and Tochikura, T. (1981) *Applied and Environmental Microbiology*, **41**, 392–5.
18 Harden, A. and Young, W. (1905) *Proceedings of the Royal Society of London, Series B*, **77**, 405–20.
19 Wakisaka, S., Ohshima, Y., Ogawa, M., Tochikura, T. and Tachiki, T. (1998) *Applied and Environmental Microbiology*, **64**, 2953–7.
20 Yamamoto, S., Wakayama, M. and Tachiki, T. (2005) *Bioscience, Biotechnology, and Biochemistry*, **69**, 784–9.
21 Horinouchi, N., Ogawa, J., Kawano, T., Sakai, T., Saito, K., Matsumoto, M., Sasaki, M., Mikami, Y. and Shimizu, S. (2006) *Bioscience, Biotechnology, and Biochemistry*, **70**, 1371–8.
22 Fujio, T. and Maruyama, A. (1997) *Bioscience, Biotechnology, and Biochemistry*, **61**, 956–9.
23 Horinouchi, N., Ogawa, J., Kawano, T., Sakai, T., Saito, K., Matsumoto, M., Sasaki, M., Mikami, Y. and Shimizu, S. (2006)

*Applied Microbiology and Biotechnology*, **71**, 615–21.

**24** Ouwerkerk, N., Steenweg, M., de Ruijter, M., Brouwer, J., van Boom, J.H., Lugtenburg, J. and Raap, J. (2002) *Journal of Organic Chemistry*, **67**, 1480–9.

**25** Ishii, K. and Shiio, I. (1972) *Agricultural and Biological Chemistry*, **36**, 1511–22.

**26** Fruya, A., Kato, F. and Nakayama, K. (1975) *Agricultural and Biological Chemistry*, **39**, 767–71.

**27** Ogata, K. (1974) *Advances in Applied Microbiology*, **19**, 209–47.

**28** Satoh, T., Ishizaki, A. and Ueda, S. (1989) *Journal of Fermentation and Bioengineering*, **68**, 92–5.

**29** Prazeres, D.M.F., Ferreira, G.N.M., Monterio, G.A., Cooney, C.L. and Cabral, J.M.S. (1999) *Trends in Biotechnology*, **17**, 169–74.

**30** Horinouchi, N., Ogawa, J., Kawano, T., Sakai, T., Saito, K., Matsumoto, M., Sasaki, M., Mikami, Y. and Shimizu, S. (2006) *Biotechnology Letters*, **28**, 877–81.

**31** Utagawa, T. (1999) *Journal of Molecular Catalysis B: Enzymatic*, **6**, 215–22.

**32** Yokozeki, K. and Tsuji, T. (2000) *Journal of Molecular Catalysis B: Enzymatic*, **10**, 207–13.

# 10
# Combination of Biocatalysis and Chemical Catalysis for the Preparation of Pharmaceuticals Through Multi-Step Syntheses

*Vicente Gotor-Fernández, Rosario Brieva, and Vicente Gotor*

## 10.1
## Introduction: Biocatalysis and Chemical Catalysis

The production of pharmaceuticals generally involves multi-step organic synthesis, commonly performed in a classical stepwise approach. In this manner, a starting material is transformed into the required product *via* one or several intermediates. In many cases each intermediate is isolated and purified prior to use of as starting material in the next step. Usually all the reaction steps are individually optimized, and in some cases tedious purification protocols are necessary with the formation of great amounts of waste and loss of part of the reaction yield. In recent decades, biocatalysis has emerged as a very useful methodology for the synthesis of intermediates in the pharmaceutical industry, especially when the objectives are the synthesis of chiral drugs [1]. Nowadays, biocatalysis is commercially applied at volumes that range from the gram scale in the pharmaceutical industry to multi-thousand-ton scales in the manufacturing of commodity chemicals. In addition, biocatalytic production of chemicals appears to be a very promising way to reduce the environmental impact of the industrial processes.

Over the past few years, interesting processes have appeared to catalyze single-step transformations using hydrolases, oxidoreductases or lyases as biocatalysts. The advantage of using biocatalysis is that this methodology offers a clean and ecological way to perform chemical processes under mild reaction conditions and with a high degree of selectivity. For these reasons, the combination of biocatalysis and chemical catalysis in multi-step synthesis using different enzymes is growing in the industrial sector, especially in the pharmaceutical industry due to the increasing demand of enantiomerically pure compounds [2].

In order to develop cost-effective and environmentally friendly chemical processes at the industrial scale, it is essential to integrate biocatalysis and modern chemical catalysis to produce manufacturing routes with fewer synthetic steps, reducing waste steams and improving overall synthetic yields. Normally, in these tandem procedures, the initial step is the preparation of the intermediate by

*Multi-Step Enzyme Catalysis: Biotransformations and Chemoenzymatic Synthesis*
Edited by Eduardo Garcia-Junceda

ISBN: 978-3-527-31921-3

biocatalysis to yield later the target drug by conventional chemical catalysis. In some cases, where the structure of the pharmaceutical is very simple, the enzymatic resolution of the racemic product can be carried out in the final step. The importance of the synthesis of single enantiomers is noteworthy because usually one enantiomer is more active than its counterpart [3].

This chapter covers some general aspects of the use of enzymes in aqueous and organic media. Although lipases are the most common biocatalysts in these processes [4], other hydrolytic enzymes such as esterases and nitrilases have also shown their utility in the manufacture of pharmaceuticals. In addition, some representative examples using oxidoreductases and lyases will be also discussed.

## 10.2 Pharmaceuticals with Hydrolases

### 10.2.1 Enzymatic Hydrolysis

Although the hydrolysis of esters with lipases and esterases represents the most common process to obtain chiral intermediates for the synthesis of pharmaceuticals, proteases and other hydrolytic enzymes such as epoxide hydrolases and nitrilases have also been used for this purpose. We show here a few representative examples of the action of these biocatalysts that have been recently published.

β-Adrenergic-blocking agents, such as propranolol, have been synthesized by different chemoenzymatic methods where the key step to introduce the chirality is an enzymatic acylation or a hydrolysis process. The main reason to prepare these amino alcohols in optically pure form is due to the fact that the activity of these pharmaceuticals resides in the (*S*)-enantiomer. In Scheme 10.1 we have represented a chemoenzymatic approach that has been carried out for the preparation of this drug where the key step is the resolution of the key intermediate 1-chloro-

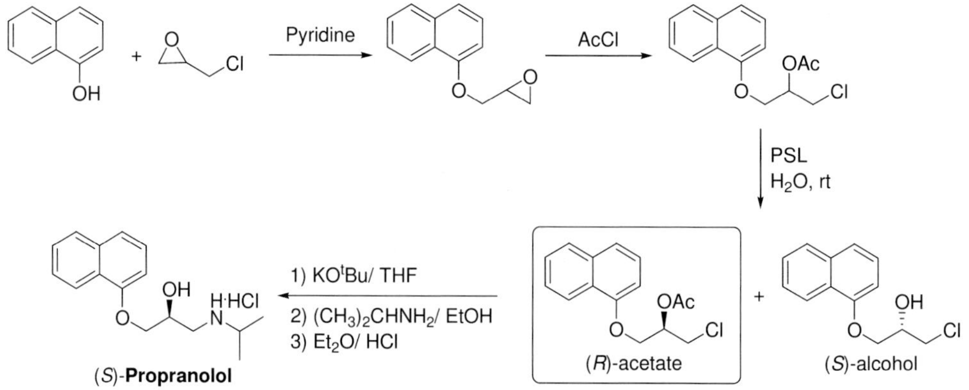

**Scheme 10.1** Chemoenzymatic preparation of (*S*)-propranolol.

2-acetoxy-3-(1-naphthyloxy)propane using *Pseudomonas cepacia* lipase (PSL) in aqueous media [5].

α-Aryl propionic acid derivatives form a major class of non-steroidal antiinflammatory drugs and several approaches have been devised using either microorganisms or isolated enzymes for the synthesis of this class of compounds [6]. For instance, racemic methyl and ethyl esters have been synthesized and later resolved using *Candida rugosa* lipase (CRL) pretreated with 2-propanol to increase the activity and enantioselectivity of the enzyme [7]. An example is shown in Scheme 10.2, where the CRL-catalyzed hydrolysis of the ester allows the recovery of the (*R*)-methyl ester that lowers serum cholesterol levels and prevents platelet aggregation, while the (*S*)-ester causes a side-effect (i.e. muscle irritability and spasms) by inhibiting the chloride channel in muscles.

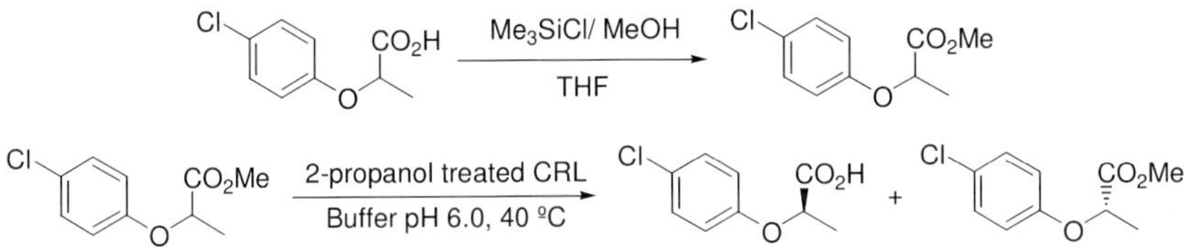

**Scheme 10.2** Synthesis of a biologically active methyl ester.

Naproxen and ibuprofen are probably the best-known aryl propionic acids, and are sold in large quantities. The sole administration of the (*S*)-enantiomer of ibuprofen is indicated because it has been demonstrated that (*R*)-ibuprofen accumulates in fatty issue as a glycerol ester whose long-term effects may be critical. Several strategies have been recently developed using different kind of hydrolytic enzymes for the synthesis of enantiomerically pure (*S*)-ibuprofen. Two examples are shown in Scheme 10.3 using in the enzymatic step the porcine pancreas lipase (PPL) [8] or *Aspergillus niger* lipase [9].

An interesting example of biocatalysis and chemical catalysis is the synthesis of a derivative of γ-aminobutyric acid (GABA) that is an inhibitor for the treatment of neuropathic pain and epilepsy (Scheme 10.4). The key intermediate is a racemic mixture of *cis*- and *trans*-diastereoisomer esters obtained by a hydrogenation following a Horner–Emmons reaction. The enzymatic hydrolysis of both diastereoisomers, catalyzed by *Candida antarctica* lipase type B (CALB), yields the corresponding acid intermediate of the GABA derivative. It is of note that both *cis*- and *trans*-diastereoisomers of the desired enantiomer of the acid intermediate can be converted into the final product in the downstream chemistry [10].

(±)-Zopiclone is a chiral cyclopyrrolone with hypnotic properties, possessing a pharmaceutical profile of high efficacy and low toxicity, similar to that of benzodiazepines. Zopiclone has been commercialized as a racemic mixture; however, the (*S*)-enantiomer is more active and less toxic than the (*R*)-enantiomer [11]. Although enzymatic hydrolysis of esters or transesterification processes of alcohols have been widely applied for enzymatic resolution or desymmetrization

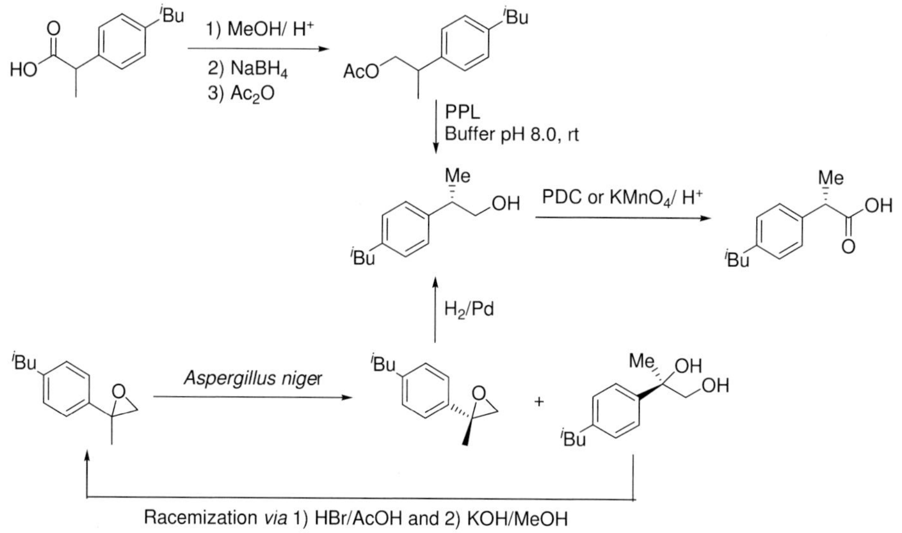

Scheme 10.3 Two different preparations of (*S*)-ibuprofen by chemoenzymatic methods.

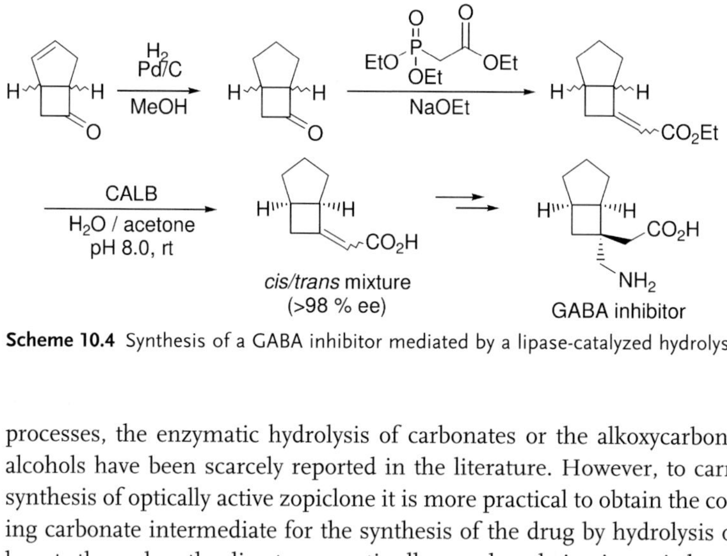

Scheme 10.4 Synthesis of a GABA inhibitor mediated by a lipase-catalyzed hydrolysis.

processes, the enzymatic hydrolysis of carbonates or the alkoxycarbonylation of alcohols have been scarcely reported in the literature. However, to carry out the synthesis of optically active zopiclone it is more practical to obtain the corresponding carbonate intermediate for the synthesis of the drug by hydrolysis of the carbonate than when the direct enzymatic alkoxycarbonylation is carried out (Scheme 10.5). The optimum process was reached when the less-expensive chloromethylchloroformate is used as starting material to prepare the corresponding carbonate ($R{=}CH_2Cl$) [12]. In this case the last step takes place with an isolated yield higher than 90%.

Epoxide hydrolases are less used than lipases; however, in recent years the synthesis of chiral diols with these biocatalysts has emerged as an excellent methodology to develop new and interesting chemoenzymatic processes [13].

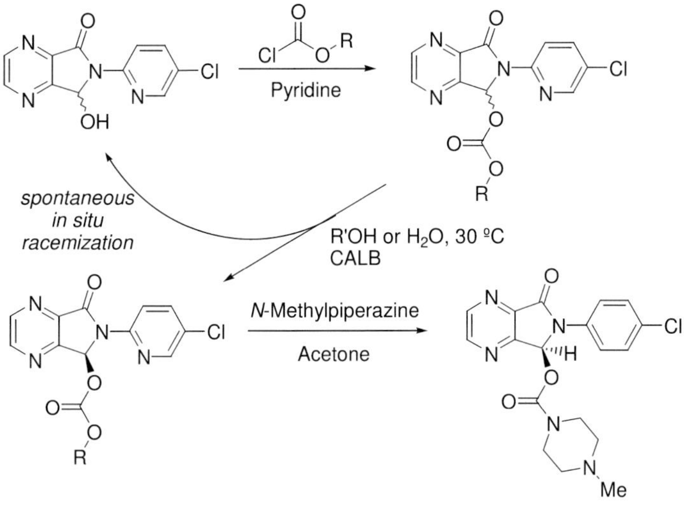

spontaneous in situ racemization

Pyridine

R'OH or $H_2O$, 30 ºC CALB

*N*-Methylpiperazine

Acetone

(*S*)-(+)-**Zopiclone**

**Scheme 10.5** Chemoenzymatic synthesis of (*S*)-(+)-zopiclone.

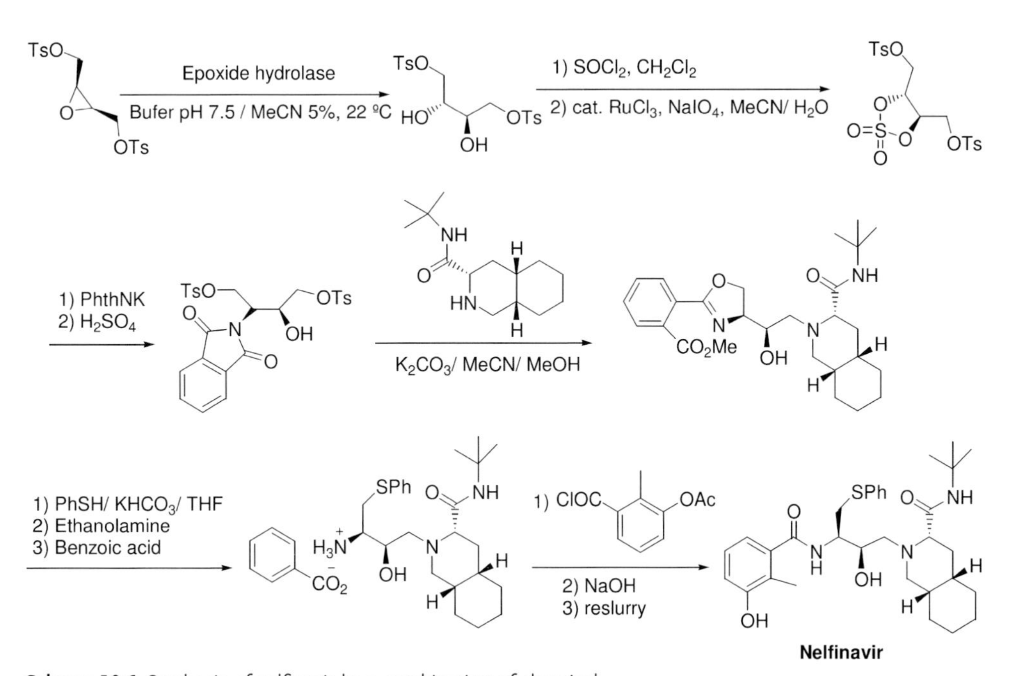


**Scheme 10.6** Synthesis of nelfinavir by a combination of chemical and enzymatic methodologies.

In addition, new some epoxide hydrolases have also shown a great utility for the desymmetrization of *meso*-epoxides. An interesting example is the synthesis of nelfinavir–the active pharmaceutical ingredient (API) of the anti-human immunodeficiency virus drug Viracept–where the (*R*,*R*)-diol obtained by opening the *meso*-epoxide is a suitable starting material. Scheme 10.6 shows a synthetic route to nelfinavir [14].

The use of microbial nitrilases has increased in recent years for the hydrolysis of nitriles to carboxylic acid under milder conditions than by chemical methods [15]. 1-Cyanoacyclohexaneacetic acid is an intermediate in the synthesis of 1-aminoethyl-cyclohexanoacetic acid (gabapentin), which has been used in the treatment of cerebral diseases [16]. Scheme 10.7 shows an easy preparation of these compounds where a regioselective biocatalytic hydrolysis followed by a catalytic chemical reduction yields gabapentin. Reactions were performed using either unimmobilized cells or alginate-immobilized cells as biocatalyst [17], where quantitative conversion of the dinitrile is obtained with total regioselectivity towards the desired product.

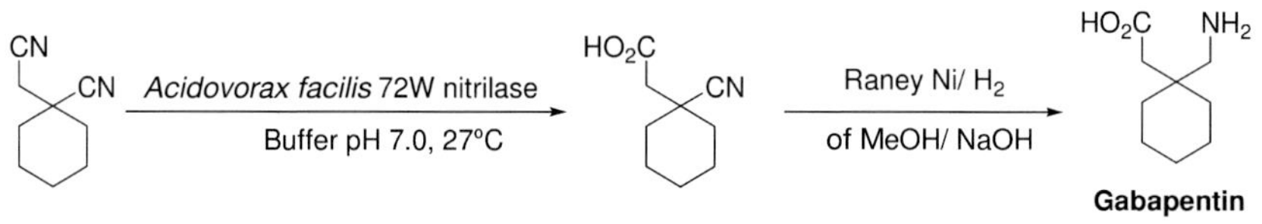

CN CN *Acidovorax facilis* 72W nitrilase, Buffer pH 7.0, 27°C → $HO_2C$ CN; Raney Ni/ $H_2$ of MeOH/ NaOH → $HO_2C$ $NH_2$ **Gabapentin**

**Scheme 10.7** Production of gabapentin by chemoenzymatic methods.

Recently, another interesting application of nitrilases has been demonstrated for the synthesis of pregabalin–the API of the neurophatic pain drug Lyrica. In this approach, the key step is the resolution of racemic isobutylsuccinonitrile (Scheme 10.8) [18], the process takes place with total regio- and stereoselectivity, and the (*S*)-acid is obtained and the (*R*)-substrate can be recycled under basic conditions. To improve the biocatalytic step, directed evolution was applied using the 'electronic polymerase chain reaction' and in the first round of evolution a single C236S mutation led to a mutant with 3-fold increase in activity [19].

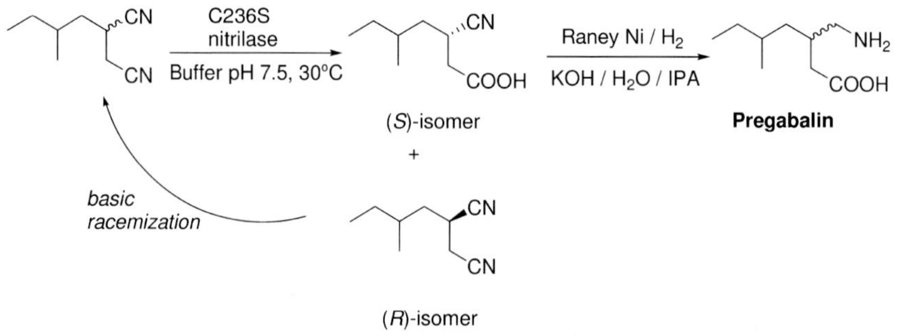

CN CN C236S nitrilase, Buffer pH 7.5, 30°C → CN COOH (*S*)-isomer + CN CN (*R*)-isomer; basic racemization; Raney Ni / $H_2$, KOH / $H_2O$ / IPA → $NH_2$ COOH **Pregabalin**

**Scheme 10.8** Synthesis of pregabalin using an evolved nitrilase.

## 10.2.2
## Enzymatic Transesterification

Hydrolytic enzymes, especially lipases, have been exploited for asymmetric synthesis transformations, led by the growing demand for enantiopure pharmaceuticals. Furthermore, lipase-catalyzed processes are normally carried out under mild reaction conditions and can be used in organic solvents. In addition, biocatalysis in non-aqueous media has been widely used for the resolution of alcohols, acids or lactones through enzymatic transesterification reactions using different lipases. Although the hydrolysis reaction is older than the acylation of alcohols or amines, since the knowledge that some lipases can work in organic solvents, this methodology has dramatically increased in the last 20 years [20]. We have chosen a few representative examples where the key step to obtain the chiral pharmaceutical is an enzymatic acylation of primary or secondary alcohols.

The potent $\beta_2$-receptor agonist formoterol is on the market as a diastereomeric mixture despite varying efficacy of the stereoisomers. The preparation of the active (*R*,*R*)-stereoisomer was carried out from the optically active (*R*)-bromohydrin, which was achieved by enzymatic resolution of the corresponding racemic bromohydrin (Scheme 10.9) [21]. Again, the key step is the biocatalytic process because later the coupling of the (*R*)-alcohol with the corresponding amine yields the chiral drug in good overall yield.

(−)-Paroxetine hydrochloride is a selective serotonin reuptake inhibitor that is used as an antidepressant. The chemoenzymatic synthesis can be carried out

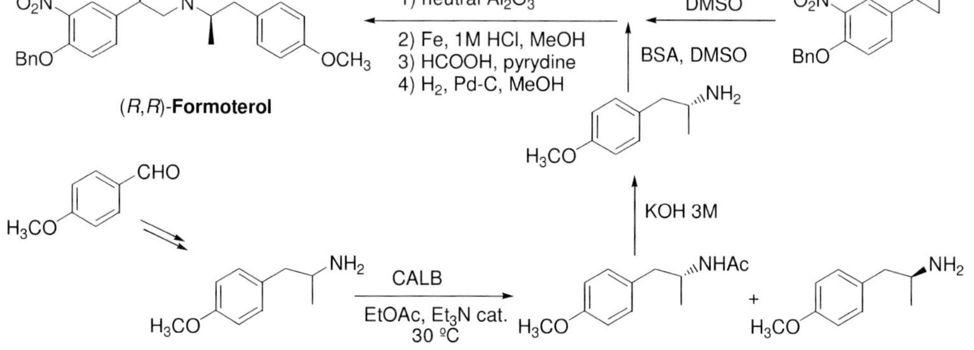

**Scheme 10.9** Synthesis of optically active formoterol using an enzymatic transesterification step.

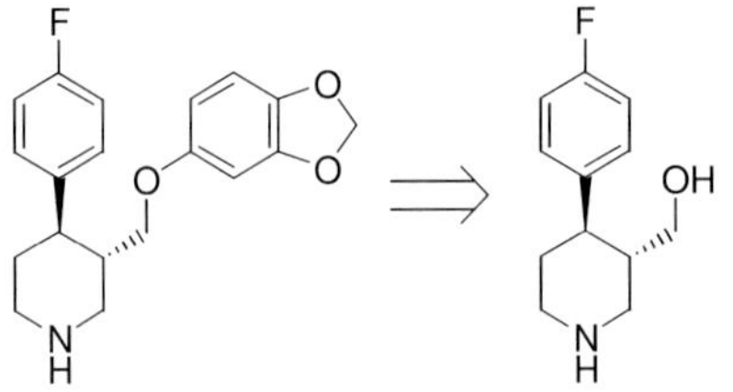
F
O
O
O
F
OH
N
H
N
H

(-)-**Paroxetine**

**Scheme 10.10** Retrosynthetic analysis of (–)-paroxetine.

through the piperidine shown in Scheme 10.10. In principle there are two possibilities – enzymatic hydrolysis of the corresponding ester or enzymatic acylation of the primary alcohol.

It is remarkable that better enantioselectivities are achieved when CALB-catalyzed acylations of the alcohol are carried out in organic solvent rather than in water. Excellent enantioselectivities are obtained when the process is carried out with vinyl esters [22]. However, in some cases the use of vinyl or alkyl esters as acyl donors has the drawback of the separation of the ester (product) and the alcohol (substrate). A practical strategy to avoid this problem is the use of cyclic anhydrides [23]. In this case an acid is obtained as product, which can be readily separated from the unreacted alcohol by a simple aqueous base–organic solvent liquid–liquid extraction. This methodology has been successfully used for the synthesis of (–)-paroxetine as indicated in Scheme 10.11 [24].

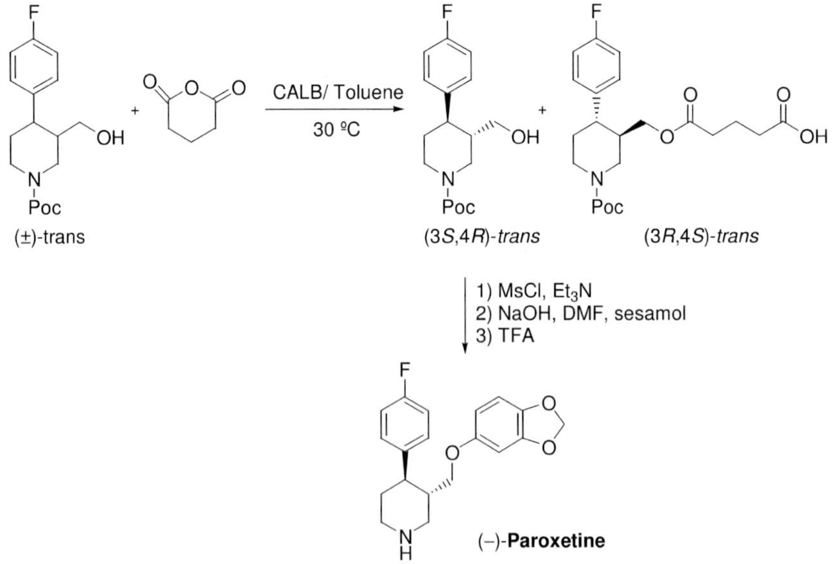
F
OH
N
Poc
(±)-trans
+
O
O
O
CALB/ Toluene
30 ºC
F
OH
N
Poc
(3S,4R)-trans
+
F
O
O
O
OH
N
Poc
(3R,4S)-trans
1) MsCl, $Et_3N$
2) NaOH, DMF, sesamol
3) TFA
F
O
O
O
N
H
(–)-Paroxetine

**Scheme 10.11** Chemoenzymatic preparation of (–)-paroxetine using cyclic anhydrides in the enzymatic step.

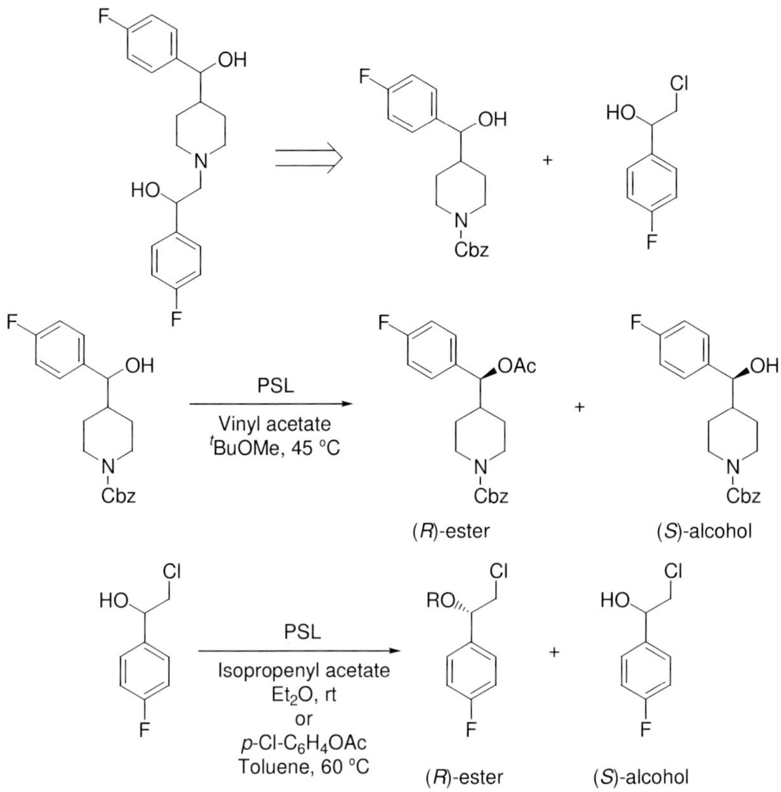

F
OH
N
HO
F
F
OH
N
Cbz
+
Cl
HO
F
F
OH
N
Cbz
PSL
Vinyl acetate
ᵗBuOMe, 45 °C
F
OAc
N
Cbz
(R)-ester
+
F
OH
N
Cbz
(S)-alcohol
Cl
HO
F
PSL
Isopropenyl acetate
Et2O, rt
or
p-Cl-C6H4OAc
Toluene, 60 °C
Cl
RO
F
(R)-ester
+
Cl
HO
F
(S)-alcohol

**Scheme 10.12** Use of lipases for the production of a pharmaceutical used as an analgesic and muscle relaxant.

Resolution of compounds with several chiral centers presents a significant challenge, so the compound shown in Scheme 10.12 that is also serotonin antagonist, which was identified as a non-narcotic analgesic and muscle relaxant, has two chiral centers and therefore exists in the form of four stereoisomers with different biological properties. It was divided in two intermediates that are secondary alcohols, which were resolved by enzymatic acylation obtaining four chiral building blocks. The simple chemical combination of these blocks gave the four different stereoisomers [25].

The resolution of primary or secondary alcohols is easy by enzymatic acylation; however, few examples have been described for the resolution of tertiary alcohols [26].

One example is the enzymatic resolution of 4-[(4-dimethylamino)-1-(4′-fluorophenyl)-1-hydroxy-1-butyl]-3-(hydroxymethyl)benzonitrile – a useful intermediate for the synthesis of optically pure citalopram (an efficient human antidepressant). It has been demonstrated that almost the entire inhibition activity resides in the (*S*)-(+)-enantiomer [27]. Its synthesis has been investigated using several lipases

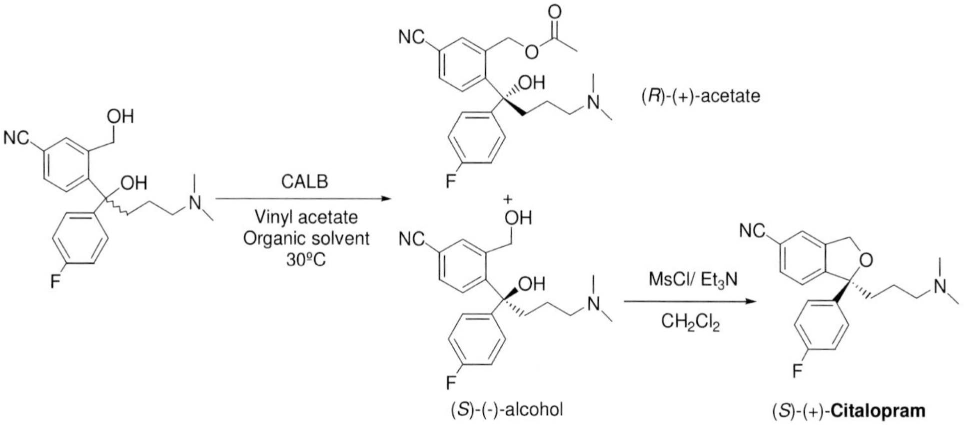

**Scheme 10.13** Chemoenzymatic synthesis of (*S*)-citalopram catalyzed by CALB.

and different acyl donors under different reaction conditions [28]. It is remarkable that CALB catalyzes the enzymatic acetylation of the primary benzylic alcohol with high enantioselectivity at the quaternary chiral centre and no acylation of tertiary alcohol was observed; the final step consists in a simple chemical cyclization with mesyl chloride (Scheme 10.13).

Although so far we have only commented on kinetic transesterification resolutions, there is another possibility of a great interest–the desymmetrization of prochiral or *meso*-compounds using hydrolytic enzymes. Enantioselective enzymatic desymmetrizations belong to the field of asymmetric synthesis and, accordingly, a maximum yield of 100% can be attained. For this reason, they constitute a very interesting alternative to kinetic resolutions for the preparation of optically active compounds, which is reflected in the increasing number of enzymatic desymmetrizations applied to asymmetric synthesis; many authors have been using this methodology in recent years [29]. For example, this methodology has been applied to the synthesis of (*R*)-4-amino-3-(4-chlorophenyl)butanoic acid, also known as (*R*)-baclofen, that is an analog of GABA used as an inhibitory neurotransmitter that regulates the control of neuronal activity in the central nervous system and several physiological mechanisms (Scheme 10.14) [30].

### 10.2.3
### Enzymatic Aminolysis

Although in recent years transesterification processes of racemic alcohols have received major attention, enzymatic acylation of amines for synthetic purposes is also being employed as a conventional tool for the synthesis of chiral amines and amides [31], using CALB as the biocatalyst in the majority of these reactions [31a]. The main difference between enzymatic acylation of alcohols and amines is the use of the corresponding acyl donor, because activated esters which are of utility

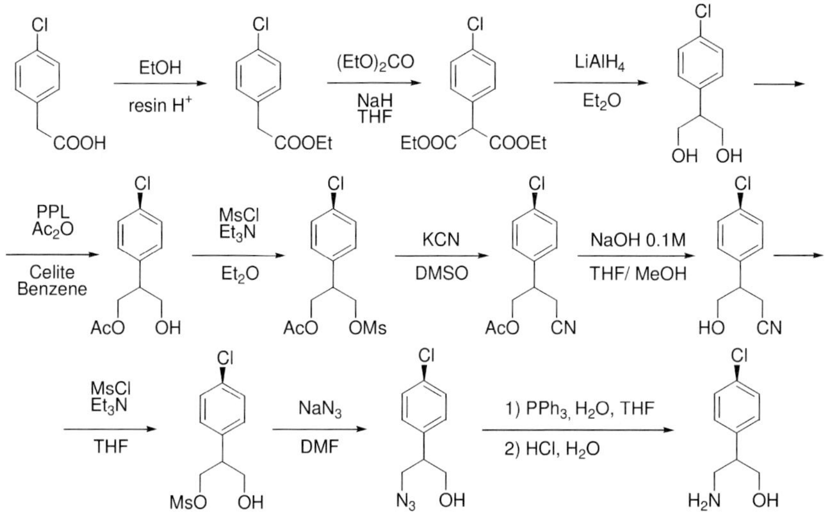

Scheme 10.14 Synthesis of (*R*)-baclofen by chemoenzymatic methods.

in acylation of alcohols react with amines in the absence of biocatalyst, so that non-activated esters must be used to carry out an enzymatic aminolysis or ammoniolysis reaction. In addition, lipases are the most efficient hydrolases to catalyze the acylation of amines and ammonia, because these hydrolytic enzymes have very low amidase activity, although in some cases the hydrolysis or alcoholysis of amides can be useful to achieve chiral amines [32].

Rasagiline and some of its derivatives have been shown to be highly selective and potent inhibitors of the B form of monoamine oxidase; in contrast, the levorotatory enantiomer is inactive. These compounds have been shown to be useful in the treatment of Parkinson's disease, memory disorders, dementia of the Alzheimer type, depression and hyperactive syndrome in children. The synthesis of (*R*)-*N*-propargyl-1-aminoindan (rasagiline) has been easily performed from propargyl chloride and (*R*)-1-aminoindan previously enzymatically resolved using Subtilisin Carslberg and 2,2,2-trifluoroethyl butyrate in a continuous reactor using 3-methyl-3-pentanol as solvent (Scheme 10.15) [33].

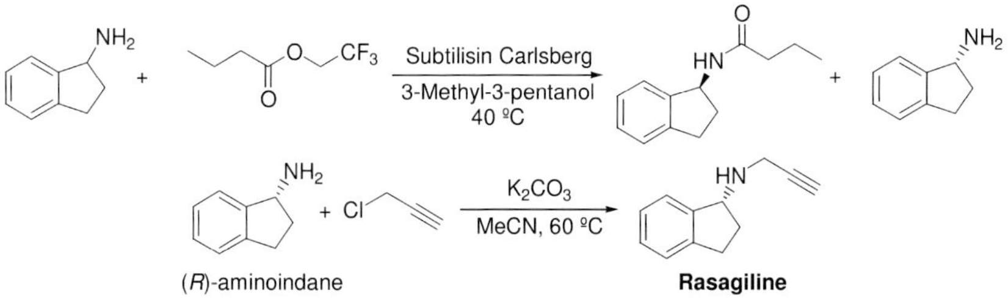

Scheme 10.15 Chemoenzymatic synthesis of rasagiline using an aminolysis process.

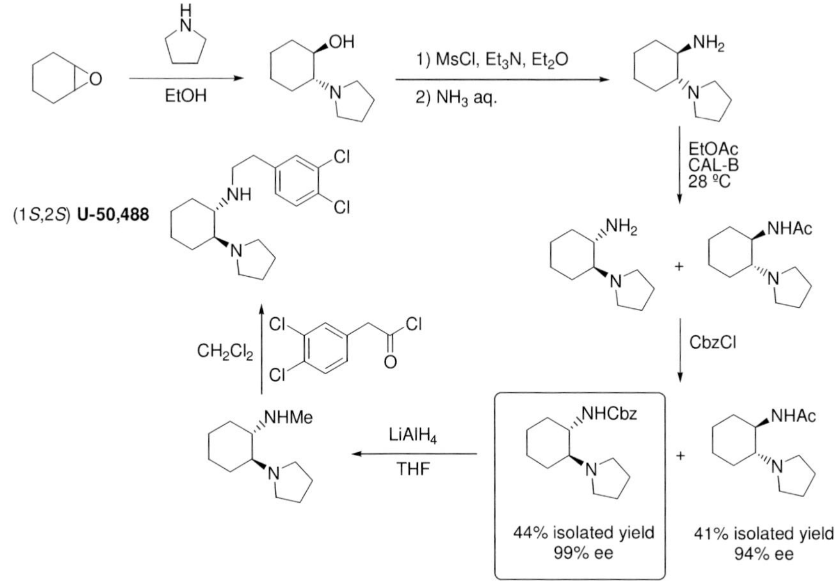

Scheme 10.16 Chemoenzymatic synthesis of compound U-50,488.

Compound U-50,488 (Scheme 10.16) and other structural analogs have been reported to be highly selective κ-opioid agonists, free from the adverse side-effects of μ-agonists like morphine. The preparation of this drug is easily achieved starting from cyclohexene oxide where the key step is the resolution of the intermediate (±)-*trans*-2-(pyrrolidin-1-yl)cyclohexanamine using CALB as biocatalyst. In the enzymatic process CALB shows very high enantioselectivity preferentially catalyzing the acetylation of the (1*R*,2*R*); the protected substrate protected with the benzyloxycarbonyl group is used in the following chemical step to yield the chiral drug with excellent yields [34].

Although the enzymatic resolution of 1,2-amino alcohols has been exhaustively investigated, the bioresolution of 1,3-amino alcohols has been scarcely reported, especially through an enzymatic aminolysis processes. Recently, the enzymatic resolution of 3-amino-3-phenylpropan-1-ol derivatives–key intermediates for the synthesis of (*S*)-dapoxetine–has been studied by an enzymatic acylation process (Scheme 10.17). *Candida antarctica* lipase type A (CALA) has been identified as the best biocatalyst for the *N*-acylation reaction of 3-amino-3-phenyl-1-*tert*-butyldimethylsilyloxypropan-1-ol using ethyl methoxyacetate as acylating agent and *tert*-butyl methyl ether as solvent [35]. This is not a surprising result as CALA is considered as an ideal biocatalyst for the resolution of sterically hindered compounds [36]. The synthesis of dapoxetine starting from benzaldehyde is shown in Scheme 10.17; the final chemical step takes place through a Mitsunobu reaction.

The potential of enzymes, especially lipases, to catalyze the aminolysis and ammonolysis of prochiral substrates has been scarcely studied, and only the

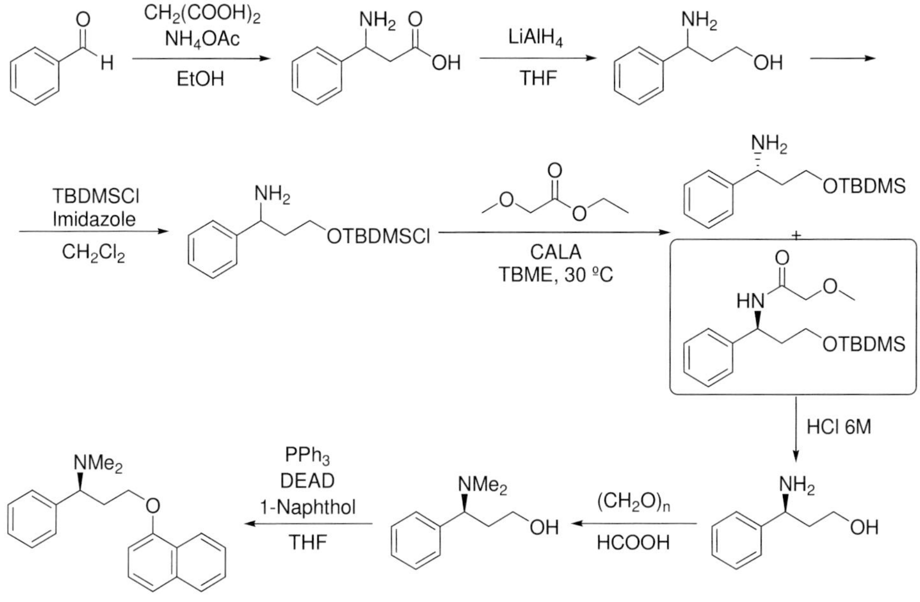

**Scheme 10.17** Synthesis of (*S*)-dapoxetine using CALA.

desymmetrization of prochiral glutarates has been reported [37]. CALB is again the only biocatalyst that yields good results and reacts the pro-(*R*)-ester group to yield the monoamidoester of (*S*)-configuration with very high yields and enantiomeric excesses (Scheme 10.18). (*R*)-4-Amino-3-hydroxybutanoic acid [(*R*)-GABOB] is a compound of great importance because of its biological function as a neuromodulator in the mammalian central nervous system. Moreover (*R*)-GABOB is a precursor of (*R*)-carnitine (vitamin BT) – a therapeutic agent for the treatment of myocardial ischemia. (*R*)-GABOB has been synthesized by a combination of biochemical and chemical catalysis. The reaction was carried out with ammonia-saturated 1,4-dioxane, the enzyme catalyzing the transformation of the pro-(*R*) group of the ester, affording the corresponding enantiopure (*S*)-monoamide.

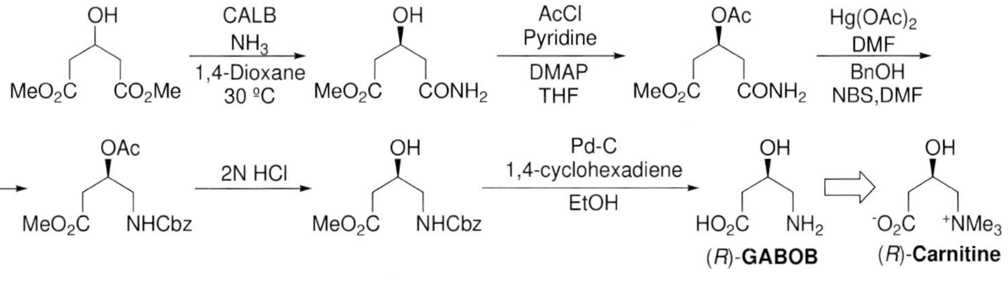

**Scheme 10.18** Chemoenzymatic synthesis of (*R*)-GABOB.

From this monoamide, (*R*)-GABOB was obtained after four sequential synthetic steps [38].

## 10.3
## Pharmaceuticals with Oxidoreductases

Oxidoreductases are, after lipases, the second most-used kinds of biocatalysts in organic synthesis. Two main processes have been reported using this type of enzymes – bioreduction of carbonyl groups [39] and biohydroxylation of non-activated substrates [40]. However, in recent few years other processes such as deracemization of amines or alcohols [41] and enzymatic Baeyer–Villiger reactions of ketones and aldehydes [42] are being used with great utility in asymmetric synthesis.

The main drawback of the processes catalyzed by oxidoreductases in comparison with hydrolytic reactions mediated by lipases resides in the necessity of cofactors; for this reason microorganisms are normally used instead of isolated enzymes. However, they present a very important advantage – the possibility to obtain only one enantiomer in the reaction – so that higher yields can be achieved than in normal kinetic resolution processes catalyzed by lipases.

We show here some representative examples of these types of reactions employing alcohol dehydrogenases or amino oxidases because biohydroxylation processes are generally used to improve the activity of a drug or in the synthesis of metabolites [43], meanwhile Baeyer–Villiger monooxigenases have emerged in recent years and only model substrates are generally studied [44].

The bioreduction of carbonyl compounds with reductases has been exploited for many years, especially in the case of ketones, with baker's yeast (*Saccharomyces cerevisiae*) being the most popular biocatalyst [45]. For instance, yeast treatment of 3-chloropropiophenone affords the expected (1*S*)-3-chloro-1-phenylpropan-1-ol, which was treated with trifluorocresol in tertrahydrofuran in the presence of triphenylphosphine and diethyl azodicarboxylate at room temperature to give (3*R*)-1-chloro-3-phenyl-3-[4-(trifluoromethyl)phenoxy]propane and the later reaction with methylamine leads to (*R*)-fluoxetine that is an important serotonin uptake inhibitor (Scheme 10.19) [46].

The reduction of other carbonylic compounds such as 1,3-diketones or β-oxo esters has also been extensively successfully described in the literature. In this manner, (*R*)-fluoxetine can be also synthesized through the bioreduction of ethyl benzoylacetate using *Geotrichum* sp. G38 that shows a complementary selectivity to the one shown by baker's yeast (Scheme 10.19) [47].

Recently, Turner *et al.* described the synthesis of the alkaloid (*R*)-(+)-crispine A, which shows cytotoxic activity against HeLa human cancer cell lines, using in the final step a deracemization procedure with the combination of an enantioselective amine oxidase obtained by directed evolution methods and a chemical non-selective reducing agent (Scheme 10.20) [48].

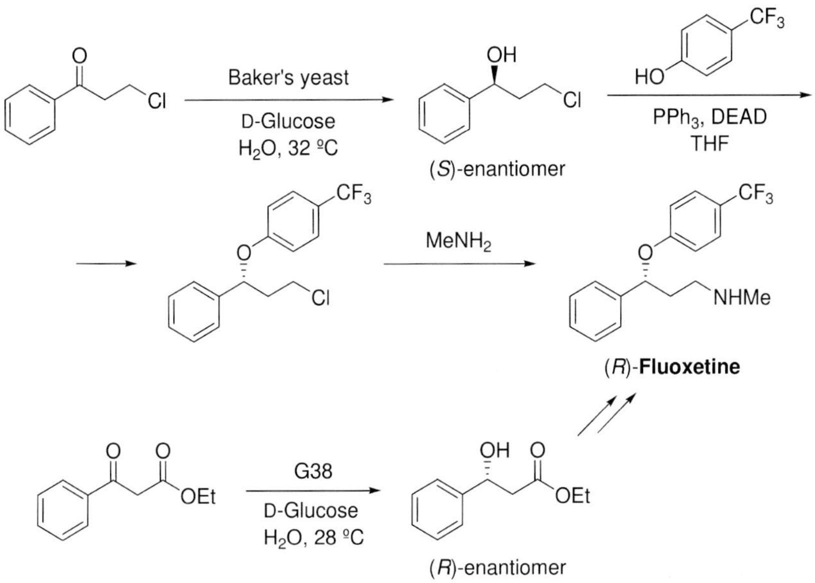

**Scheme 10.19** Chemoenzymatic synthesis of (*R*)-fluoxetine.

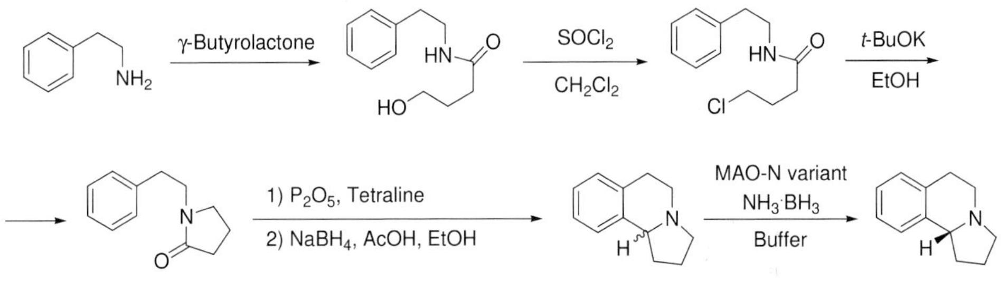

**Figure 10.20** Production of (+)-crispine A using an amino oxidase obtained by directed evolution methods.

## 10.4
## Pharmaceuticals with Lyases

Although lyases represent only about 5% of the biocatalytic processes, aldolases and oxynitrilases show great utility in the synthesis of compounds of physiological interest.

In the last few years new aldolases have emerged as useful biocatalysts in the enantioselective formation of C–C bonds. For instance, the value of enzyme discovery has been illustrated in the synthesis of a precursor for the synthesis of

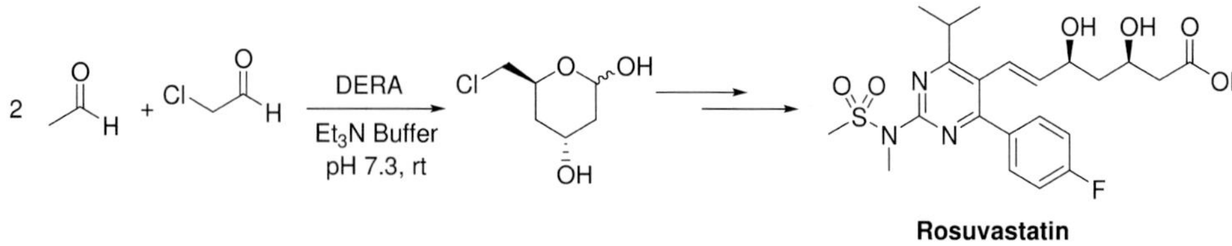

Scheme 10.21 Chemoenzymatic synthesis of rosuvastatin.

statin-type 3-hydroxy-3-methyl-glutaryl-CoA reductase inhibitors as, for example, rosuvastatin, the API of Crestor, and atorvastatin, the API of Lipitor, to reduce low-density lipoprotein cholesterol (Scheme 10.21). The most efficient way to introduce this type of dihydroxy acid is using deoxyribose-5-phosphate aldolase that catalyzes the sequential aldol condensation between one equivalent of 2-chloroacetaldahyde and two equivalents of acetaldehyde [49]. In the initial study, a recombinant *E. coli* 2-deoxy-D-ribose-5-phosphate aldolase (DERA) was used and the reaction performed with high enzyme loading (20% w/w) with low volumetric productivity (2 g/l/day) [49a]. Later a novel DERA was discovered allowing the improvement of the volumetric productivity in the aldol reaction (720 g/l/day) and also reducing the enzyme loading to 2% w/w [49b].

Oxynitrilases are enzymes that catalyze the formation and cleavage of cyanohydrins through the stereoselective addition of hydrogen cyanide to aldehydes or methyl ketones giving enantiopure α-hydroxynitriles. The use of (*R*)-oxynitrilases for the preparation of chiral cyanohydrins has dramatically grown in the last decade because of their possibilities as precursors for the synthesis of many compounds with physiological properties [50].

The use of this kind of lyases has been exploited in organic synthesis by several groups using (*R*)- and (*S*)-oxynitrilases for the synthesis of a great variety of organic compounds. A representative example of the utility of this reaction in the synthesis of pharmaceuticals is the preparation of adrenergic bronchodilators [51]. Scheme 10.22 shows a chemoenzymatic route to (*R*)-terbutaline.

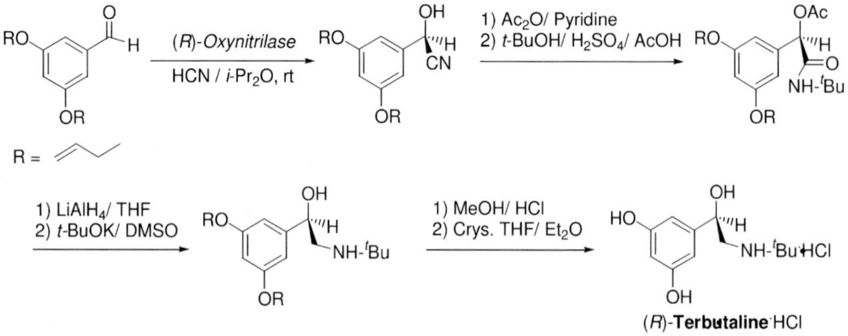

Scheme 10.22 Synthesis of (*R*)-terbutaline using an oxynitrilase.

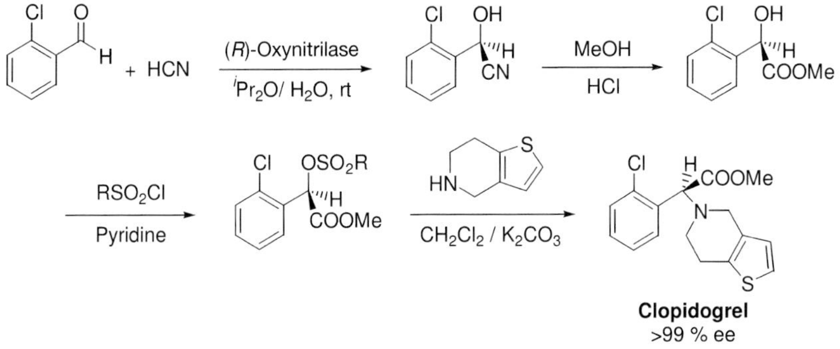

**Scheme 10.23** Chemoenzymatic preparation of (*S*)-clopidogrel.

Other interesting applications of the optically active cyanohydrins obtained by (*R*)-oxynitrilase-catalyzed processes are the production of the blockbuster clopidogrel (Scheme 10.23) [52], and the angiotensin-converting enzyme inhibitors enalapril and lisinopril (Scheme 10.24) [53, 54].

In these synthesis, the optically active (*R*)-cyanohydrin is transformed into the corresponding α-hydroxy carboxylic ester and the hydroxyl function is activated by sulfonylation. The treatment of the corresponding intermediate with tetrahydrothieno[3,2-*c*]pyridine stereoselectively yields the (*S*)-configured clopidogrel (Scheme 10.23). In the second case, a mutant of the recombinant almond (*Prunus amigdalus*) (*R*)-oxynitrilase isoenzyme 5 catalyzes the formation of enantiopure (*R*)-2-hydroxy-4-phenylbutyronitrile [54]. Reaction of the sulfonylated hydroxyester derivative with the corresponding dipeptide leads to the formation of enalapril or lisinopril (Scheme 10.24).

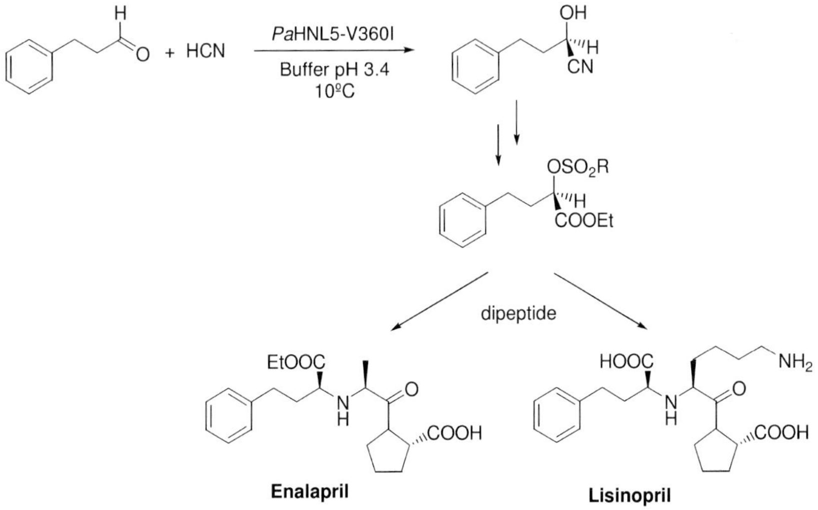

**Scheme 10.24** Preparation of enalapril or lisinopril catalyzed by an (*R*)-oxynitrilase.

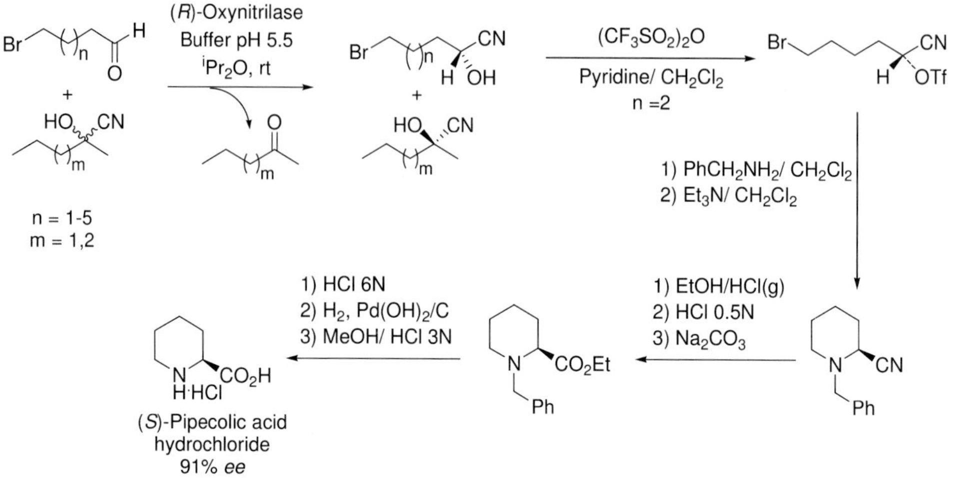

**Scheme 10.25** Chemoenzymatic synthesis of (*S*)-pipecolic acid.

The double process of cyanation/transcyanation of ω-bromoaldehydes and racemic cyanohydrins as a source of HCN is a really interesting process (Scheme 10.25). Thus, using this reaction it is possible to obtain optically active (*S*)-ketone- and (*R*)-aldehyde-cyanohydrins in one pot [55]. The reaction is carried out in diisopropyl ether using a crude extract of almond containing (*R*)-oxynitrilase as biocatalyst. The optically active ω-bromocyanohydrins prepared by this method is used as starting materials for the synthesis of valuable compounds such as (*R*)-2-cyanotetrahydrofuran and (*R*)-2-cyanotetrahydropyran – common structural components of interesting biologically active compounds. These cyanohydrins are also used for the preparation of piperidine ring forms [56], an integral feature of many compounds and synthetic derivatives that exhibit interesting biological activities, and 2,3-disubstituted piperidines [57], which constitute the basic skeleton of many physiologically interesting compounds, and occurs, for example, in special non-peptidic tachykinin receptor antagonists and cardiovascular agents [58]. As an example of these chemoenzymatic routes, we show the synthesis of (*S*)-pipecolic acid in Scheme 10.25 [58a].

The enzyme-catalyzed cyanohydrin reaction offers new and interesting perspectives for the synthesis of different kinds of chiral cyanohydrins, because over the next few years the continuous development of new genetically modified oxynitrilases will be without any doubt of great utility for the preparation of pharmaceuticals.

## 10.5 Conclusions

Nowadays, biocatalysis in combination with chemical catalysis has emerged in the pharmaceutical industry as a routine tool for the preparation of enantiopure drugs.

The possibility of carrying out these enzymatic processes under mild conditions and their high selectivity make the biocatalytic step very attractive to perform some transformations that are difficult to achieve only by chemical procedures. We have shown here the versatility of some biocatalysts for the preparation of a great variety of pharmaceuticals. Among the enzymes tested for organic synthesis, lipases have demonstrated a great versatility and around 50% of biotransformations take place using hydrolytic enzymes, especially lipases. Oxidoreductases have also been used for a long time in the pharmaceutical industry and over the last few years lyases have been found to be of great utility for this preparation of important building blocks through enantioselective formation of C–C bonds. In recent years, directed evolution of enzymes has allowed the isolation of new biocatalysts, which will offer broad new possibilities for the organic chemist. This enzymatic methodology allows the synthesis of products of great added value, and we have shown here that biocatalysis and chemical catalysis is an excellent combination in synthetic organic chemistry for the production of pharmaceuticals and other manufacturing processes.

## References


1 Gotor, V. (2002) *Organic Process Research and Development*, **6**, 420–6.

2 Patel, R.M. (ed.) (2006) *Biocatalysis in the Pharmaceutical and Biotechnology Industry*, Taylor & Francis, New York.

3 Csuk, R. (2006) *Biocatalysis in the Pharmaceutical and Biotechnology Industry* (ed. R.M. Patel), Taylor & Francis, New York, pp. 699–716.

4 Gotor-Fernández, V., Brieva, R. and Gotor, V. (2006) *Journal of Molecular Catalysis B: Enzymatic*, **40**, 111–20.

5 (a) Bevinakatti, H.S. and Banerji, A.A. (1991) *Journal of Organic Chemistry*, **56**, 5372–5.
(b) Ader, U. and Schneider, M.P. (1992) *Tetrahedron: Asymmetry*, **3**, 521–4.

6 Alcántara, A.R., Sánchez-Montero, J. and Sinisterra, J.V. (2000) *Stereoselective Biocatalysis* (ed. R.M. Patel), Marcel Dekker, New York, pp. 713–40.

7 Colton, I.J., Ahmed, S.N. and Kazlauskas, R.J. (1995) *Journal of Organic Chemistry*, **60**, 212–17.

8 Basak, A., Nag, A., Battacharya, G., Mandal, S. and Nag, S. (2000) *Tetrahedron: Asymmetry*, **11**, 2403–7.

9 Cleij, M., Archelas, A. and Furtoss, R. (1999) *Journal of Organic Chemistry*, **64**, 5029–35.

10 Yazbeck, D., Derrick, A., Panesar, M., Desse, A., Gujral, A. and Tao, J. (2006) *Organic Process Research and Development*, **10**, 655–60.

11 Blaschke, G., Hempel, G. and Muller, W. (1993) *Chirality*, **5**, 419–21.

12 Fernández-Solares, L., Díaz, M., Brieva, R., Sánchez, V.M., Bayod, M. and Gotor, V. (2002) *Tetrahedron: Asymmetry*, **13**, 2577–82.

13 Zhao, L., Han, B., Huang, Z., Millar, M., Huang, H., Malashock, D., Zhu, Z., Milan, A., Robertson, D.E., Weiner, D.P. and Burk, M.J. (2004) *Journal of the American Chemical Society*, **126**, 11156–7.

14 (a) Inaba, T., Birchler, A.G., Yamada, Y., Sagawa, S., Yokota, K., Ando, K. and Uchida, I. (1998) *Journal of Organic Chemistry*, **63**, 7582–3.
(b) Albizati, K.F., Babu, S., Birchler, A., Busse, J.K., Fugett, M., Grubbs, A., Haddach, A., Pagan, M., Potts, B., Remarchuk, T., Rieger, D., Rodriguez, R., Shanley, J., Szendroi, R., Tibbetts, T., Whitten, K. and Borer, B.C. (2001) *Tetrahedron Letters*, **42**, 6481–5.
(c) Albizati, K. and Babu, S. (2005) US 2005250949, CAN 143:46002, and references cited therein.

(d) Burk, M.J., Barton, N., DeSantis, G., Greenberg, G., Weiner, D. and Zhao, L. (2006) *Handbook of Chiral Fine Chemicals* (eds D. Ager and I. Fotheringham), 2nd edn, Marcel Dekker, New York, pp. 405–18.

**15** Martinkova, L. and Mylerova, V. (2003) *Current Organic Chemistry*, **7**, 1279–95.

**16** Jennings, R.A., Johnson, D.R., Seamans, R.E. and Zeler, J.R. (1994) US 5319135, CAN: 121:108049.

**17** Burns, M.P. and Wong, J.W. (2004) WO 2004111256, CAN 142:54867.

**18** Burns, M.P., Weaver, J. and Wong, J. (2005) WO 2005100580, CAN 143:404598.

**19** (a) Xie, J., Feng, J., García, E., Bernett, M., Yazbeck, D. and Tao, J. (2006) *Journal of Molecular Catalysis B: Enzymatic*, **41**, 71–80.
(b) Yazbeck, D., Durao, P., Xie, J. and Tao, J. (2006) *Journal of Molecular Catalysis B: Enzymatic*, **39**, 156–9.

**20** Ghanem, A. and Aboul-Enein, H.Y. (2004) *Tetrahedron: Asymmetry*, **15**, 3331–51.

**21** Campos, F., Bosch, M.P. and Guerrero, A. (2000) *Tetrahedron: Asymmetry*, **11**, 2705–17.

**22** de Gonzalo, G., Brieva, R., Sánchez, V.M., Bayod, M. and Gotor, V. (2001) *Journal of Organic Chemistry*, **66**, 8947–53.

**23** Bouzemi, N., Debbeche, H., Aribi-Zouiou, L. and Fiaud, J.-C. (2004) *Tetrahedron Letters*, **45**, 627–30.

**24** de Gonzalo, G., Brieva, R., Sánchez, V.M., Bayod, M. and Gotor, V. (2003) *Journal of Organic Chemistry*, **68**, 3333–6.

**25** (a) Carr, A.A., Nieduzak, T.R., Miller, F.P. and Sorensen, S.M. (1989) EP 317997, CAN 111:232581.
(b) Nieduzak, T.R. and Margolin, A.L. (1991) *Tetrahedron: Asymmetry*, **2**, 113–22.
(c) Pàmies, O. and Bäckvall, J.-E. (2002) *Journal of Organic Chemistry*, **67**, 9006–10.

**26** Henke, E., Pleiss, J. and Bornscheuer, U.T. (2002) *Angewandte Chemie (International Edition in English)*, **41**, 3211–13.

**27** Boegeso, K.P. (1989) EP 0347066, CAN 113:78150.

**28** Fernández-Solares, L., Brieva, R., Quirós, M., Llorente, I., Bayod, M. and Gotor, V. (2004) *Tetrahedron: Asymmetry*, **15**, 341–5.

**29** García-Urdiales, E., Alfonso, I. and Gotor, V. (2005) *Chemical Reviews*, **105**, 313–54.

**30** Chênevert, R. and Desjardins, M. (1994) *Canadian Journal of Chemistry*, **72**, 2312–17.

**31** (a) Gotor Fernández, V., Busto, E. and Gotor, V. (2006) *Advanced Synthesis and Catalysis*, **348**, 797–812.
(b) Gotor-Fernández, V. and Gotor, V. (2006) *Current Organic Chemistry*, **10**, 1125–43.

**32** Kato, K., Saito, Y., Gong, T. and Yokogawa, Y. (2004) *Journal of Molecular Catalysis B: Enzymatic*, **30**, 61–8.

**33** (a) Youdim, M.B.H., Finberg, J.P.M., Levy, R., Sterling, J., Lerner, D. and Berger-Paskin, T. (1991) EP 436492, CAN 115:158747.
(b) Gutman, A.L., Meyer, E., Kalerin, E., Polyak, F. and Sterling, J. (1992) *Biotechnology and Bioengineering*, **40**, 760–7.

**34** González-Sabín, J., Gotor, V. and Rebolledo, F. (2004) *Chemistry – A European Journal*, **10**, 5788–94.

**35** Torre, O., Gotor-Fernández, V. and Gotor, V. (2005) *Tetrahedron: Asymmetry*, **16**, 860–6.

**36** Domínguez de María, P., Carboni-Oerlemans, C., Tuin, B., Bargeman, G., van der Meer, A. and van Gemert, R. (2005). *Journal of Molecular Catalysis B: Enzymatic*, **37**, 36–46.

**37** (a) López-García, M., Alfonso, I. and Gotor, V. (2003) *Journal of Organic Chemistry*, **68**, 648–51.
(b) López-García, M., Alfonso, I. and Gotor, V. (2003) *Tetrahedron: Asymmetry*, **14**, 603–9.

**38** Puertas, S., Rebolledo, F. and Gotor, V. (1996) *Journal of Organic Chemistry*, **61**, 6024–7.

**39** Nakamura, K., Yamanaka, R., Matsuda, T. and Harada, T. (2003) *Tetrahedron: Asymmetry*, **14**, 2659–81.

**40** Holland, H.J. (2000) *Stereoselective Biocatalysis* (ed. R.M. Patel), Marcel Dekker, New York, pp. 131–52.

**41** Turner, N.J. (2004) *Current Opinion in Chemical Biology*, **8**, 114–19.

**42** Hilker, I., Alphand, V., Wohlgemuth, R. and Furstoss, R. (2004) *Advanced Synthesis and Catalysis*, **346**, 203–14.

**43** Gotor, V., Quirós, M., Liz, R., Frigola, J. and Fernández, R. (1997) *Tetrahedron*, **53**, 6421–32.

**44** (a) van Berkel, W.J.H., Kamberbeek, N.M. and Fraaije, M.W. (2006) *Journal of Biotechnology*, **124**, 670–89.
(b) Mihovilovic, M.D. (2006) *Current Organic Chemistry*, **10**, 1265–87.

**45** Csuk, R. and Glanzer, B.I. (1991) *Chemical Reviews*, **91**, 49–97.

**46** Fronza, G., Fuganti, C., Grasselli, P. and Mele, A. (1991) *Journal of Organic Chemistry*, **56**, 6019–23.

**47** Jian-Xin, G., Zu-Yi, L. and Guo-Qiang, L. (1993) *Tetrahedron*, **49**, 5805–16.

**48** Bailey, K.R., Ellis, A.J., Reiss, R., Snape, T.J. and Turner, N.J. (2007) *Chemical Communications*, 3640–2.

**49** (a) Machajewski, T.D. and Wong, C.-H. (2000) *Angewandte Chemie (International Edition in English)*, **39**, 1352–74.
(b) Hanson, S.R., Wong, K., Huang, H., Chen, P. and Burk, M.J. (2004) *Proceedings of the National Academy of Sciences of the United States of America*, **101**, 5788–93.
(c) Greenberg, W., Wong, K., Varvak, A. and Swanson, R.V. (2004) WO 2004027075, CAN 140:302423.
(d) Jennewein, S., Shuermann, M., Wolberg, M., Hilker, I., Luiten, R., Wubbolts, M. and Mink, D. (2006) *Biotechnology Journal*, **1**, 537–48.

**50** (a) Griengl, H., Schwab, H. and Fechter, M. (2000) *Trends in Biotechnology*, **18**, 252–6.
(b) Daußmann, T., Rosen, T.C. and Dünkelmann, P. (2006) *Engineering in Life Sciences*, **6**, 125–9.

**51** Effenberger, F. and Jaeger, J. (1997) *Journal of Organic Chemistry*, **62**, 3867–73.

**52** Bousquet, A. and Musolino, A. (1999) WO 9918110 A1, CAN 130:296510.

**53** Sheldon, R.A., Zeegers, H.J.M., Houbiers, J.P.M. and Hulshof, L.A. (1991) *Chimica Oggi*, **9**, 35–47.

**54** Weis, R., Gaisberger, R., Skranc, W., Gruber, K. and Glieder, A. (2005) *Angewandte Chemie (International Edition in English)*, **44**, 4700–4.

**55** Menendez, E., Brieva, R., Rebolledo, F. and Gotor, V. (1995) *Journal of the Chemical Society, Chemical Communications*, **10**, 989–90.

**56** (a) Nazabadioko, S., Perez, R.J., Brieva, R. and Gotor, V. (1998) *Tetrahedron: Asymmetry*, **9**, 1597–604.
(b) Monterde, M.I., Nazabadioko, S., Rebolledo, F., Brieva, R. and Gotor, V. (1999) *Tetrahedron: Asymmetry*, **10**, 3449–55.

**57** Monterde, M.I., Brieva, R. and Gotor, V. (2001) *Tetrahedron: Asymmetry*, **12**, 525–8.

**58** (a) Carlier, P., Simond, J.A.L. and Monteil, A.J.C. (1989) FR 2608603, CAN 110:38903.
(b) Harrison, T., Williams, B.J., Swain, C.J. and Ball, R.G. (1994) *Bioorganic and Medicinal Chemistry Letters*, **4**, 2545–50.
(c) Baker, R., Macleod, A.M., Seward, E.M. and Swain, C.J. (1996) WO 9521819, CAN 124:8636.

# Index


*a*

acetaldehyde, 2′-deoxyribonucleoside synthesis 129, 130, 199–209
acetyloxyphenyl ketones 7, 8
actinomycetes, glycosylation pattern modification 159–198
actinomycin 144
actinorhodin 142, 143
acylation 116, 117, 123, 181, 214, 215, 220–224
adenine 204–208
adenosine triphosphate (ATP) 202, 203, 206, 207
*S*-adenosylmethionine (SAM) 150–152
aglycone 176, 178, 185, 189, 192
air-stable racemization catalysts 13, 14
D-alanine 24, 25
L-alanine 31, 36, 37, 119, 120
alcalase 24, 25
alcohols
– chiral 112–114
– secondary 1, 6–16, 112
aldol reactions
– aza sugar synthesis 69
– 2′-deoxyribose-5-phosphate production 129, 130, 203, 204, 228
– natural/unnatural sugars 71, 72
– stereochemistry 61, 62, 127–130
aldolases
– C–C bond formation 127–130
– classes 61, 62
– DHAP-dependent 61–81, 128
– FBPA-mediated 70, 71, 73–76
– pharmaceuticals preparation 227, 228
D-allo-isoleucine 24, 25
allylic alcohols 9, 10, 15, 16
aluminum catalysts 14, 15
amidase 25, 26, 122, 123
amine oxidase 118, 119, 226
amines
– chiral 21, 22, 33–37, 117–121
– secondary 35, 36
– tertiary 35, 36
amino acid dehydrogenases 27–29, 118–120
amino acids
– β-2-amino acids 30
– cyclic 28, 29
– deracemization 31, 32, 47, 48, 119
– enantiomerically pure 21–33
β-amino alcohols 36, 37
γ-aminobutyric acid (GABA) 215, 216
aminocoumarins 193, 194
aminocyclitol analogs of valiolamine 70, 71
aminocyclopentadienyl ruthenium chloride complex 11, 12
(*R*)-4-amino-3-hydroxybutanoic acid ((*R*)-GABOB) 225, 226
aminolysis 222–226
α-amino nitriles 26
aminotransferases 31, 32, 119, 120, 163, 164
ammonia lyases 30, 31
ammoniolysis 121, 122, 223, 224
analapril 229
angucyclines 181, 183–186
anthracyclines 186–191
antibiotics production
– cascade reactions 123
– glycosylated derivatives 167–186, 193, 194
– multi-modular synthases 140–146
– multi-step processes 46, 75, 76
antitumor drugs 142, 175–181, 182, 186–191
aromatization 142–144
2-aryl-4-pentenenitriles 27


*Multi-Step Enzyme Catalysis: Biotransformations and Chemoenzymatic Synthesis*
Edited by Eduardo Garcia-Junceda

ISBN: 978-3-527-31921-3

asymmetric hydrogenation 7, 8, 32, 125–127
asymmetric reductive acetylation 7, 8
asymmetric synthesis
– chiral amines 119, 120
– DHAP-dependent aldolases 61–81
– pharmaceuticals 222
asymmetric transformations 1–19
atorvastatin 113, 130, 228
aureolic acid group 175–181, 182
aza sugar synthesis 68–71, 77, 78

***b***
(*R*)-baclofen 222, 223
bacterial coupling 98, 99
benz[*a*]anthraquinone ring system 181, 183–186
benzylic amines 34
biocatalysis
– amino acids and amines 21–39
– combinatorial 46, 48, 85–97
– combined chemical/biocatalysis 213–233
– coupled oxidizing enzymes 48–51, 53–57
– industrial processes 46, 47
– oxidizing enzymes 41
– redox systems 41, 43–45, 48–51
– whole-cell biotransformations 55
biochemical retrosynthesis, 2′-deoxyribonucleosides 199–201, 204–207
bioelectrocatalysis 56
biofuel cells 56, 58
biological oxidation mechanisms 44, 45
biomimetic heterocyclization 141, 142
biosensors 56, 57
biosynthesis, combinatorial 86–88, 97–102, 159–198
biotransformations
– multi-step 41–60
– whole-cell 55, 56, 110, 125, 126
bisindolylmaleimides 191
brevicomin 73

***c***
cancer
– anticancer drugs 142, 175–182, 186–191
– glycopeptides 96, 97
carbon–carbon bonds
– double bond reduction 125, 126
– formation 61, 70, 127–130
carboxylic acid derivatives
– chiral 121–127, 141, 142
cascade reactions 109–131
catalase 119, 128, 130
catalysts, racemization 3–5, 16
cephalexin 123
chemical transformations, excised domains 139–155
chemoenzymatic routes
– amino acids and amines 21–39
– iminocyclitol 69, 70, 77, 78
– macrocyclization 144–147
– natural products 73–77
– oxidizing enzymes 46–48, 52–56, 58, 60
– pharmaceuticals preparation 21, 138–156, 213–233
chiral alcohols 112–114
chiral amines
– chemoenzymatic routes 21, 22, 33–37
– enzymatic cascade reactions 114–121
chiral amino acids 21, 22, 47
chiral carboxylic acid derivatives
– enzymatic cascade reactions 121–127
– heterocyclic 141, 142
chlorinase 149
chloroalanine 31, 32
chlorobiocin 193, 194
chloroperoxidase (CPO) 46, 47, 131
chromomycin 175–176, 179–181, 182
citalopram 221, 222
clavulanic acid 154
cloning
– genes 165, 167, 168, 170, 176, 181, 188
– oxidizing enzymes 55, 56
(*S*)-clopidogrel 229
Codexis process 113, 114
cofactors
– aldolases 61
– NAD(P)H 42, 44, 52, 54, 56, 112, 118, 119, 125, 126
– oxidoreductases 226
– oxygenases 153–155
– recycling/regeneration 51, 56, 112, 113, 131
combinatorial biocatalysis 46, 48, 85–97
combinatorial biosynthesis
– actinomycete glycosylation pattern modification 159–198
– glycoconjugates 86–88, 97–100
coronatine 149
coumarin 154
(+)-crispine A 226, 227

cyanohydrins 123, 124, 229, 230
cyclic amines 35, 36
cyclic amino acids 28, 29
cymene–ruthenium catalysts 9–11
Cytochrome P450 42, 56, 153, 154

*d*
dapoxetine 224, 225
daunorubicin 187, 188
deacylation, easy-on/easy-off process 116
Degussa (now Evonik) process 118
dehydratases 162, 163, 169
dehydrogenases 43, 44, 49, 118, 226
deoxy sugars 87, 88–94, 161, 167–175
deoxygenation 163, 164
deoxyhexoses 159–194
2′-deoxyinosine 204, 205, 207, 208
deoxyriboaldolase 129, 130, 199–205, 207, 208
2′-deoxyribonucleoside synthesis 199–211
– one-pot process 206–209
– three-step process 204–206, 207
2′-deoxyribose-5-phosphate (DR5P) 129, 130, 200–205
2′-deoxyribosyl groups 200, 201
deracemization
– amines 35–36
– amino acids 31, 32, 47, 48, 119
– chiral secondary alcohols 114, 115
desosamine 170, 172–175
desosaminyl transferase 173–175
desymmetrization 222, 225
diamino acids 29
dihydroxyacetone (DHA) 64
dihydroxyacetone phosphate (DHAP)-dependent aldolases 61–81, 128
– aza sugar synthesis 68–71, 77, 78
– DHAP preparation 64–68
– natural product synthesis 73–76, 77
– natural/unnatural sugar synthesis 73, 74, 128, 129
– stereospecificity 63–68, 78, 128
dioxygenases 42, 43, 154, 155
DNA building blocks
– glycosylated derivatives 165, 189, 191
– microbial production 129, 130, 199–210
dTDP-β-L-rhamnose pathway 88, 89, 91–94
dynamic kinetic resolution (DKR)
– chemoenzymatic routes 23, 33–35
– enzymes 2, 3, 5
– enzymo-metallic catalysis 1–19
– (*R*)- and (*S*)-selective 5, 12, 13, 16
– racemic chiral amines 33–35, 114, 116, 117, 119, 120, 122
– secondary alcohols 1, 6–16, 112

*e*
elloramycin 189–191
enantiomerically pure compounds
– amino acids and amines 21, 22
– enzymatic cascade processes 111–127
– pharmaceuticals 213–231
– synthesis 1, 2, 5, 16
enoate reductases (EREDs) 26, 125
enzymatic aminolysis 222–226
enzymatic hydrolysis 214–218
enzymatic resolution 1, 5
enzymatic transesterification 219–222
enzyme–metal combination 5, 16
enzymes
– cascade reactions 109–131
– chemoenzymatic routes 21–39
– compartmentalization 109, 110
– immobilization 109–111
– kinetic parameters 49, 50, 88, 89
– oxidizing 48, 49, 52, 53, 55, 56, 58
– sugar biosynthesis 161–165
epimerases 163, 164
epirubicin 189
epoxide hydrolases 216, 217
erythromycin 146, 147, 150, 167–169
Evonik (formerly Degussa) process 118
excised domains for chemical transformations 139–155
– aromatization 142–144
– glycosylation 150, 151
– halogenation 147–150
– heterocyclization 139–142
– macrocyclization 155–157
– methyltransferases 151–153
– oxidation 153–155

*f*
fatty acids (FAs) 137
fermentation 203, 204
flavin-dependent enzymes
– cascade reactions 125
– halogenation 147, 148
– multi-step biotransformation processes 42, 44, 45, 56
– sugar biosynthesis 163
fluorinase 148, 150
(*R*)-fluoxetine 226, 227

formate dehydrogenase (FDH) 113, 118–121, 126
formoterol 219
fructose-1,6-bisphosphate aldolase (FBPA) 70, 71, 74–76
fructose-1,6-disphosphate (FDP) 202–204, 206
fructose-6-phosphate aldolase (FSA) 76–78
fusion proteins 99, 102

***g***

gabapentin 217, 218
galactose oxidase/catalase 130, 131
gene clusters 161, 162, 165–182, 185–193
gene expression 166, 170–174, 179, 184, 185, 188–190, 194
gene inactivation 165, 166, 168–174, 176–182, 188, 193
glucose, 2′-deoxyribonucleoside synthesis 199–209
glucose dehydrogenase (GDH) 113, 126
glucose oxidase/catalase 130, 131
glutamate racemase 120, 121
D-glyceraldehyde-3-phosphate (G3P) 199, 200, 202
glycoconjugate synthesis, multi-enzyme systems 83–100, 102
glycolipid-related oligosaccharides 96
glycolysis, D-glyceraldehyde-3-phosphate generation 199, 200, 202, 205–207
glycopeptides 96, 97, 150, 151
glycosidases
– glycoconjugate synthesis 84, 86, 97
– inhibitors 68, 70
glycosidic oligosaccharide derivatives 96
glycosylation
– actinomycetes 159–198
– bioactive compounds 165–194
– excised domains for chemical transformations 150, 151
– gene expression 166, 170–174, 179, 184, 185, 188–190, 194
– gene inactivation 165, 166, 168–174, 176–182, 188, 193
glycosyltransferases 150, 151
– actinomycetes biosyntheses 166, 169, 176, 179, 180, 184–186, 188, 191
– fused proteins 99
– Leloir glycosyltransferases 86–88, 97
– nucleotide sugar synthesis 91–94, 96
– oligosaccharide synthesis 99, 100
griseorhodin 142, 143

***h***

halogenases 147–150
halogenation 147–150
halohydrin dehalogenase (HHDH) 114
haloperoxidases 43, 147, 148
hepatitis C virus (HCV) protease inhibitors 21
heterocyclization 139–142
Hofmann rearrangement 25
hybrid peptides 144, 145
hydrocyanation 123–125
hydrogen peroxide 130, 131
hydrogenation
– asymmetric 7, 8, 32, 125–127
– transfer-hydrogenation 6
hydrolases 2, 3, 213–226
hydrolysis 214–216, 218
α-hydroxy acids 119, 120, 123–125
hydroxylases 42
hydroxynitrile lyase 123

***i***

ibuprofen 215, 216
iminium ion 35, 36
iminocyclitol 68–70, 77, 78
immobilization of enzymes 109–111
*in vitro/in vivo* multi-enzyme systems 85, 86
indenyl ruthenium complex 8, 9
indoles 148
indolocarbazoles 148, 149, 191–193
intramolecular reductive amination 69, 70
ionic liquids 9, 10
iridium catalysts 35
L-isoleucine 24, 25, 149
isomerases 70, 71, 72, 74
isopenicillin 154
isopropenyl acetate 12–14
isopropyl butyrate 14

***k***

α-keto amino acids 27–29, 119
ketones, asymmetric reductive acetylation 7, 8
ketoreductase (KRED) 112–114, 125, 162, 164, 176
kinetic parameters 49, 50, 88, 89

***l***

laccases 46, 47, 49–51
lactate dehydrogenase 37, 120, 121

lactones 146, 147, 167–175
landomycins 184–186
Leloir glycosyltransferases 86–88, 97
*tert*-leucine 32, 33, 118
leucine dehydrogenase 33, 119
lipases
– amine acylation 114–117
– chemoenzymatic synthesis 23, 24, 33–35
– enzymo-metallic catalysis 2, 3, 5, 15
– pharmaceuticals preparation 214–216, 219–225, 231
lisinopril 229
living-cell processes, oligosaccharide synthesis 98–102
lyases 30, 31, 123, 213, 227–230, 231
lysine cyclodeaminase gene 29, 30
L lysine oxidase 28, 29

***m***
macrocyclization 155–157
macrolides 167–175
malononitriles 25, 26
(*R*)-mandelic acid 123, 124
Meerwein–Ponndorf–Verley–Oppenauer reaction 14
metabolic engineering
– 2′-deoxyribonucleoside synthesis 199–205
– gene clusters 165
– multi-modular synthases 137, 139, 149, 150, 156
metabolically engineered cells 97–100
metal-catalyzed racemization 1, 2, 5, 6
– catalysts 3–5, 16
*N*-methyl-L-amino acid dehydrogenases (NMAADH) 28, 29
β-methyl-β-aryl alanine 32
methyl esters 24, 215
*N*-methyl-L-phenylalanine 28, 29
methylation 151–153, 163–165, 179
*N*-methylpyrrolidine 35, 36
methyltransferases 151–153, 163–165
methymycin 171–175
microbial production, DNA building blocks 129, 130, 199–211
mithramycin 175–179
modular biosynthetic systems 137–158
molecular oxygen 118, 119, 153–155
momensin 140, 141
monoamine oxidase (MAO) 35, 36
monosaccharides 86, 97, 161
multi-enzyme systems
– bioelectrocatalysis 56
– cascade reactions 110
– chemoenzymatic routes 21
– 2′-deoxyribonucleoside synthesis 209
– design and development 48, 49
– glycoconjugate synthesis 83–103
– *in situ* regeneration of nucleotide sugars 88–94
– *in vitro/in vivo* 85, 86
– natural/unnatural sugar synthesis 71–73
– one-pot processes 206–208
– redox systems 41, 43–45, 55, 56
– whole-cell biotransformations 55, 56
multi-modular synthases 137–158
multi-step biotransformation processes
– cofactor recycling 51, 52, 55, 56
– combinatorial biocatalysis 46, 48
– DHAP-dependent aldolases 61–81
– enzymatic cascade reactions 109, 110
– examples 52, 53
– FSA-mediated synthesis 77, 78
– industrial processes 46, 47
– oxidation steps 45–48
– oxidizing enzymes 48–60
– pharmaceuticals preparation 217, 218
mycaminose 170
D-mycarose 163–165, 168–170, 176, 179
myxothiazol 152, 153

***n***
NAD(P)H cofactors 42, 44, 52, 54, 56, 112, 118, 119, 125, 126
nanotechnology
– multi-enzyme systems 53, 57
– Pd catalysts 116, 117
natural products
– actinomycetes synthesis 159
– chemoenzymatic routes 73–76, 77
– multi-modular synthases 137–158
natural/unnatural sugar synthesis 71–73, 77, 78, 128, 129
nelfinavir 217
nitrilases 26, 122–124, 218, 228, 229
nitrile hydratase 25, 26, 122, 123
nitriles, chiral carboxylic acid synthesis 122–125
non-ribosomal peptides (NRPs) 137–154
novobiocin 152–154
nucleobases, 2′-deoxyribonucleoside synthesis 199, 200, 204–209

nucleoside phosphorylase 200, 204–207
nucleotide deoxy sugars 87–94
nucleotide sugar synthesis 85–94

***o***

Old Yellow Enzyme 30, 124
oligoesters 10, 11
oligosaccharide synthesis
– deoxyhexoses 161
– metabolically engineered cells 97–102
– multi-enzyme systems 85, 87, 88, 92, 94–96
D-olivoses 162, 165, 170, 173, 175, 176, 179, 181
optically active polyesters 10, 11
oxazoles 140, 141
oxazolines 140
oxidases
– biocatalytic systems 42–47, 50, 51
– biosensors 57
– chemoenzymatic routes 31, 32, 35, 54, 55
– coupling with non-redox enzymes 53
– enzymatic cascade reactions 130, 131
oxidation
– biological mechanisms 44, 45
– enzymatic cascade reactions 118, 119, 130, 131
– excised domains for chemical transformations 153–155
– molecular oxygen 118, 119, 153–155
oxidizing enzymes
– biocatalysis 41–43
– classes 41–43
– cloning 55
– coupling 48, 53, 54, 56
– heterologous expression 55, 58
– kinetic parameters 49
– multi-step biotransformation processes 41–60
oxidoreductases 47, 213, 226, 227, 231
oxovanadium(V) complex 15, 16
oxygenases 42, 43, 45–48, 153–155
oxynitrilases 228, 229

***p***

palladium catalysts 34, 117
pancratistatin analogs 74, 75
paroxetine hydrochloride 219, 220
penicillin acylase 36, 37, 116, 117, 123
penicillin G amidase 116, 124, 125
pentamycin 75, 76
pentoses 164
peroxidases
– biocatalytic systems 42, 43, 46, 47
– biosensors 57
– cofactor regeneration 52–55
– enzymatic cascade reactions 130, 131
pharmaceuticals 21, 138–156, 213–233
phenylalanine ammonia lyases (PAL) 30, 31
phenylalanine dehydrogenase (PDH) 27, 28
phenylalanines 28, 29, 31, 32
1-phenylethanol 6, 7, 12–15
(*R*)-1-phenylethylamine 36, 37
phenylglycoside moiety 184–186
phenylpyruvic acid 28
phosphopentomutase 204, 205, 207, 208
phosphorylation 128, 129
phytase 128
pikromycin 172–174
pipecolic acid 29, 30, 230
polyesters, optically active 10, 11
polyketides (PKs) 46, 137–155
polysaccharides 97, 161, 175
pregabalin 218
prejadomycin 185, 186
premithramycin 177–179
propranolol 214, 215
proteins, transition metal/protein catalysts 30, 31
pyrophosphorylase 92, 93
pyruvate 37

***r***

racemization
– chemoenzymatic routes 23, 33
– metal-catalyzed 1, 2, 5, 6
– room temperature 11–14, 16
rasagiline 223
rebeccamycin 148, 149, 191
redox systems
– bioelectrocatalysis 56
– coupling enzymes 48–51, 54
– enzyme biocatalysis 41
– heterologous expression 55, 58
– redox potentials 43–45, 48–50, 52
– whole-cell biotransformations 55
reductases 48, 54, 163, 226
reductive amination 27, 28, 69, 70, 117–121
regioselectivity, methylation 152, 153
retrosynthesis, 2′-deoxyribonucleosides 199–201, 204–207
D-rhodinoses 181, 183, 184, 188

rhodium catalysts 31
room temperature racemization 11–14, 16
rosuvastatin 228
ruthenium catalysts
– air-stable 13, 14
– amine DKR 33
– racemization 3–5
– secondary alcohol DKR 6–14

**s**
Schiff's bases 24, 61
secondary alcohols, dynamic kinetic resolution 1, 6–16, 112
secondary amines 35, 36
sphydrofuran 74
spinosyn 171, 172
staurosporine 191–193
stereochemistry
– aldol reactions 61, 62, 127–130
– C–C bond formation 127, 128
– DHAP-dependent aldolases 63–76, 78, 128
stereoinversion, amino acids 31, 32
subtilisin
– chemoenzymatic synthesis 24
– enzymo-metallic catalysis 3, 5, 12, 13, 16
sugar synthesis
– actinomycetes 159–175
– aza sugars 68–71, 77, 78
– enzymes 161–165
– glycoconjugates 85, 86
– glycosylation 150, 151
– natural/unnatural sugars 73, 74, 77, 78, 128, 129
– nucleotide deoxy sugars 87–94
– nucleotide sugars 85–94
– sugar exchange 150, 151
surfactin 147
syringolides 76, 77

**t**
tandem heterocycles 140–142
TDP-4-keto-6-deoxy-D-glucose 162
telomerase inhibitors 140, 142
telomestatin 140
(*R*)-terbutaline 228
tertiary amines 35, 36
tetrahydroisoquinoline-1-carboxylic acid ethyl ester 23
thiazoles 140–142
thioesterase (TE) 141, 142, 144–147
thioesters 23, 24
thiyl radicals 34, 35
threonine aldolase 36, 37
transaminases 36, 37, 119–121
transesterification 219–222
transfer-hydrogenation 6
transition metal/protein catalysts 30, 31
transketolase 129
trifluoroethyl butanoate 12, 13
tylosin 162–164, 170, 171
tyrocidine A 144–146
tyrosinase 50, 51
L-tyrosine decarboxylase 36, 37

**u**
U-50,488 224
urdamycin 181, 183, 184

**v**
valiolamine, amincyclitol analogs 70, 71
vanadium catalysts 15, 16
vinyl esters 8

**w**
whole-cell biotransformations
– enzymatic cascade reactions 110, 125, 126
– multi-enzyme systems 55, 56

**y**
yeast treatment
– DNA building blocks 202–204, 206–208
– fusion proteins 99, 102
– pharmaceuticals preparation 226, 227

**z**
zopiclone 215–217